Pathways to Math Literacy

Second Edition

Dave Sobecki
Miami University Hamilton

Brian Mercer
Parkland College

PATHWAYS TO MATH LITERACY, SECOND EDITION

Some ancillaries, including electronic and print components, may not be available to customers outside the United States.

This book is printed on acid-free paper.

6 7 8 9 LMN 21 20 19

ISBN: 978-1-260-40493-7 (Bound Edition)
MHID: 1-260-40493-5

ISBN: 978-1-259-98560-7 (Loose Leaf Edition)
MHID: 1-259-98560-1

ISBN: 978-1-260-18930-8 (Annotated Instructor's Edition)
MHID: 1-260-18930-9

Product Developer: *Luke Whalen*
Marketing Manager: *Noah Evans*
Content Project Manager: *Peggy Selle*
Buyer: *Sandy Ludovissy*
Design: *Tara McDermott*
Content Licensing Specialist: *Shannon Manderscheid*
Cover Image: *©Alfred Pasieka/Science Photo Library/Getty Images RF.*
Compositor: *SPi Global*

Library of Congress Cataloging-in-Publication Data

Names: Sobecki, Dave, author. | Mercer, Brian A., author.
Title: Pathways to math literacy / Dave Sobecki, Associate Professor of Mathematics, Miami University Hamilton, Brian Mercer, Professor of Mathematics, Parkland College.
Description: Second edition. | New York, NY : McGraw-Hill Education, [2019] | Includes index.
Identifiers: LCCN 2017034142| ISBN 9781259985607 (alk. paper) | ISBN 9781260189308 (annotated instructor's edition)
Subjects: LCSH: Mathematics—Textbooks.
Classification: LCC QA39.3 .S63 2019 | DDC 510—dc23
LC record available at https://lccn.loc.gov/2017034142

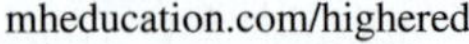
mheducation.com/highered

A Letter from the Authors

When we set out to write the first edition of this book, Math Literacy wasn't a particularly well-defined subject. During the planning stages, we did our best to combine our own classroom experiences with information gathered from instructors across the country who were interested in this new non-STEM pathways movement. We're very proud of the result, but over the last four years we've learned a whole lot more about a field that is far more developed than it was in the early days. The result is the second edition of *Pathways to Math Literacy.*

For those who are new to math literacy, let's first address what this whole thing is about. In our view, we as college math faculty have spent too many years teaching developmental courses as if the students were all headed to calculus. We know that most of them aren't, so our approach is based on answering this question: **what is it really useful for non-STEM students to learn?** And what if we agreed to move past the "this is important because it's important" mentality, and thought about the topics and activities that will best serve a group of students that are, for the most part, poorly served by traditional developmental algebra?

Our project is the result of attempting to do just that. Most importantly, it's not about watering down the curriculum in an attempt to pass more students. It's about providing non-STEM students with an alternate but challenging pathway that will get them into the college-credit math courses they need without getting trapped behind the roadblocks that have emerged within the traditional developmental math track. But more importantly, **it's about focusing on context and critical thinking, and showing these students why the math they've struggled with for so many years is relevant in their lives.**

A big part of our approach was developed after studying the results of research that's been done on what exactly employers are looking for in college graduates. Working productively in teams, being able to solve problems without hand-holding, and being able to use technology efficiently are consistently at the top of the list, so we made these the cornerstones of *Pathways to Math Literacy.* Surely we want our students to become better educated, but we also want them to become gainfully employed.

What we've discovered along the way is how much richer the experience can be for non-STEM students when they stop trying to memorize and mimic, and start to really think and learn. By using a workbook format, focusing on active learning, incorporating technology, and approaching every single topic from an applied standpoint, we've been able to build a course (and a book) that elicits our favorite response from students: **"This doesn't feel like a math course. We're kind of using math. . . ." YES. Yes we are.**

This is where we started, and it won't ever change. Based on feedback from far too many gracious instructors to count, we had three main goals for this second edition. The first was to redesign the order of the lessons so that the mathematics flows more logically, and to include additional topics that the market was asking for. Second, we aimed to place more emphasis on the fact that algebra doesn't have to be an abstract, frightening topic, and that having to struggle productively isn't just okay; it's GREAT! The third goal was to add new features and exercises that make the book and its online supplements more flexible and comprehensive so that instructors can adapt our basic ideas to their own needs and philosophies easily. We surely don't expect that everyone, or maybe even anyone, will cover every single lesson in the book. We think we know a lot about math literacy, but you're still the most important expert on what is best for your students. Think of the book as a suggested curriculum that's fully intended to be customized for your needs. No matter which lessons you choose to cover, you'll be fully supported by our entire program. The specifics of these revisions can be found on page v of the book.

If you look very carefully, you'll find many of the topics that typically make up the core of the developmental algebra curriculum. We like to think of it as giving your children medicine they don't want by mixing it into a bowl of ice cream. By making everything contextual, and liberally mixing in important study skills and a variety of topics

that are usually in the province of liberal arts math, we're making the process of learning useful problem-solving skills through algebra more palatable for students, which at the end of the day is a significant part of the battle. When your students open their minds to the possibility of really understanding a math course, and really seeing how math can be useful, they blossom into the learner that we try to bring out in all of our students.

Many thanks for looking at our materials, and ALWAYS feel free to reach out to us with questions, comments, suggestions, or support.

Dave and Brian
davesobecki@gmail.com
bmercer@parkland.edu

What's New in the Second Edition?

For those that are familiar with the first edition of *Pathways,* you may have noticed that we seem to have put on some weight. (The book, not Dave and Brian.) We thought it would be a good idea to discuss the variety of reasons for the increase in size.

First, we improved the layout of the book in a couple of subtle but important ways. Many instructors contacted us to say that there wasn't enough space for students to write the kind of thoughtful responses that we hope for, so we spaced out the questions quite a bit more where appropriate. We also found the layout of the **Portfolio** to be very awkward for our students: when they turned in their **Applications,** the Portfolio came along with it and they no longer had access to the **Technology, Reflections,** or **Looking Ahead** questions. That's been rectified by expanding the Portfolio to a second page.

Speaking of expansion, we've added a few new features based on four years' worth of feedback from our wonderful users. Probably the most common request we got was for extra problems that students could practice on, so the **Did You Get It?** feature was born. Whenever a key concept has been covered, we insert an extra question or two to help students reinforce that concept. Answers are included at the end of each lesson, and solutions videos can be found in the online resources. These aren't intended to be in-class activities: they're reinforcement for students who feel that some extra practice would be helpful.

The next new feature is **Prep Skills** for each lesson. In the first edition, the online skills problems were divided into "Prep" and "Practice" categories, but in this edition we took prep to a whole new level. Each lesson begins with a list of specific skills that are needed for that lesson, along with some exposition, solved sample questions, and a list of problems for students to work on. Again, the answers are provided at the end of each lesson, and solutions videos can be found online. (The problems can be worked online as well, of course.) We've found that students with many different academic backgrounds and levels of preparation are taking Math Literacy courses, so a mechanism for letting both students and instructors know exactly what's needed for success in each lesson should help to level the playing field.

The third new feature is end of unit materials. In order to wrap up units and help students prepare for unit exams, we've written extensive summary materials. This starts with an interactive review of all key terms and formulas: rather than just asking students to read yet another list of definitions, we're asking them to test their knowledge of key terms with accessible fill-in-the-blank questions. Next comes a summary of all the learning objectives from the lesson, along with pages of review problems that are similar to those covered within the unit. For units in which new technology skills were particularly important, a tech review is included as well.

Finally, the most obvious way to expand the size of a book is to add new content. This is probably least responsible for our expansion, but we did add some new topics that were requested over the last few years. These topics include gathering and organizing data, expected value and weighted averages, and margin of error in polling. On the algebraic side, there are new expanded sections covering inequalities, systems of equations, and finding linear equations using the point-slope form.

We also did a pretty substantial reorganization of the original topics, again based on feedback from too many generous users to count. While we still believe that a solid foundation in numeracy should come first, we made an effort to incorporate more algebraic skills a bit earlier. This allows much more flexibility in the types of problems students can solve, and also helps to build their algebra skills gradually over the course of the semester.

If you have a moment, please drop us a line and let us know how you feel about the new edition of *Pathways.* Improvement is a never-ending process, and your feedback will allow us to continue on that pathway. Pun intended.

Pathways to Math Literacy Worktext Features

- **NEW Prep Skills:** Appearing directly before nearly every lesson in this book, this new feature provides every unique learner with the knowledge and skills he or she will need to successfully complete the next lesson in the course.
- **NEW Did You Get It?:** Allowing students the opportunity to perform frequent self-checks for understanding and mastery, this new feature is sprinkled throughout lessons to cover the most critical mathematical concepts in the book.
- **Portfolio:** This 2-page section, found near the end of every lesson in the book and printed strategically so that it can be removed by students as needed without affecting other book content, gives students an excellent record of their knowledge and achievements within each lesson. It's also a useful item for instructors to collect as desired, and can be a wonderful study tool at the end of the term.
- **Technology:** The use of technology does not have to be crutch for math students and act as a barrier to success; it should instead *enhance* their mastery of mathematical content, provide opportunity for enrichment and further exploration, and help prepare them for success beyond the classroom. The Technology boxes, videos, and assignments littered throughout *Pathways* lessons aim to achieve all of these objectives.
- **Online Practice:** An enormous bank of adaptive, algorithmic exercises on McGraw-Hill Education's powerful digital learning platforms gives students almost endless opportunities for skills practice as they work on problems that are carefully programmed to mesh seamlessly with the writing and instructional style of the authors.
- **Applications:** In a context-based course, application problems are paramount among homework problems. The problems in the Applications portion of each lesson can be completed on paper and handed in, or can be done online. In either case, the focus is on why the mathematical skills students have studied are useful in their lives.
- **Reflections:** Successful math students—and indeed, successful people—get in the habit of constantly engaging in self-reflection. These critical open-ended questions, located in the Portfolio section of each lesson, help students review mathematical content from the lesson, articulate the purpose of learning that content, and determine what concepts they might still need to continue reviewing and working on to master moving forward.
- **Looking Ahead:** This feature, which appears at the end of the Portfolio section within each lesson, helps ensure continuity and alignment for students as they transition from one lesson to the next throughout the program.

Pathways to Math Literacy Additional Resources for Instructors

- **NEW Quick Start Guide:** This concise tool gives instructors everything they need to successfully implement *Pathways to Math Literacy* with outstanding results starting immediately Day 1, even with minimal prep time available.
- **Annotated Instructor Edition (Includes teaching tips and exercise answers)**
- **Instructor Notes for each lesson:** Created by Brian Mercer, these notes walk an instructor through each lesson sharing an overview, best practices, and common student challenges.
- **Solutions videos:** Solutions videos for every problem in the book are available for the instructor to distribute as needed. (Solutions PowerPoints are also available as needed.)
- **"Math Literacy in action!":** Since not everyone is able to make the trip to Parkland College, Brian Mercer has recorded several of his classes to help instructors get an idea for what a typical class looks like.
- **First Day of Class Presentation:** This resource assists instructors in helping to set up expectations and goals of the course.
- **Unit Exams/TestGen test bank software**
- **Other valuable resources:** These include group projects, evaluations forms, and sample rubrics.

Pathways to Math Literacy Additional Resources for Students

- **NEW SmartBook with Learning Resources:** This powerful digital resource provides students with an assignable, adaptive eBook and study tool that directs them to the content they don't know and helps them study more efficiently.
- **NEW Over 500 new algorithmic skills practice and homework exercises available through ConnectMath Hosted by ALEKS**
- **NEW Dozens of new online videos—in addition to the over 300 such videos that already exist—in a comprehensive online video series, all of which are tied directly to problems from *Pathways to Math Literacy***

- **Expanded ALEKS Math Literacy Pie:** In support of the expanded content in the book, we've expanded content in the math literacy pie as well, which now includes up to 788 topics that students can learn. ALEKS is able to identify what students know and don't know upon entering the course and puts them on a personalized path to success that is aligned with the material you include in the course.
- **Excel Support:** The consistent integration of technology activities is supported by student resources. Author-created video tutorials walk students through all new Excel skills learned in the course, and Excel templates help students get started on Technology assignments.
- **Unit Exam reviews**
- **eBook (rich in media features)**

**Digital homework and practice, along with all resources, are available through ConnectMath Hosted by ALEKS and ALEKS*

Lesson 1-5 A Coordinated Effort 69

1-5 Portfolio

Name ______________________________

Check each box when you've completed the task. Remember that your instructor will want you to turn in the portfolio pages you create.

Technology

1. ☐ Use Excel to create two different scatter diagrams for the ordered pairs in the table on page 71. The first should just have the points; the second should connect the points with curves. A template to help you get started can be found in the online resources for this lesson.

Online Practice

1. ☐ Include any written work from the online assignment along with any notes or questions about this lesson's content.

Applications

1. ☐ Complete the Applications problems.

Reflections

Type a short answer to each question.

1. ☐ If someone says that the point of graphing is plotting points and connecting the dots, how would you explain to him how very, very wrong he is? It'll be tough, but try to be nice.
2. ☐ Why do you think we use the word "ordered" in "ordered pair"?
3. ☐ Explain the advantages of graphed data over data in table form.
4. ☐ Name one thing you learned or discovered in this lesson that you found particularly interesting.
5. ☐ What questions do you have about this lesson?

Looking Ahead

1. ☐ Complete the Prep Skills for Lesson 1-6.
2. ☐ Read the opening paragraph in Lesson 1-6 carefully and answer Question 0 in preparation for that lesson.

70 Unit 1 Numeric Data: Visualize and Organize

Answers to "Did You Get It?"

1.

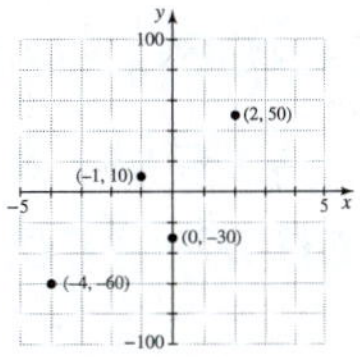

2. It looks a little above \$3, maybe \$3.10 or so. This isn't as precise as the \$3.09 we can get from the table.

3.

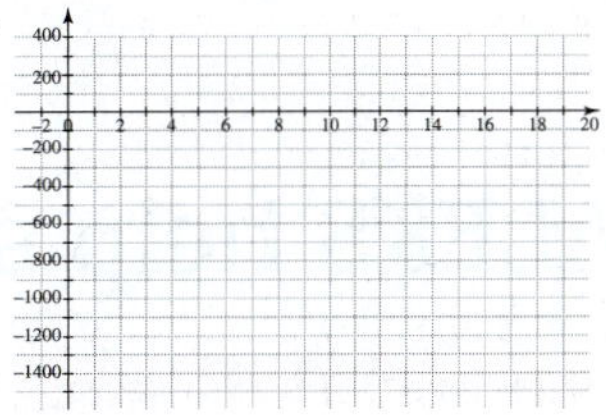

Answers to Prep Skills

1. **a.** 12–19 **b.** Anyone in the 35–44 range

   c. From ages 12–19 to 20–34 the rate increases, but then the rate decreases as age goes up.

2.

Hours since 12 P.M.	0	1	2	3	4	5	6	7	8

3. **a.** / **b.**

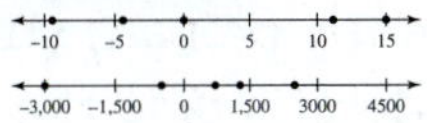

The improved 2-page portfolio in each lesson helps reinforce student learning.

Looking to motivate and engage students? Problem solved!

ALEKS® uses artificial intelligence to precisely map what each student knows, doesn't know, and is most ready to learn in a given course area. The system interacts with each student like a skilled human tutor, delivering a cycle of highly individualized learning and assessment that ensures mastery. Students are shown an optimal path to success, and instructors have the analytics they need to deliver a data-informed, impactful learning experience.

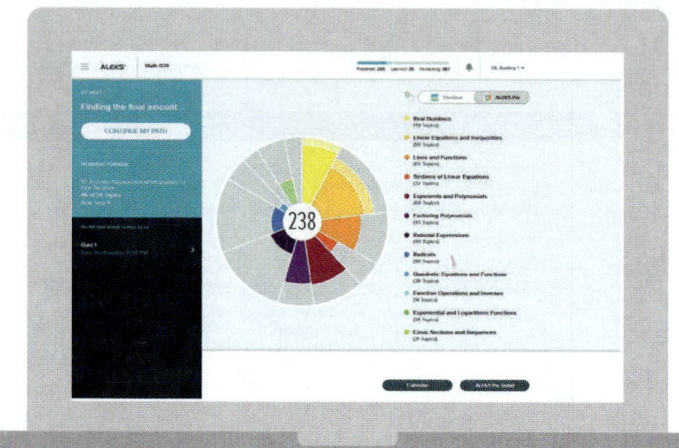

> "ALEKS has helped to create the best classroom experience I've had in 36 years. I can reach each student with ALEKS."
> – *Tommy Thompson, Cedar Valley College, TX*

How can ALEKS help solve your students' challenges?

I did all my homework, so why am I failing my exams?

The purpose of homework is to ensure mastery and prepare students for exams. ALEKS is the only adaptive learning system that ensures mastery through periodic assessments and delivers just-in-time remediation to efficiently prepare students. Because of how ALEKS presents lessons and practice, students learn by understanding the core principle of a concept rather than just memorizing a process.

I'm too far behind to catch up. - OR - I've already done this, I'm bored.

No two students are alike. So why start everyone on the same page? ALEKS diagnoses what each student knows and doesn't know, and prescribes an optimized learning path through your curriculum. Students only work on topics they are ready to learn, and they have a proven learning success rate of 93% or higher. As students watch their progress in the ALEKS Pie grow, their confidence grows with it.

How can ALEKS help solve your classroom challenges?

I need something that solves the problem of cost, time to completion, and student preparedness.

ALEKS is the perfect solution to these common problems. It provides an efficient path to mastery through its individualized cycle of learning and assessment. Students move through the course content more efficiently and are better prepared for subsequent courses. This saves both the institution and the student money. Increased student success means more students graduate.

My administration and department measure success differently. How can we compare notes?

ALEKS offers the most comprehensive and detailed data analytics on the market. From helping the student in the back row to monitoring pass rates across the department and institution, ALEKS delivers the data needed at all levels.

The customizable and intuitive reporting features allow you and your colleagues to easily gather, interpret, and share the data you need, when you need it.

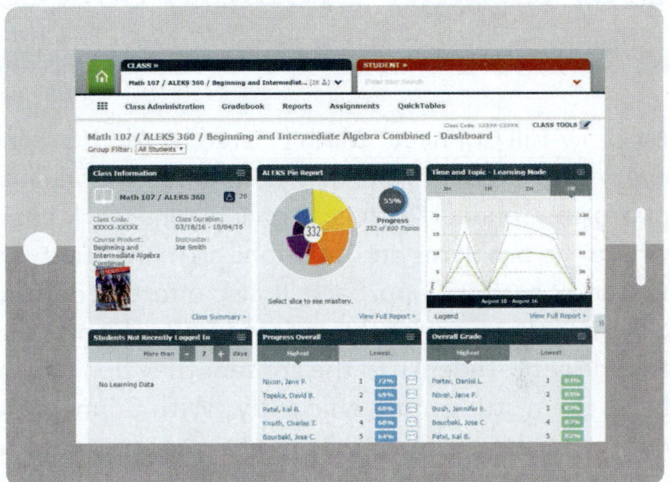

The ALEKS Instructor Module offers a modern, intuitive interface to manage courses and track student progress.

Trusted Service and Support

A unique solution requires unique support. A dedicated team of specialists and faculty consultants ensure that your ALEKS implementation is seamless and painless . . . from start to finish.

ALEKS Service and Support Offers:

- LMS integration that provides single sign-on capability and gradebook synchronization
- Industry-leading technical support and 99.97% uptime
- Flexible courses that can align with any textbook and/or resources, for any classroom model
- Resources for implementation, first day of class orientation, how-to videos and more
- Onsite seminars/worshops and webinars with McGraw-Hill and faculty consultants

www.aleks.com/highered

Looking for a consistent voice between text and digital? Problem solved!

McGraw-Hill Connect® Math Hosted by ALEKS® offers everything students and instructors need in one intuitive platform. ConnectMath is an online homework engine where the problems and solutions are consistent with the textbook authors' approach. It also offers integration with SmartBook, an assignable, adaptive eBook and study tool that directs students to the content they don't know and helps them study more efficiently. With ConnectMath, you get the tools you need to be the teacher you want to be.

©Steve Debenport/Getty Images

> "I like that ConnectMath reaches students with different learning styles . . . our students are engaged, attend class, and ask excellent questions."
> *– Kelly Davis, South Texas College*

Trusted Service and Support

A dedicated team of specialists and faculty consultants ensure that your ConnectMath implementation is seamless and painless . . . from start to finish.

ConnectMath Service and Support Offers:

- LMS integration that provides single sign-on capability and gradebook synchronization
- Industry-leading technical support and 99.97% uptime
- Resources for implementation, first day of class orientation, how-to videos and more
- Onsite seminars/worshops and webinars with McGraw-Hill and faculty consultants

How can ConnectMath help solve your students' challenges?

I like to learn by __________.

Whether it's reading, watching, discovering, or doing, ConnectMath has something for everyone. Instructors can create assignments that accommodate different learning styles, and students aren't stuck with boring multiple-choice problems. Instead they have a myriad of motivational learning and media resources at their fingertips. SmartBook delivers an interactive reading and learning experience that provides personalized guidance and just-in-time remediation. This helps students to focus on what they need, right when they need it.

I still don't get it. Can you do that problem again?

Because the content in ConnectMath is author-developed and goes through a rigorous quality control process, students hear one voice, one style, and don't get lost moving from text to digital. The high-quality, author-developed videos provide students ample opportunities to master concepts and practice skills that they need extra help with . . . all of which are integrated in the ConnectMath platform and the eBook.

How can ConnectMath help solve your classroom challenges?

I need meaningful data to measure student success!

From helping the student in the back row to tracking learning trends for your entire course, ConnectMath delivers the data you need to make an impactful, meaningful learning experience for students. With easy-to-interpret, downloadable reports, you can analyze learning rates for each assignment, monitor time on task, and learn where students' strengths and weaknesses are in each course area.

We're going with the ________ (flipped classroom, corequisite model, etc.) implementation.

ConnectMath can be used in any course setup. Each course in ConnectMath comes complete with its own set of text-specific assignments, author-developed videos and learning resources, and an integrated eBook that cater to the needs of your specific course. The easy-to-navigate home page keeps the learning curve at a minimum, but we still offer an abundance of tutorials and videos to help get you and your colleagues started.

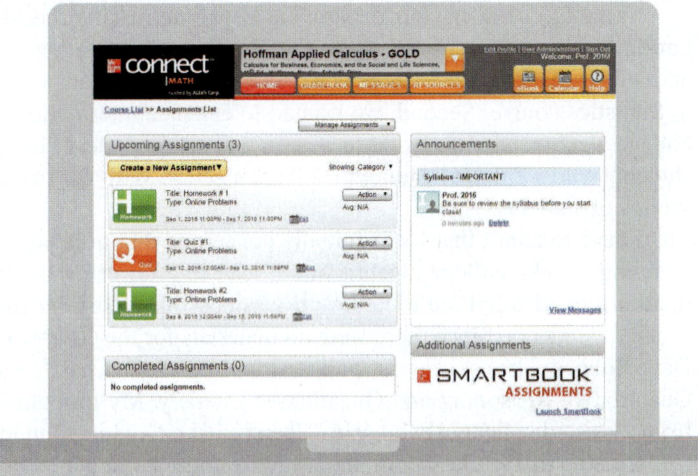

About the Authors

Dave Sobecki

©McGraw-Hill Education

I was born and raised in Cleveland, and started college at Bowling Green State University majoring in creative writing. Eleven years later, I walked across the graduation stage to receive a PhD in math, a strange journey indeed. After two years at Franklin and Marshall College in Pennsylvania, I came home to Ohio, accepting a tenure-track job at the Hamilton campus of Miami University. I've won a number of teaching awards in my career, and while maintaining an active teaching schedule, I now spend an inordinate amount of time writing textbooks and course materials. I've written or co-authored either nine or sixteen textbooks, depending on how you count them, as well as a wide variety of solutions manuals and interactive CD-ROMS. I've also worked on an awful lot of the digital content that accompanies my texts, including Connect, LearnSmart, and Instructional videos.

I'm in a very happy place right now: my love of teaching meshes perfectly with my childhood dream of writing. (Don't tell my publisher this—they think I spend 20 hours a day working on textbooks—but I'm working on my first novel in the limited spare time that I have.) I'm also a former coordinator of Ohio Project NExT, as I believe very strongly in helping young college instructors focus on high-quality teaching as a primary career goal. I live in Fairfield, Ohio, with my lovely wife Cat and fuzzy dogs Macleod and Tessa. I'm a recovering sports fan, still rooting for Ohio State and the Cleveland teams in a saner manner. Other passions include heavy metal music, travel, golf, collecting fine art, and visiting local breweries.

Brian Mercer

©McGraw-Hill Education

I can say without a doubt that I was made to be in a classroom. I followed the footsteps of my father, a 35-year middle school math teaching veteran, into this challenging yet rewarding career. My college experience began as a community college student at Lakeland College in Mattoon, Illinois. From there, I received a Bachelor of Science in Mathematics from Eastern Illinois University and a Master of Science in Mathematics from Southern Illinois University. I accepted a tenure-track faculty position at Parkland College, where I have taught developmental and college-level courses for 20 years. I had the opportunity to begin writing textbooks shortly after I started teaching at Parkland. My then department chair and mentor, James W. Hall, and I co-authored several textbooks in Beginning and Intermediate Algebra.

In the fall of 2011, our department began discussing the idea of creating two tracks through our developmental math sequence. The idea stemmed from two issues. First, most of our beginning and intermediate algebra students were headed to either our Liberal Arts Math or our Introduction to Statistics course. Second, we wanted to enhance intermediate algebra to better prepare those students who were headed to college algebra. Obviously, these were two competing ideas! Increasing the algebraic rigor of these courses seemed to "punish" students who were not heading to college algebra. With the two track system, we implemented a solution that best serves both groups of students.

I have to admit that I was initially concerned that offering an alternate path through developmental math for students not planning to take college algebra would lead to a lowering of standards. However, my participation in our committee investigating this idea led me to believe it was possible to offer a rigorous course that was exceedingly more appropriate for this group of students. Since there were no materials for the course, I began creating my own and was paired by McGraw-Hill with Dave Sobecki. The successful partnership that resulted has led to two editions of this book, along with follow-up projects in Quantitative Reasoning and Quantitative Literacy. My thoughts on all three of these projects are summed up by a comment from a trusted colleague who said, "This is just a long overdue idea."

Outside of the classroom and away from the computer, I am kept educated, entertained and ever-busy by my wonderful wife Nikki, and our two children, Charlotte, 10 and Jake, 9. I am an avid St. Louis Cardinals fan and enjoy playing recreational softball and golf in the summertime with colleagues and friends.

Acknowledgments

First, we'd like to thank the following individuals who reviewed or class tested *Pathways to Math Literacy*. Some of you helped make the original version great, others among you contributed to making the second edition even better, and many of you have been helping to enhance *Pathways* along every step of the journey from the very beginning.

Maria Andersen, CEO, Coursetune, Inc.
Amber Anderson, Danville Area Community College
Rachel Anschicks, College of DuPage
Abbey Auxter, Community College of Philadelphia
Jack Bennett, Ventura College
Gale Brewer, Amarillo College
Andrea Buettner, Hennepin Technical College
Cindy Burns, McLennan Community College
Elizabeth Cannis, Pasadena City College
Edie Carter, Amarillo College
Dorothy Marie Carver, Portland Community College
Lori Chapman, Macomb Community College
Mark Chapman, Lansing Community College
Carl Clark, Indian River State College
Brandi Cline, Lone Star College—Tomball
James Condor, State College of Florida
Trey Cox, Chandler-Gilbert Community College
Awilda Delgado, Broward College
Nicole Duvernay, Spokane Community College
Hope Essien, Malcolm X College
Asha Hill, Georgia Highlands College
Jennifer Hill, College of DuPage
Linda Hintzman, Pasadena City College
Brandon Huff, Lewis and Clark Community College
Laura Iossi, Broward College
Gretta Johnson, Amarillo College
Gizem Karaali, Pomona College
Michelle Kershner, Elgin Community College
Taylor Kilman, Indian River State College
Tamela Kostos, McHenry County College
Gayle Krzemien, Pikes Peak Community College
Brian Leonard, Southwestern Michigan College
Christine Mac, Front Range Community College
Caren McClure, Santa Ana College
David Miller, Black Hawk College
Faith Miller, Macomb Community College
Melissa Morgan, Waubonsee Community College
Catherine Moushon, Elgin Community College
Nicole Munden, Lewis and Clark Community College
Bette Nelson, Alvin College
Peter Nodzenski, Black Hawk College
Dawn Peterson, Illinois Central College
Karey Pharris, Pikes Peak Community College
Pat Rhodes, Treasure Valley Community College
Lisa Rombes, Washtenaw Community College
Martin Romero, Santa Ana College
Doug Roth, Pikes Peak Community College
Jack Rotman, Lansing Community College
Cynthia Schultz, Illinois Valley Community College
Jo Lynn Marie Sedgwick, Waubonsee Community College
Mary Sheppard, Malcolm X College
Nigie Shi, Bakersfield College
Craig Slocum, Moraine Valley Community College
Lindsey Small, Pikes Peak Community College
Robin Stutzman, Southwestern Michigan College
Kelly Thannum, Illinois Central College
Ria Thomas, Southwestern Michigan College
Cassonda Thompson, York Technical College
Diane Veneziale, Rowan College at Burlington County
Carol Weideman, St. Petersburg College
Karen White, Amarillo College
Erin Wilding-Martin, Parkland College

Much of the information we needed to bring this vision to life was provided through a variety of focus groups. So next, we thank our focus group participants, whose valuable insights helped focus our efforts. Guess that's why they call them focus groups.

Patricia Anderson, Arapahoe Community College
Beth Barnett, Columbus State Community College
Ratan Barua, Miami Dade College
Jonathan Brucks, University of Texas at San Antonio
Robert Cantin, MassBay Community College
Billye Cheek, Grayson College
Diana Coatney, Clark College
Mahshid Hassani, Hillsbourough Community College

Jessica Lickeri, Columbus State Community College
Rita Lindsay, Indian River State College
Faun Maddux, West Valley College
Tanya Madrigal, San Jacinto College
Teri Miller, Clark College
Jeff Morford, Henry Ford Community College
Arumugam Muhundan, State College of Florida
Bill Parker, Greenville Technical College
Betty Peterson, Mercer Community College
Paul Stephen Prueitt, Atlanta Metro State College
Leslie Sterrett, Indian River State College
Pat Rhodes, Treasure Valley Community College
Wendy Pogoda, Hillsborough CC, South Shore
Saisnath Rickhi, Miami Dade College
Cynthia Roemer, Union County College
Arlene Rogoff, Union County College
Mark Roland, Dutchess Community College
Jorge Sarmiento, County College of Morris
Cathy Schnakenburg, Arapahoe Community College
Pat Suess, St. Louis Community College
Mel Taylor, Ridgewater Community College
Robyn Toman, Anne Arundel Community College
LuAnn Walton, San Juan College
Keith White, Utah Valley University
Valerie Whitmore, Central Wyoming College
Latrica Williams, St. Petersburg College
Mina Yavari, Allan Hancock College

We'd also like to take a moment to thank some of the people at McGraw-Hill Education who helped to make this new edition of *Pathways* a reality and get out the good word about it.

Kathleen McMahon—Managing Director, Math & Physical Sciences
Caroline Celano—Director of Mathematics
Kim Moreno—Director, ALEKS Implementation
Tami Hodge—Director of Marketing
Noah Evans—Marketing Manager, Developmental Mathematics & Non-STEM Pathways
Annie Clarke—Marketing Coordinator
Mary Ellen Rahn—Market Development Manager, Mathematics & Statistics
Robin Reed—Lead Product Developer
Luke Whalen—Product Developer, Developmental Mathematics
Marisa Dobbeleare—Product Development Coordinator
Megan Platt—Product Development Coordinator
Cynthia Northrup—Director of Digital Content, Mathematics
Adam Fischer—Digital Product Analyst
Ruth Czarnecki-Lichstein—Digital Product Analyst
Lora Neyens—Program Manager
Peggy Selle—Content Project Manager (Core)
Rachael Hillebrand—Content Project Manager (Assessment)
Tara McDermott—Designer
Shannon Manderscheid—Content Licensing Specialist

And finally, a few additional people merit above and beyond thanks:

Erin Wilding-Martin, who contributed to this product in so many ways that I'm not even sure she can count all of them.

Jack Rotman, a national leader in the Pathways movement, whose thoughtful and in-depth reviews went about eight miles above and beyond.

Jim Hall, who's almost solely responsible for getting Brian into the wild and crazy world of textbook writing.

The students of Parkland College, who contributed an incredible amount of insight by providing feedback on the earliest incarnations of this book.

Cat Sobecki and Nikki Mercer, our wives. We still can't figure out how two math nerds managed to snag these two beauties.

Detailed Table of Contents

Unit 1
Numeric Data: Visualize and Organize

Outline

Lesson 1-1 Where Does the Time Go?

LEARNING OBJECTIVES

- ☐ 1. Analyze personal time management for a week of activities.
- ☐ 2. Solve problems involving percentages.
- ☐ 3. Create and interpret pie charts.
- ☐ 4. Create and interpret bar graphs.

Time is what we want most, but what we use worst.
—William Penn

©Corbis/agefotostock RF

One of the most important aspects of success in college is very underrated: learning how to manage time. This is a skill that you're unlikely to acquire by chance: In order to understand how to use your time most effectively, you have to first become aware of how you're using your time. Then you'll have to develop a plan that will help you best take advantage of your valuable time. This can help you to do better in your classes, and can also help you to have more time to do the things you enjoy.

0. Do you think you do a good job of budgeting your time? Explain.

1-1 Class

Here's a sample time chart put together by a student who was interested in identifying the amount of time she spent for one week. Each hour is marked with C (time in class), H (time spent on homework), S (sleep), W (work), or O (other time commitments). In the Portfolio portion of this lesson, you'll be asked to fill out a similar time chart of your own. For now, we'll analyze this chart.

	Sunday	Monday	Tuesday	Wednesday	Thursday	Friday	Saturday
12am–1:00	O	S	S	S	S	S	O
1:00–2:00	O	S	S	S	S	S	O
2:00–3:00	S	S	S	S	S	S	S
3:00–4:00	S	S	S	S	S	S	S
4:00–5:00	S	S	S	S	S	S	S
5:00–6:00	S	S	S	S	S	S	S
6:00–7:00	S	S	O	S	O	S	S
7:00–8:00	S	S	W	S	W	S	S
8:00–9:00	S	O	W	O	W	O	S
9:00–10:00	O	C	W	C	W	C	S
10:00–11:00	O	C	W	C	W	C	S
11:00–Noon	H	H	H	H	H	H	H
Noon–1:00	H	O	C	O	C	O	H
1:00–2:00	W	C	C	C	C	C	O
2:00–3:00	W	H	C	H	C	H	O
3:00–4:00	W	O	H	O	H	O	O
4:00–5:00	W	O	H	O	H	O	H
5:00–6:00	W	O	O	O	O	O	H
6:00–7:00	W	O	O	O	O	O	H
7:00–8:00	W	H	H	H	H	O	O
8:00–9:00	H	H	H	H	H	O	O
9:00–10:00	H	O	H	O	H	O	O
10:00–11:00	H	O	O	O	O	O	O
11:00–12am	O	S	O	S	O	O	O

1. Count the number of spaces containing each letter in the time chart (C, H, S, W, O). Use the results to fill in the following chart, writing the total number of hours devoted to each activity during one week.

	Hours
Class	
Homework	
Sleep	
Work	
Other	

2. Without adding the hours in the table, how can you decide what the total number of hours should be? Do your values from the table add to the right number of hours?

3. Most college advisors will tell you that a good rule of thumb is to allow 2 hours of study time outside of class for each hour spent in class. Is the student that filled out the table in Question 1 following that advice?

One really good way to analyze the amount of time spent on each activity is to compute percentages. The word "percent" literally means "per hundred." So if you found that you spent 42% of your time playing a trendy new game on your phone, that would mean (1) you should consider psychological help, and (2) for every 100 hours of time, you spend 42 of them playing the game.

4. Write a fraction with the number of hours spent by our hypothetical student in class in the numerator, and the total number of hours in the denominator. You'll find those numbers in your answers to Questions 1 and 2.

5. To convert your fraction from Question 4 to a percent, first use a calculator to perform the division. This gives you a percentage in decimal form. To convert to percent form, move the decimal point two places to the right. The result is the percentage of time that particular student spent in class in one week.

6. Use the steps described in Questions 4 and 5 to fill in the next chart with the percentage of hours in a week devoted to each activity. (Round to the nearest percent.)

	Percentage
Class	
Homework	
Sleep	
Work	
Other	

Math Note

If you need more practice on working with percentages, have a look at the review box on page 7, and the online prep resources for this lesson.

7. Studies have shown that college students get about 6½ hours of sleep per night on average, but that academic achievement improves when students average 8 hours of sleep. What percentage of the hours in a week would be devoted to sleep if you get 6½ hours of sleep per night? What if you get 8 hours? How do you feel our friend from Question 1 is doing when it comes to sleep? Explain your answer.

Did You Get It

Try this problem to see if you understand the concepts we just studied. The answers can be found at the end of the Portfolio section.

1. Of 118 new enrollees in a college nursing program, 68 are planning on pursuing a four-year degree, 32 a two-year degree, and the rest are undecided. Find the percentage of enrollees that fall into each category.

8. A circle can be divided into 360 equal units of measure, which we call **degrees**. In other words, an entire circle is made up of 360° (the ° symbol represents degrees). Multiply the decimal form of each percentage in Question 6 by 360°, and put the results into the next chart. Round to the nearest whole degree.

	Degrees
Class	
Homework	
Sleep	
Work	
Other	

How Often Do You Wash Your Hands After Using a Public Restroom?

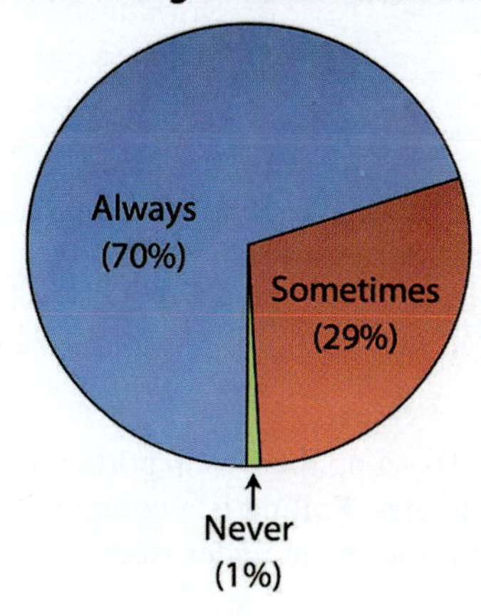

A **pie chart** is a diagram used to compare the relative sizes of different parts of a whole. For example, the pie chart to the right, adapted from a *USA Today* survey, describes people's hand-washing tendencies when using a public restroom. In this case, the whole is the number of people surveyed, and each of the parts is the percentage of folks that gave a certain response. Since 70% of respondents said "Always," the "Always" category fills 70% of the circle. This represents $0.7 \times 360° = 252°$. (See page 7 for a quick review of percents.) The "Sometimes" category fills 29%, which corresponds to $0.29 \times 360° \approx 104°$, and the "Never" category just 1%, corresponding to $0.01 \times 360° \approx 4°$.

Now let's see if we can build a pie chart that illustrates the amount of time spent on various activities for the student survey that began this lesson.

9. Build a pie chart for the information in Question 8 by marking off angles on the circle to the right, starting at the 0° mark, that correspond to the numbers of degrees in Question 8. In this case, the "whole" refers to the total number of hours in a week, and each part is the portion of that time spent on one of the activities. Each of the light gray lines on the graph represents 10°.

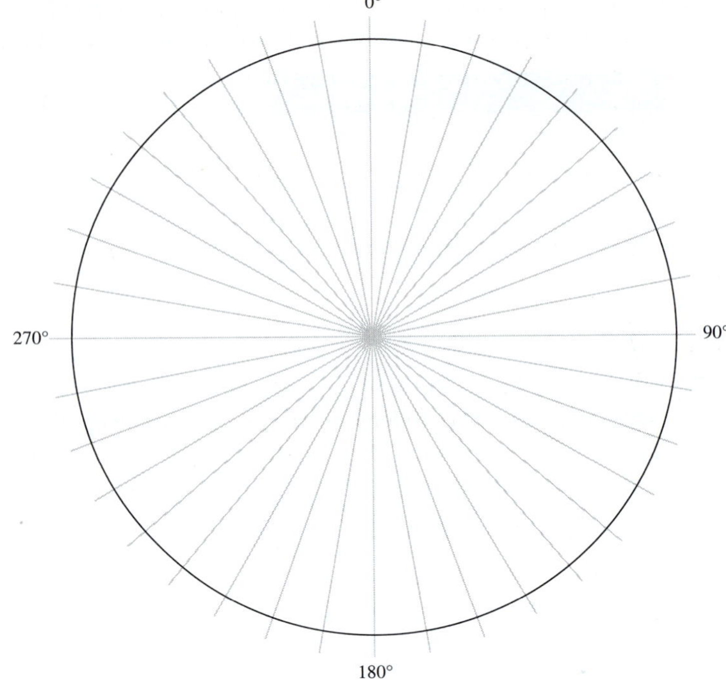

10. Does the pie chart make it easy to analyze the amount of time this student spends on each activity? Explain.

Did You Get It

2. Make a pie chart using the data provided in Did You Get It 1.

Drawing a Pie Chart

1. Find the percentages that fall into each category among the data your chart will describe.
2. Multiply those percentages by 360° to find the number of degrees that needs to be covered by each portion of the chart.
3. Draw an initial boundary line from the center of the chart to the 0° mark. Then, starting at the 0° mark, measure the number of degrees corresponding to your first portion, then draw a boundary. Label this first portion of your chart.
4. Starting at this new boundary, measure the number of degrees corresponding to your second portion, draw in a new boundary, and label that portion.
5. Continue until all categories have been accounted for.

To set up the Group portion of this lesson, we're going to conduct a short class poll to help you learn a little bit about your classmates. Polling is a common and useful way to gather some **data**, which are measurements or observations that are gathered for an event under study.

If we wanted to learn about characteristics of all the students at your college, the ideal approach would be to poll every single one of them. But in most cases, that's not particularly realistic. So instead, we'd likely choose a **sample** of students from the larger **population** of all students at your school. In this case, the sample will be the students in your class. As we learn about statistical studies in this course, we'll need to think about whether or not a chosen sample of subjects provides a good representation of the overall population we're wondering about.

To gather some data, everyone in your class should respond to the two poll questions below. You'll compile and analyze the results in the Group portion of the lesson.

11. Do you think the sample for this poll (students in your class) would or would not provide a good representation of the population in question (all the students at your school)? Explain your reasoning.

How do you feel about math in general?	
I love it	☐
I like it	☐
I can take it or leave it	☐
I don't like it	☐
I hate it	☐

Check all of the boxes that apply to you.	
I live with a parent.	☐
I work and go to school.	☐
I have children.	☐
I'm taking more than 10 credit hours.	☐
I started college right after high school.	☐
At least one of my parents attended college.	☐
I drive to class.	☐
I was born in a year that begins with 1.	☐
I know what my major is.	☐

Math Note

The word "data" is plural, so we say "data are" not "data is." The singular version is "datum."

Review of Percents

1. The word "percent" literally means "per hundred." When we read that 70 percent of respondents to a survey always wash their hands in a public restroom, it means that 70 people per hundred do so.
2. Translating a percent as "per hundred" makes it easy to convert percents to fraction form for using them in calculations: 70% means 70 per hundred, or $\frac{70}{100}$. And there's your fractional form for 70%!
3. This also shows us how to convert percents to decimal form: Dividing any number by 100 moves the decimal two places to the left. That makes 70% equal to 0.70, 5% equal to 0.05, and 23.5% equal to 0.235.
4. If you forget to convert a percent into fractional or decimal form when doing a calculation, you can usually catch your mistake with a bit of thought. In the pie chart on page 5, if we used $70 \times 360°$ instead of $0.7 \times 360°$, we'd have found an angle of 25,200°, which is pretty silly given that a full circle is 360°.

1-1 Group

Numerous studies have shown that one of the best ways to do better in college classes is to study in pairs or groups. To help you get started, if you feel comfortable sharing contact information, exchange the information in the table below. The group you're in now will be your small group for the first unit of this course. When you get used to meeting in class, you'll likely find that meeting outside of class to study and work on homework is a good idea as well, so include some study times that would be convenient for you to meet.

Name	Phone number	Email	Available times

1. Use data from the first class poll on page 7 about what students think about math to complete the table. Then use your results to draw a pie chart comparing the responses.

Feelings toward math	Number	Percent	Degrees
Love it			
Like it			
Take or leave			
Don't like			
Hate it			

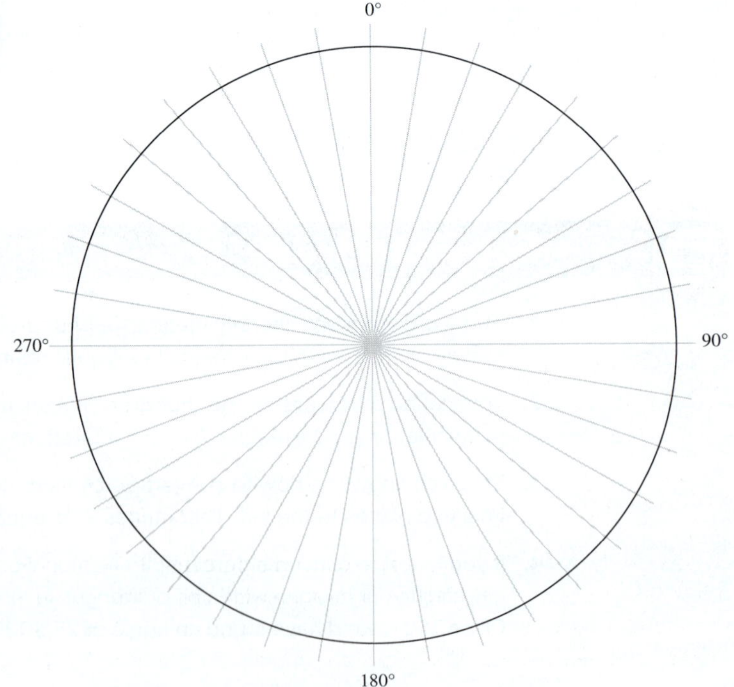

2. Describe how the pie chart allows you to analyze the result of the poll differently than looking at the raw numbers in the chart does.

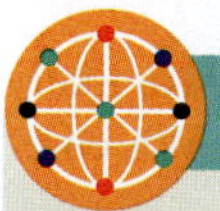

Using Technology: Creating a Pie Chart

To create a pie chart in Excel:

1. Type the category names in one column (or row).
2. Type the category values in the next column (or row).
3. Use the mouse to drag and select all the data in those two columns (or rows).
4. With the appropriate cells selected, click the **Insert** tab, then choose **Chart** and Pie chart. There are a few different styles you can experiment with, but starting with the simplest is a good idea.

You can add titles, change colors and other formatting elements by right-clicking on certain elements, or using the options on the **Charts** menu. Try some options and see what you can learn!

See the Lesson 1-1-1 Using Tech video in class resources for further instruction.

Pie charts are ideal for representing information where a whole amount is divided into distinct categories. Notice that was the case for the poll on attitudes toward math; every individual could choose only one of the responses. This is reflected in the fact that the percentages for each response add up to 100%. But the second poll is different: It's likely that many respondents will choose more than one response.

3. Use the next table to tabulate the results of that second poll question. Round to the nearest percent again.

Response	Number	Percent
Live with parent		
Work and school		
Have children		
More than 10 hours		
Straight to college		
A parent went to college		
Drive to class		
Born before 2000		
Know major		

As we expected, the percentages add up to far more than 100%, so a pie chart would be a really bad choice for illustrating these data. Instead, we'll use a different way to visualize the data. A **bar graph** is a visual way to compare the sizes of different values. Pie charts are great for comparing parts to a whole; bar graphs are effective for comparing parts to other parts. In a bar graph, the length of each bar gives an indication of how the size of a quantity compares to other related quantities.

Let's look at an example first, then get back to our class polling data. In a telephone poll conducted in October 2016, likely voters were asked two questions about the four candidates for U.S. president: Who are you planning to vote for? and Which of the candidates do you think are qualified to be president? The results are summarized in the following two graphs.

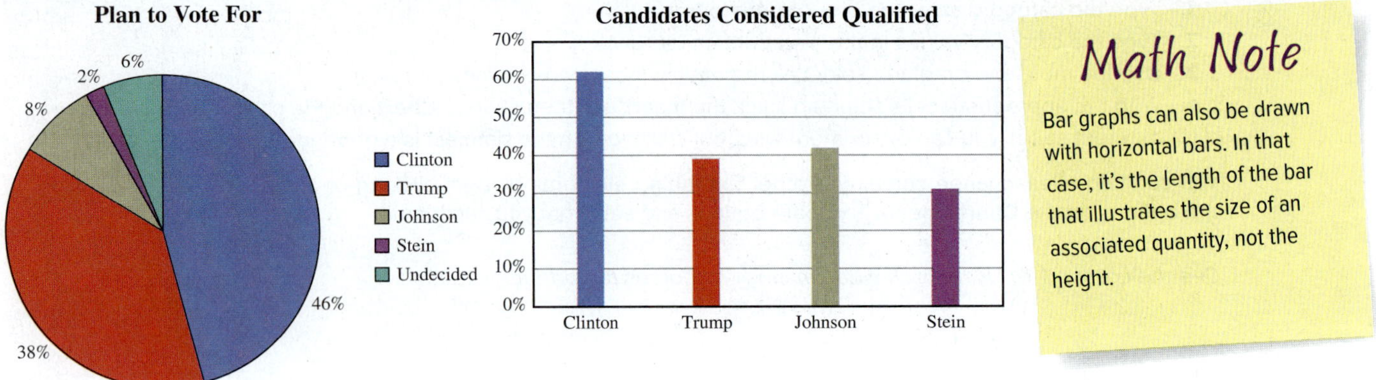

Math Note

Bar graphs can also be drawn with horizontal bars. In that case, it's the length of the bar that illustrates the size of an associated quantity, not the height.

4. Why did we choose a pie chart for the first question and a bar graph for the second?

5. Use the information in the graphs to fill in the table here.

Candidate	Plan to vote for	Qualified
Clinton		
Trump		
Johnson		
Stein		
Undecided		

6. Illustrate the results of the second class poll question with a bar graph. Don't forget to label the responses along the bottom, and label appropriate heights along the left side.

Drawing a Bar Graph

1. Label the name of each category in your data, equally spaced along the bottom of your graph (or along the left if you want to draw horizontal bars).
2. Label an appropriate scale along the left of your graph (or along the bottom for horizontal bars). Make sure your scale starts at zero, and extends out a little bit further than the largest number in your data set.
3. Draw in rectangular bars whose heights or lengths correspond to the number representing each category in your data set.

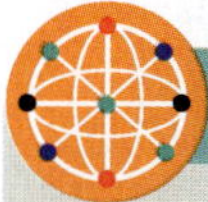

Using Technology: Creating a Bar Graph

To create a bar graph in Excel:

1. Type the category names in one column (or row).
2. Type the category values in the next column (or row).
3. Use the mouse to drag and select all the data in those two columns (or rows).
4. With the appropriate cells selected, click the **Insert** tab, then choose **Chart** and Column or Bar graph. ("Column" gives you vertical bars, "Bar" gives you horizontal.) Again, there are a few different styles you can experiment with, but starting with the simplest is a good idea.

You can add titles, change colors and other formatting elements by right-clicking on certain elements, or using the options on the **Charts** menu. Try some options and see what you can learn!

See the Lesson 1-1-2 Using Tech video in class resources for further information.

7. Write at least three observations you can make from your bar graph. What do you find interesting or notable about what it tells you?

Did You Get It

3. Of the 118 nursing enrollees from Did You Get It 1, 96 are taking a biology class this semester, 59 a math class, 85 an English class, 16 a psychology class, and 38 a phys ed class. Draw a bar graph illustrating this data.

1-1 Portfolio

Name ____________________

Check each box when you've completed the task. Remember that your instructor will want you to turn in the portfolio pages you create.

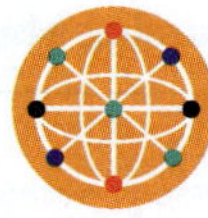

Technology

1. ☐ Fill in a time chart like the one at the beginning of this lesson based on your schedule. Then compile the results and use Excel to create a pie chart and a bar graph showing where your time is spent. For the pie chart, use percentages of time spent on each activity. For the bar graph, use the number of hours spent. There's a sample template in the online resources for this lesson to help you get started. Copy and paste your charts into a Word document, and write a couple of sentences below the charts describing any thoughts you have about what you learned.

Online Practice

1. ☐ Include any written work from the online assignment along with any notes or questions about this lesson's content.

Applications

1. ☐ Complete the Applications problems.

Reflections

Type a short answer to each question.

1. ☐ Do you spend your time efficiently? Explain.
2. ☐ Why are you in college? If you've never thought about a good answer to this question, you certainly should have!
3. ☐ What changes do you plan to make in how you budget your time?
4. ☐ List some benefits of working/studying in groups rather than alone.
5. ☐ Give an example of a survey question that would best be represented by a pie chart, and another that would best be represented by a bar graph. No using any questions from this lesson! Explain your reasoning.
6. ☐ Name one thing you learned or discovered in this lesson that you found particularly interesting.
7. ☐ What questions do you have about this lesson?

Looking Ahead

1. ☐ Complete the Prep Skills for Lesson 1-2.
2. ☐ Read the opening paragraph in Lesson 1-2 carefully and answer Question 0 in preparation for that lesson.

Answers to "Did You Get It?"

1. Four year: 57.6%; two year: 27.1%; undecided: 15.3%

2. **Number of Enrollees**

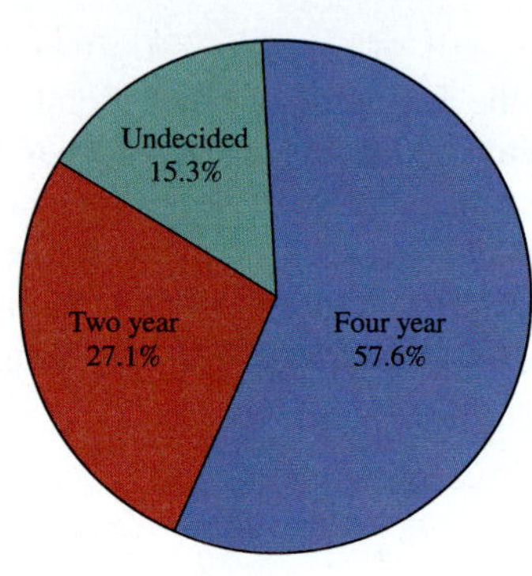

3. **Classes Being Taken by Nursing Majors**

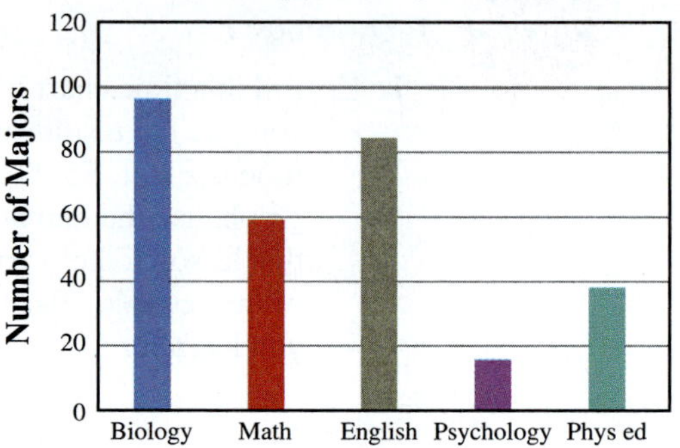

1-1 Applications

Name ______________________________

According to the 2015–2016 National Pet Owners Survey conducted by the American Pet Products Association, 65% of households in the United States have at least one pet. The first table below shows the percentage of households that have each of the most popular pets.

The second table shows the results of a poll done by Public Policy Polling in 2013, in which respondents were asked, "Which of the following exotic animals would you most like to have as a pet: an alligator, a dinosaur, an elephant, a giraffe, a polar bear, or a tiger?"

Percentage of households having various types of pets

Bird	6.1
Cat	42.9
Dog	54.4
Horse	2.5
Fish	13.6
Reptile	4.9

Which exotic animal would you most like to have as a pet?

Alligator	6%
Dinosaur	18%
Elephant	16%
Giraffe	20%
Polar bear	14%
Tiger	26%

1. Which data set is best illustrated with a pie chart? Explain why you feel that way, then draw a pie chart representing the data.

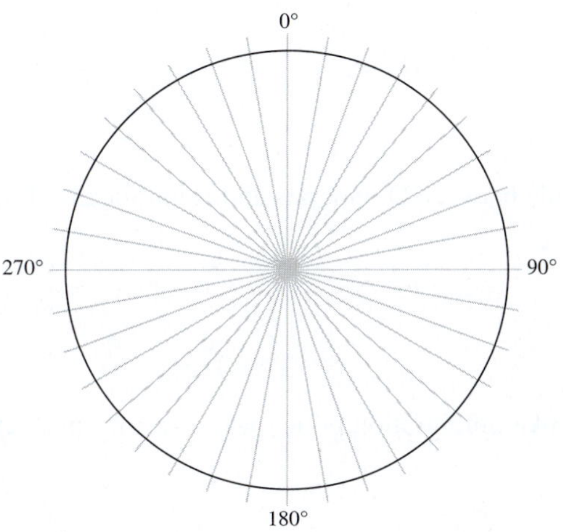

2. Which data set is best illustrated with a bar graph? Explain why you feel that way, then draw a bar graph representing the data.

1-1 Applications

Name ____

In May 2015, the Panetta Institute for Public Policy released the results of an extensive survey of college students. One of the questions was, "Do you think the following issues are or are not serious problems facing our country?" Results were divided by political party. Data for those self-reporting as either Democrat or Republican are illustrated by the following bar graph.

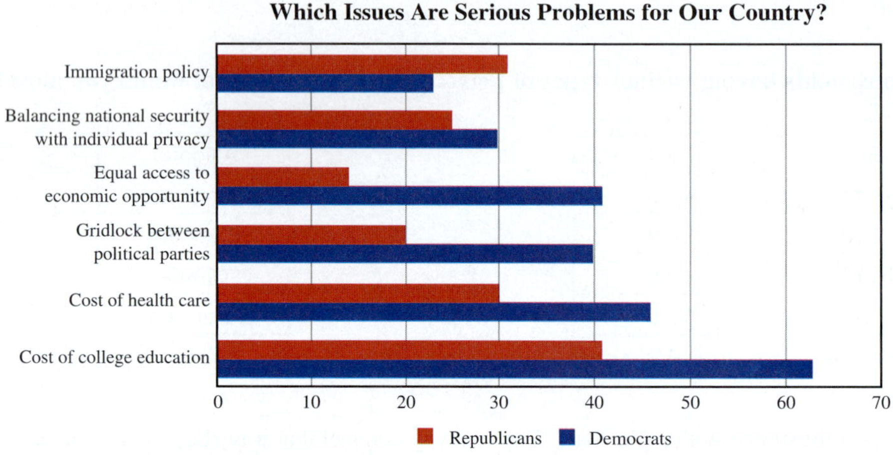

3. What issue was seen as serious by most respondents overall?

4. What issue showed the greatest divide between Democrats and Republicans? Explain how you decided.

5. What percentage of Democrats felt like immigration policy was a serious problem? What about Republicans?

6. Of the two political parties, which would you say had the most positive outlook on the current state of the country? Explain.

1-1 Applications

Name ____

On the show *How I Met Your Mother,* Marshall demonstrated the true versatility of pie charts and bar graphs in the most brilliant way possible: by making a pie chart describing his favorite bars, and a bar graph describing his favorite pies. The charts are reproduced below.

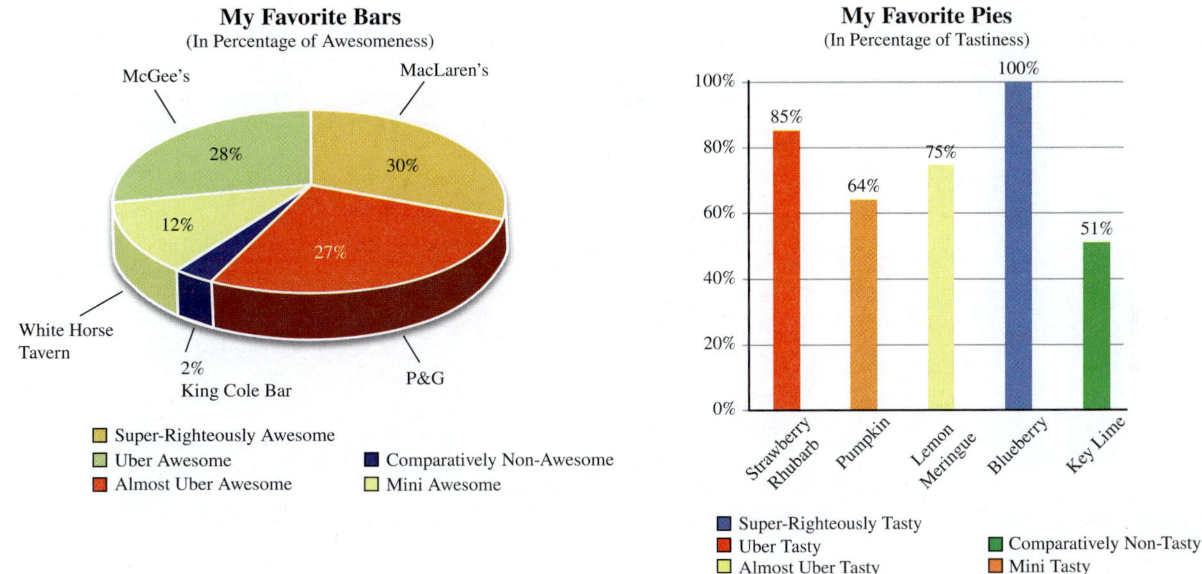

7. Comment on how well you think Marshall did in terms of appropriately using each type of graph based on the data illustrated.

8. What's different about this bar graph that makes it easier to read the data?

9. Look at Marshall's pie chart and try to ignore the percentages for a moment. Which slice appears to be the biggest? Which actually should be?

10. What does your answer to Question 9 tell you about pie charts drawn in a three-dimensional format? Explain.

Lesson 1-2 Prep Skills

SKILL 1: PERFORM ARITHMETIC OPERATIONS ON INTEGERS

- $12 + (-3) = 12 - 3 = 9$ — Adding a negative is the same as subtracting.
- $15 - (-8) = 15 + 8 = 23$ — Subtracting a negative results in addition.
- $12 \times (-5) = -60$ — Multiplying or dividing opposite signs results in a negative.
- $-84 \div (-7) = 12$ — Multiplying or dividing same signs results in a positive.

SKILL 2: ADD OR SUBTRACT FRACTIONS

- $\frac{3}{7} + \frac{2}{7} = \frac{5}{7}$ — If denominators are the same, rewrite the common denominator and add/subtract the numerators.
- $\frac{8}{3} - \frac{2}{3} = \frac{6}{3} = 2$
- $\frac{2}{3} \cdot \frac{4}{4} = \frac{8}{12}$ — To rewrite a fraction with a new denominator, multiply numerator and denominator by the same number.
- $\frac{3}{4} - \frac{5}{6}$ — Find least common denominator: 12 is the smallest number that's a multiple of both 4 and 6.
  $= \frac{9}{12} - \frac{10}{12}$ — Rewrite each fraction with the least common denominator.
  $= -\frac{1}{12}$ — Rewrite common denominator and subtract numerators.

SKILL 3: ADD DECIMALS

- $2.3 + 0.5$
  $= 2.8$ — Add digits in the tenths place to get 8; add digits in the ones place to get 2.
- $3.5 + 5.9$
  $= 9.4$ — Add digits in the tenths place to get 14; carry the 1 over to the ones place; add that 1 to 3 + 5 in the ones place to get 9.

PREP SKILLS QUESTIONS

Perform each operation.

1. a. $12 + 4$ b. $8 + (-10)$ c. $-5 - (-7)$
2. a. 4×6 b. -8×4 c. $(-12) \times (-5)$
3. a. $\frac{8}{5} + \frac{3}{5}$ b. $\frac{2}{9} - \frac{23}{36}$ c. $\frac{1}{3} + \frac{4}{5}$
4. a. $4.8 + 3.1$ b. $0.3 + 1.9$ c. $0.36 + 0.77$

Lesson 1-2 Do You Have Anything to Add?

LEARNING OBJECTIVES

☐ 1. Identify circumstances where addition or subtraction is possible.

☐ 2. Add or subtract quantities.

We are what we repeatedly do. Excellence, therefore, is not an act, but a habit.
—Aristotle

©Comstock RF

We've all heard statistics describing the collective weight problem that we as Americans have. So it's not at all surprising that counting calories has become something of a national pastime. Nutritional information labels are now mandated by law on food packaging, and more and more restaurants are including calorie and fat information on their menus. In order to keep track of what you're putting in your body, you need to be able to add up amounts of calories, fat, carbohydrates, protein, and others. This requires identifying and adding the amounts that correspond to like ingredients. This is a skill that will come in handy throughout this course (and beyond).

0. What mathematical topic do you think we'll cover in this lesson, and what does it have to do with nutrition?

1-2 Class

My coauthor's daughter Charlotte is in fourth grade, and her favorite lunch consists of a turkey sandwich, carrot slices, yogurt, and juice. The nutrition facts for the food in her lunch are given here. Use these facts to answer questions that follow. A sandwich is made from two servings of bread, one and a half servings of turkey, and one serving of cheese. Charlotte also gets a half-serving of carrots and one serving of yogurt along with one juice box.

Wheat Bread

Nutrition Facts
Serving Size 25g

Amount Per Serving	
Calories 66	Calories from Fat 8
	% Daily Value*
Total Fat 1g	1%
Saturated Fat 0g	1%
Trans Fat 0g	
Cholesterol 0mg	0%
Sodium 130mg	5%
Total Carbohydrate 12g	4%
Dietary Fiber 1g	4%
Sugars 1g	
Protein 3g	
Vitamin A 0% • Vitamin C	0%
Calcium 4% • Iron	5%

*Percent Daily values are based on a 2,000 calorie diet. Your daily values may be higher or lower depending on your calorie needs.

Sliced Turkey

Nutrition Facts
Serving Size 33g

Amount Per Serving	
Calories 34	Calories from Fat 5
	% Daily Value*
Total Fat 1g	1%
Saturated Fat 0g	1%
Trans Fat 0g	
Cholesterol 14mg	5%
Sodium 335mg	14%
Total Carbohydrate 1g	0%
Dietary Fiber 0g	1%
Sugars 1g	
Protein 6g	
Vitamin A 0% • Vitamin C	3%
Calcium 0% • Iron	3%

*Percent Daily values are based on a 2,000 calorie diet. Your daily values may be higher or lower depending on your calorie needs.

American Cheese

Nutrition Facts
Serving Size: 1 slice (19g)

Amount Per Serving	
Calories 60	Calories from Fat 40
	% Daily Value*
Total Fat 4g	6%
Saturated Fat 2.5g	12%
Trans Fat 0g	
Cholesterol 15mg	5%
Sodium 200mg	8%
Total Carbohydrate 2g	1%
Dietary Fiber 0g	0%
Sugars 1g	
Protein 3g	
Vitamin A 2% • Vitamin C	0%
Calcium 20% • Iron	0%

*Percent Daily values are based on a 2,000 calorie diet. Your daily values may be higher or lower depending on your calorie needs.

Carrots

Nutrition Facts	
Serving Size 61g	
Amount Per Serving	
Calories 25	Calories from Fat 1
	% Daily Value*
Total Fat 0g	0%
Saturated Fat 0g	0%
Trans Fat 0g	
Cholesterol 0mg	0%
Sodium 42mg	2%
Total Carbohydrate 6g	2%
Dietary Fiber 2g	7%
Sugars 3g	
Protein 1g	
Vitamin A 204% • Vitamin C	6%
Calcium 2% • Iron	1%

*Percent Daily values are based on a 2,000 calorie diet. Your daily values may be higher or lower depending on your calorie needs.

Yogurt

Nutrition Facts	
Serving Size 125g	
Amount Per Serving	
Calories 119	Calories from Fat 2
	% Daily Value*
Total Fat 0g	0%
Saturated Fat 0g	1%
Trans Fat 0g	
Cholesterol 2mg	1%
Sodium 72mg	3%
Total Carbohydrate 24g	8%
Dietary Fiber 0g	0%
Sugars 24g	
Protein 6g	
Vitamin A 0% • Vitamin C	1%
Calcium 19% • Iron	0%

*Percent Daily values are based on a 2,000 calorie diet. Your daily values may be higher or lower depending on your calorie needs.

Juice Box

Nutrition Facts	
Serving Size 1 Juice Box	
Amount Per Serving	
Calories 60	Calories from Fat 0
	% Daily Value*
Total Fat 0g	0%
Saturated Fat 0g	0%
Trans Fat 0g	
Cholesterol 0mg	0%
Sodium 10mg	1%
Total Carbohydrate 15g	5%
Dietary Fiber 0g	0%
Sugars 14g	
Protein 0g	
Vitamin A 10% • Vitamin C	100%
Calcium 10% • Iron	0%

*Percent Daily values are based on a 2,000 calorie diet. Your daily values may be higher or lower depending on your calorie needs.

For Questions 1-4, find the total amount of each in Charlotte's lunch.

1. Total calories

2. Total calories from fat

3. Total milligrams (mg) of sodium

4. Total grams (g) of carbohydrates

5. The Food and Drug Administration recommends limiting sodium to 2,300 mg per day for a person with a 2,000-calorie diet. What percentage of the recommended amount of sodium is Charlotte getting in her lunch? What percentage of a 2,000-calorie diet is contained in her lunch? What can you conclude?

Did You Get It

Try this problem to see if you understand the concepts we just studied. The answers can be found at the end of the Portfolio section.

1. How much dietary fiber is Charlotte getting in her lunch? What percentage of the recommended daily allowance of Vitamin C is she getting?

6. If we use the letter P to stand for the phrase "grams of protein," what is the result of the calculation below, and what is the significance of the answer?

$$6P + 9P + 3P + \frac{1}{2}P + 6P + 0P$$

7. Write the number of grams of sugar in Charlotte's yogurt, and the number of milligrams of cholesterol. What happens if you put a plus sign between those two quantities? Can you add them?

8. Suppose that in addition to the amount of carbohydrates in Charlotte's lunch, she also ate some of her friend's mini-pretzels, which added another 3,000 mg of carbs to her intake. At that point, she would have eaten 69.5 g + 3,000 mg of carbs. Can you perform that addition? Why or why not?

9. There are 1,000 milligrams in one gram. Change 3,000 mg into grams. Now can you add up the total number of carbs? If so, do it.

When Can You Add Two Quantities?

Two quantities can only be added when they are **like quantities**: that is, they represent a certain number of the same thing. For example, 3 g of protein and 8 g of protein can be added because each is a number of grams of protein, but 69.5 g of carbs and 3,000 mg of carbs can't be added because grams and milligrams are different things. In some cases, you can convert one quantity to a different unit in order to make the quantities like: 69.5 g + 3,000 mg can be written as 69.5 g + 3 g. In other cases, you simply can't add.

1-2 Group

Here are some addition problems for you to work out. Some can be done very quickly; for some you'll need to do some rewriting; others can't be done at all. Perform the additions that you can. For those that can't be done, explain why.

1. 12 g cholesterol + 5 g cholesterol

2. $\frac{3}{11} + \frac{7}{11}$

3. 8 yd + 2 ft

4. 10 g sodium + 17 g carbs

5. $5 + 75¢

6. $8x + 5x$

7. $\frac{2}{5} + \frac{7}{20}$

8. 5 days + 5 hours

9. $3y + 7y^2$

10. 0.5 + 0.4

Math Note

You probably recognize addition problems like Questions 6 and 9 from the study of **algebra**, where we use symbols (often letters) called **variables** to represent quantities that can VARY.

Did You Get It

2. Perform each addition, if possible.

   a. $\frac{2}{3} + \frac{5}{3}$

   b. 125 ft + 230 ft

   c. 6 ft + 18 in.

   d. 12 days + 19 mi

A bank statement gives you a detailed look at the activity in your account over a period of time. All you need to know to understand a bank statement is a few words of vocabulary, and how to add and subtract positive and negative numbers. Use the sample statement below to answer each question. (Amounts in parentheses are negative.)

Bank Statement

Statement date		
5/17/17	**Previous balance**	$2,358.25
	11 debits	($1,121.17)
	4 credits	$1,215.20
	Ending balance	$2,452.28

Date	Description	Debits	Credits	Balance
4/17/17	Rent	($400.00)		$1,958.25
4/19/17	Utilities	($120.39)		$1,837.86
4/20/17	Paycheck		$620.34	$2,458.20
4/23/17	Groceries	($79.24)		$2,378.96
4/24/17	New clothes	($93.18)		$2,285.78
4/26/17	Return shirt		$22.01	$2,307.79
5/2/17	Cell phone bill	($78.44)		$2,229.35
5/4/17	Pizza	($11.02)		$2,218.33
5/4/17	Paycheck		$572.29	$2,790.62
5/8/17	Withdraw cash	($200.00)		$2,590.62
5/12/17	Check #278	($40.00)		$2,550.62
5/13/17	Groceries	($48.23)		$2,502.39
5/15/17	Gas	($38.68)		$2,463.71
5/16/17	Netflix streaming fee	($11.99)		$2,451.72
5/17/17	Interest		$0.56	$2,452.28

11. What does a <u>credit</u> refer to?

12. What does a <u>debit</u> refer to?

13. What is meant by a <u>balance</u>?

14. How did this account holder do this month if her goal was to spend less than she made?

15. What would the ending balance be with an additional debit of $50?

16. What would the ending balance have been if the account holder hadn't returned the shirt on 4/26?

17. What would the ending balance be if we remove the $200 cash withdrawal from 5/8?

Did You Get It

3. For each transaction, find what the ending balance would have been had that single transaction taken place. Think of these as three separate questions.
   a. The account holder deposited a birthday check from her mom in the amount of $50.
   b. The account holder withdrew $40 from an ATM.
   c. Check #278 (May 12) was never cashed by the recipient.

At some point in elementary school, you were taught to remember rules for adding and subtracting positive and negative numbers. But did you ever try to understand why those rules make sense? In previous classes, you might have been so focused on *remembering* rules that you never really *thought* about them. Our goal is to *understand* why rules make sense. Let's see if our study of bank statements can shed some light on this important subject.

18. We know that credits add to the balance in an account, while debits subtract from it. If you add two credits to your account, what does that do to the balance?

19. Based on Question 18, fill in the blank: Adding two positive numbers results in a ________________ positive number.

20. If you add two debits to your account, what does that do to the balance?

21. Based on Question 20, fill in the blank: Adding two negative numbers results in a larger (further from zero) ________________ number.

22. If you add a credit, then follow that with a debit, what happens to your balance?

23. Based on Question 22, fill in the blank: Adding a negative to a positive is the same thing as ________________ from the positive.

24. If you remove a debit from your account, what happens to your balance?

25. Based on Question 24, fill in the blank: Subtracting a negative is the same thing as ________________ the positive of that number.

Rules for Adding and Subtracting Signed Numbers

1. Adding two positive numbers results in a larger positive number.
2. Adding two negative numbers results in a larger negative number.
3. Adding a negative to a positive is the same thing as subtracting from the positive.
4. Subtracting a negative number is the same thing as adding the positive of that number.

26. Suppose you have a balance of $1,200 in checking, and you make a deposit of $210. What's the new balance?

27. What if you have a balance of $210, and make a deposit of $1,200?

28. Based on Questions 26 and 27, when adding two numbers the order ________________ matter. Choose either "does" or "does not."

29. Suppose you have a balance of $1,200 in checking, and you make a withdrawal of $300. What's the new balance?

30. What if you have a balance of $300, and make a withdrawal of $1,200?

31. Based on Questions 29 and 30, when subtracting a number from another, the order ________________ matter. Choose either "does" or "does not."

The Significance of Order When Adding or Subtracting

Addition is **commutative**: The order in which you add doesn't matter.

Subtraction is not commutative: The order does matter.

In each phrase in Questions 32–34, pick out the word that's telling you to do a certain arithmetic operation, write the operation it's describing, then perform the calculation.

32. The sum of $4 and $2.50

33. My GPA was 3.1, but it increased by 0.4.

34. The difference between $3.50 per gallon and $2.05 per gallon

35. The colored box summarizes some words that are commonly used to describe either addition or subtraction. Complete each list by writing in the appropriate word from Questions 32–34.

Words and Phrases That Represent Addition or Subtraction

Addition

Plus, is added to, in addition, ________________, ________________

Subtraction

Minus, is decreased by, is taken away from, ________________

Using Excel to Add and Subtract

To add the values in cells A2 and B2, we can use the addition symbol.

In cell C2, we type "=A2+B2". Note that the value in C2 is what you get if you add the values in A2 and B2. The equal sign is a signal to Excel that the stuff you're entering after it is a formula, rather than text, so always begin a formula calculation in Excel with =.

To subtract the values in cells A5 and B5, we can use the subtraction symbol.

In cell C5, we type "=A5–B5".

To add several values listed in multiple cells, we use the SUM command. In the screen below, cell F7 is selected and it reveals the formula =SUM(F2:F6), which has been typed in cell F7 to compute the sum of the values in cells F2, F3, F4, F5, and F6. The colon (:) is used to indicate that we want to include all of the cells between the first one, F2, and the last one, F6. You can type in F2:F6, or you can drag across those cells and Excel will put it in for you.

F7 | fx =SUM(F2:F6)

	A	B	C	D	E	F
1			Sum			Calories
2	$4	$3	$7			120
3						50
4			Difference			95
5	$3.50	$2.05	$1.45			200
6						35
7					Total	500

1-2 Portfolio

Name ____

Check each box when you've completed the task. Remember that your instructor will want you to turn in the portfolio pages you create.

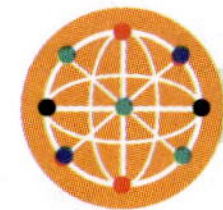

Technology

1. ☐ Design a spreadsheet to find the ending balance for the bank statement on page 24. If you enter the beginning balance and credits as positive numbers and debits as negative numbers, you can use the SUM command in one column. If you want to try something fancier, you can put the credits in one column, the debits in another, then sum each column and decide how to use each to calculate the ending balance. A template to help you get started can be found in the online resources for this lesson.

Online Practice

1. ☐ Include any written work from the online assignment along with any notes or questions about this lesson's content.

Applications

1. ☐ Complete the Applications problems.

Reflections

Type a short answer to each question.

1. ☐ In your own words, describe the importance of like terms in addition.
2. ☐ Analyze a recent bank statement for your own account, or one for a family member. Describe the significance of the credits and debits.
3. ☐ How do debits to a bank account help us to understand why subtracting a negative results in addition?
4. ☐ Name one thing you learned or discovered in this lesson that you found particularly interesting.
5. ☐ What questions do you have about this lesson?

Looking Ahead

1. ☐ Complete the Prep Skills for Lesson 1-3.
2. ☐ Read the opening paragraph in Lesson 1-3 carefully and answer Question 0 in preparation for that lesson.

Answers to "Did You Get It?"

1. 3 g; 108.5% **2. a.** $\frac{7}{3}$ **b.** 355 ft **c.** 7½ ft, or 90 in. **d.** Can't be added

3. a. $2,502.28 **b.** $2,412.28 **c.** $2,492.28

Answers to "Prep Skills"

1. a. 16 **b.** −2 **c.** 2

2. a. 24 **b.** −32 **c.** 60

3. a. $\frac{11}{5}$ **b.** $-\frac{5}{12}$ **c.** $\frac{17}{15}$

4. a. 7.9 **b.** 2.2 **c.** 1.13

1-2 Applications

Name ______________________________

1. Dwayne and his buddy Sea Bass are over-the-road truckers, and at their weekly chess match on Sundays, they compare the number of miles they drove during the week. The data for last week are illustrated by the bar graph. How far did Dwayne drive last week?

2. Who drove the most overall during the week? By how much?

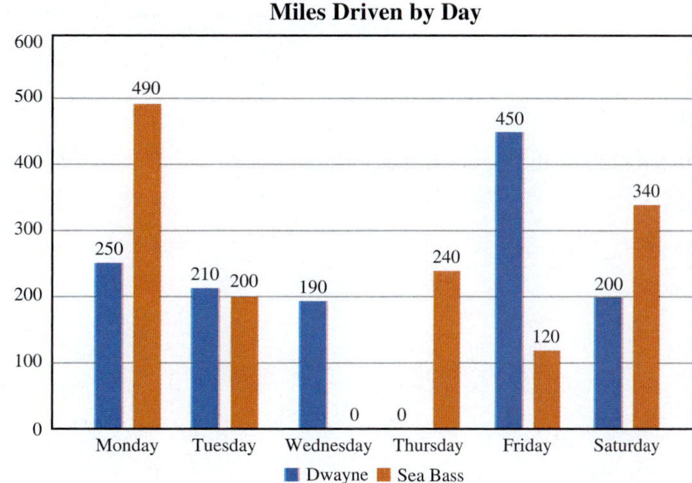

The next bar graph shows the annual net profit or loss for the Ford Motor Company for 2007 to 2015. Use it for Questions 3–5.

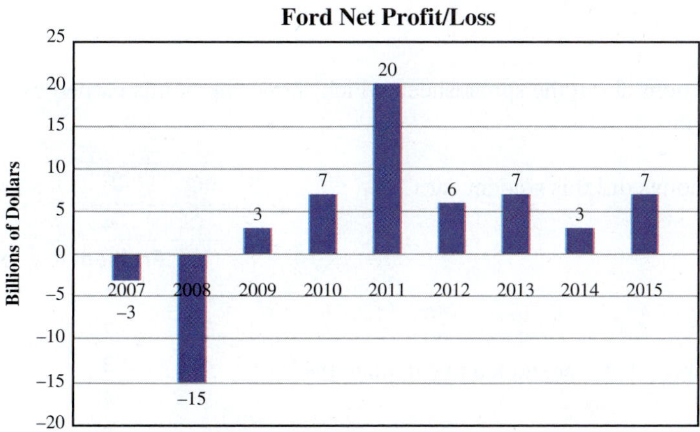

3. What was the difference between net income in 2008 and 2009? Does that represent an increase or a decrease?

4. What was the company's total net income for the years 2007 to 2010?

5. What was the total net income for all years shown in the graph?

1-2 Applications

Name ____

The **perimeter** of a figure is found by taking the sum of the lengths of each side of that figure. Find the perimeter of each figure in Questions 6 and 7.

6.

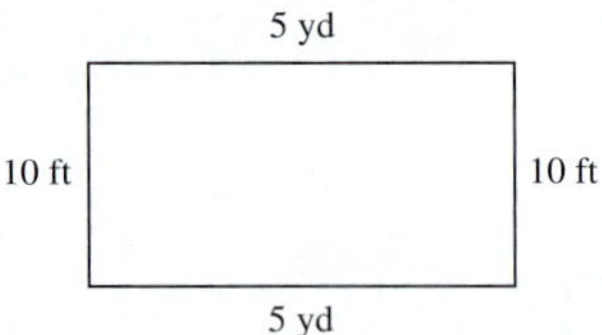

7.

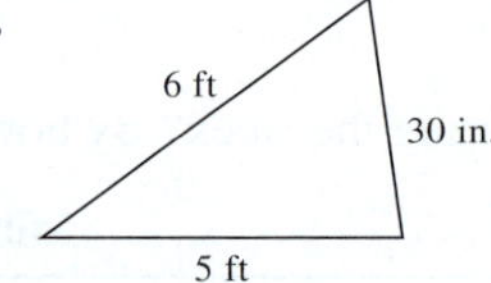

8. Find the length of the unlabeled side of the figure if the perimeter is 86 in.

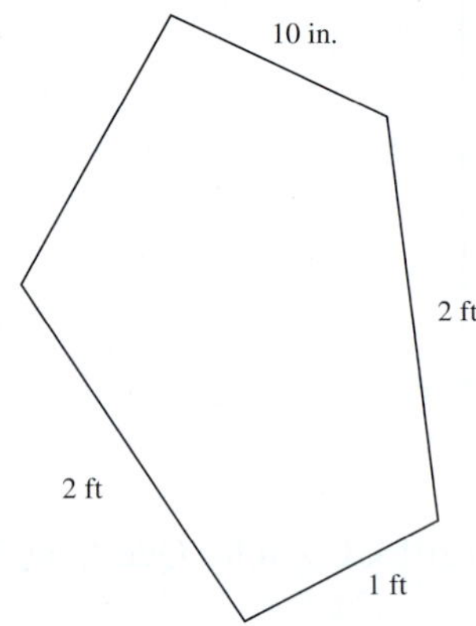

Answer the following questions about the spreadsheet, which shows the points earned by a student on a 10-question quiz worth a total of 50 points.

9. What percent of possible points did this student earn?

10. Write down all of the numbers that were added to calculate the value in cell C12.

11. A formula was entered in cell B12, not the number 50. Write the formula that was entered.

C12 fx =SUM(C2:C11)

	A	B	C
1	Problem #	Possible	Points
2	1	4	4
3	2	4	4
4	3	6	5
5	4	8	7
6	5	4	2
7	6	5	5
8	7	4	4
9	8	3	3
10	9	5	0
11	10	7	7
12		50	41

Lesson 1-3 Prep Skills

SKILL 1: RECOGNIZE MULTIPLICATION AS REPEATED ADDITION

- $7 + 7 + 7 + 7 + 7$ is the same thing as 5×7 and is equal to 35.
- $(-4) + (-4) + (-4) + (-4)$ is the same thing as $4 \times (-4)$, and is equal to -16.

SKILL 2: COMPUTE PERCENTAGE OF A WHOLE

To compute a percentage of a whole, compute the given percentage to decimal form by moving the decimal two places to the left. Then multiply that decimal by the whole amount.

- 40% of 100 is $0.4 \times 100 = 40$. 40% in decimal form is 0.40, or just 0.4.
- 5% of 30 is $0.05 \times 30 = 1.5$. 5% in decimal form is 0.05.

PREP SKILLS QUESTIONS

1. Write each repeated addition as multiplication and find the result.

   a. $10 + 10 + 10 + 10 + 10 + 10$

   b. $(-3) + (-3) + (-3) + (-3) + (-3)$

2. Find each requested value.

   a. 50% of 84 b. 35% of 200 c. 4% of 1,000 d. 11.3% of 550

Lesson 1-3 It's About Accumulation

LEARNING OBJECTIVES

- ☐ **1.** Interpret multiplication as repeated addition.
- ☐ **2.** Multiply or divide quantities.

©Design Pics/Craig Tuttle RF

A journey of a thousand miles begins with a single step.
—Lao-Tzu

The best time to plant a tree is twenty years ago. The second best time is now.
—Chinese proverb

In my experience as both a student and a professor, maybe the single biggest obstacle to succeeding in college is procrastination. As the work you have to do mounts, it's easy to feel overwhelmed at times. We're all familiar with being faced with so much to do that we just don't know where to start. Unfortunately, too many people take the default path, which is to do nothing constructive and put the important stuff off. Sometimes, though, you just have to force yourself to take that first step on the thousand-mile journey. The cumulative effect of doing small bits of work can accomplish a lot!

0. We studied the effects of addition in the last lesson. What do you think we'll study in this one? Justify your response.

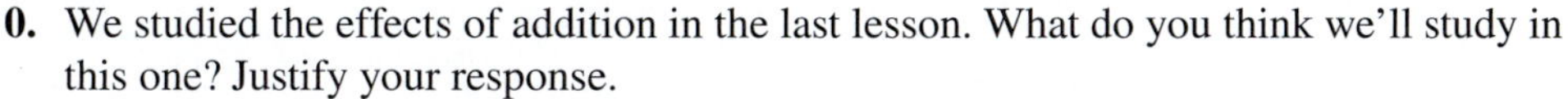

1-3 Group

1. Write about a time you put off something that you had to do, and then regretted it later.

2. Write down an activity that you're good at, like a sport, playing an instrument, a video game, or something else. How good were you when you started, compared to now? How much time went into building your skill at that activity?

3. Write about a task you've performed that required many different small steps to accomplish something bigger.

4. What do you do when you're overwhelmed with work and need to develop a plan?

Some of the problems in the remainder of this activity require calculations. Feel free to use a calculator or computer, but make sure you write down enough work to explain your results.

5. I don't do coffee, but according to an online Starbucks menu, a "Venti" cup of regular coffee costs $2.45. (Evidently, "Venti" is Italian for "really expensive coffee.") Suppose that a regular patron buys two of those every day. How much would she spend in one full year?

6. There are two ways to calculate the amount spent on coffee in Question 5. First, you can add $4.90 + $4.90 + $4.90 + $4.90 + . . . until you've written $4.90 ______ times. Second, you can multiply ______ by ____.

7. Use Question 6 to describe the relationship between multiplication and addition.

8. Going back to our coffee-buying friend in Question 5, complete the table below, based on buying two $2.45 cups of coffee every day. Don't just write an answer: Write the entire calculation.

   How much will she spend . . .

. . . after 1 day?	
. . . after 3 days?	
. . . after 10 days?	
. . . after 180 days?	

9. How much money would she save over the course of the year if she buys coffee at her friendly neighborhood convenience store, where it costs $1.49 per cup?

Did You Get It

Try these problems to see if you understand the concepts we just studied. The answers can be found at the end of the Portfolio section.

1. I also don't drink soda—hello, empty calories!—but I noticed that the soda machine down the hall from my office offers 16-ounce bottles of Coke for $1.50. If a student attends class four days a week during two 16-week semesters and has two bottles of Coke each day, how much money is he spending on Coke per year?
2. You can also get a bottle of water for $1.00. The bottle of Coke contains 185 calories, while the water has none. How much money would he save by going with the water, and how many calories?

10. Tommy (who used to work on the docks) and Gina (who works the diner all day) wisely plan to start saving for retirement. They deposit \$1,000 into an account to start out. The plan is to deposit \$100 into the account each month for the next 20 years. How much money will they have deposited into the account at the end of 20 years?

11. When you're trying to save money, it can help motivate you to calculate how much money accumulates little by little. Help Tommy and Gina with some of these calculations by filling in the table below, assuming that they stick with their plan of depositing \$100 each month after starting with the initial deposit of \$1,000. Again, include all calculations, not just the answer.

How much will they have saved . . .

. . . after 1 month?	
. . . after 1 year?	
. . . after 5 years?	
. . . after 10 years? (They're halfway there.)	

12. How much more can Tommy and Gina save in 20 years if they deposit \$125 each month rather than \$100?

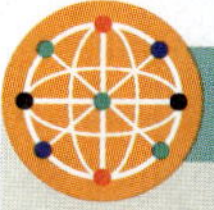

Using Technology: Using the Fill Handle to Auto-Fill Cells

Spreadsheets have the ability to complete cells automatically with either a formula or a value from a selected cell. This is done using the fill handle, which is the bottom-right corner of a selected cell.

	A	B
1	Months	Value
2	0	\$1,000.00
3	1	\$1,100.00
4		

To auto-fill the cells below B3, click on the little square (fill handle) in the bottom right corner. Then drag to auto-fill cells with either the value or the formula that is contained in cell B3.

See the Lesson 1-3 Using Technology Video in class resources for further instruction.

1-3 Class

Questions 1–5 use the pie charts below. The first illustrates the percentage of energy consumption in the United States by energy source for 2014. The second illustrates the specific breakdown of different renewable energy sources.

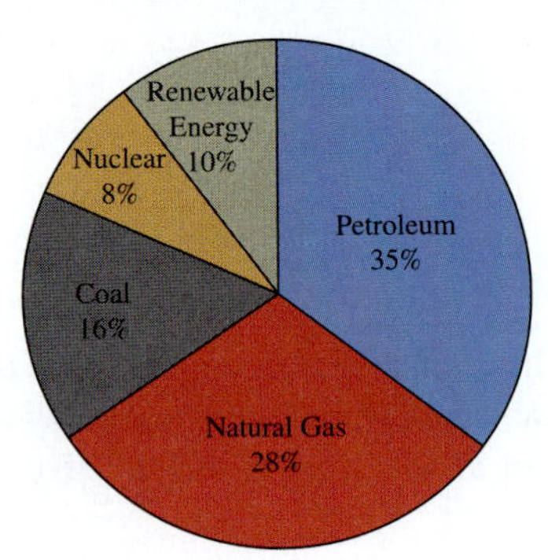

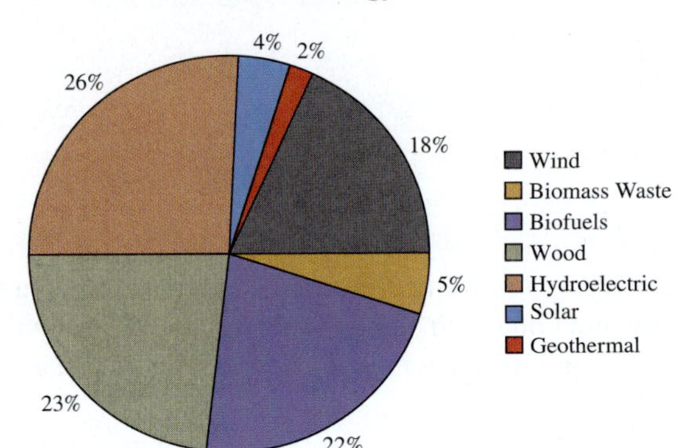

1. The first chart shows that 35% of our energy came from petroleum. The total energy usage for the United States in 2014 was 98.3 quadrillion BTU (British thermal units). What does the result of the calculation 0.35×98.3 tell us about energy usage? Include the result of that calculation in your answer.

2. Does the calculation 98.3×0.35 have the same result? What does this tell you about multiplication?

3. How many quadrillion BTU were generated using renewable energy sources?

4. How much of that energy was generated using wind power? Solar power?

Math Note

In case you're wondering, a quadrillion is almost unimaginably big: It's a million times bigger than a billion. For some perspective, it would take almost 700,000 years to count to 1 quadrillion by ones.

5. Back to the first pie chart: Looking at the natural gas slice, what is the significance of the calculation $100.8 \div 360$? Perform the calculation and describe what it tells us.

6. Does the calculation 360 ÷ 100.8 have the same result? What does this tell you about division?

Did You Get It

3. How many quadrillion BTU were generated by coal in 2014?
4. What is the significance of the calculation 0.26×360 in the renewable energy pie chart?

The Significance of Order When Multiplying or Dividing

Multiplication is **commutative**: The order in which you multiply doesn't matter.

Division is not commutative: The order does matter.

In each phrase in Questions 7–9, pick out the word that's telling you to do a certain arithmetic operation, write the operation it's describing, then perform the calculation.

7. The product of 20 and $2.50

8. Last year I made $2,000 working part time, but this year I made three times as much.

9. The quotient of $24 and 12 gallons

10. The accompanying box summarizes some words that are commonly used to describe either multiplication or division. Complete each list by writing in the appropriate word from Questions 7–9.

Words and Phrases That Represent Multiplication or Division

Multiplication

Multiplied by, ____________________, ____________________

Division

Over, divided by, ______________

1-3 Portfolio

Name ______________________________

Check each box when you've completed the task. Remember that your instructor will want you to turn in the portfolio pages you create.

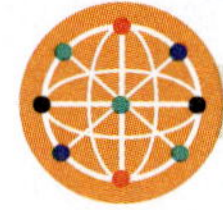

Technology

1. ☐ Create a spreadsheet that keeps track of the amount of money accumulated by Tommy and Gina (page 37). In column A, do it the long way: Enter every monthly deposit over a 10-year period, as well as the $1,000 initial deposit. You don't have to enter each amount separately: Use the auto-fill feature. See a template to help you get started in the online resources for this lesson. At the top of columns C and E, create cells in which a user can enter the monthly deposit and the number of years. At the top of column F, enter a formula to calculate the total amount deposited (including the initial $1,000).

Online Practice

1. ☐ Include any written work from the online assignment along with any notes or questions about this lesson's content.

Applications

1. ☐ Complete the Applications problems.

Reflections

Type a short answer to each question.

1. ☐ Describe a situation outside of school where multiplying was useful.
2. ☐ Describe a plan you could follow to keep from getting too far behind in class work.
3. ☐ How can thinking about the cumulative effects of small things help you either in or outside of the classroom?
4. ☐ Name one thing you learned or discovered in this lesson that you found particularly interesting.
5. ☐ What questions do you have about this lesson?

Looking Ahead

1. ☐ Complete the Prep Skills for Lesson 1-4.
2. ☐ Read the opening paragraph in Lesson 1-4 carefully and answer Question 0 in preparation for that lesson.

Answers to "Did You Get It?"

1. $384
2. $128 and 47,360 calories
3. 15.728 quadrillion BTU
4. It provides the number of degrees in the hydroelectric slice of the pie chart.

Answers to "Prep Skills"

1. **a.** $6 \times 10 = 60$ **b.** $5 \times (-3) = -15$
2. **a.** 42 **b.** 70 **c.** 40 **d.** 62.15

1-3 Applications

Name ______________________________

1. If you spend 30 minutes a day goofing off on the Internet during time you've set aside for studying, how much study time (in minutes, and in hours) will you waste over the course of a 16-week semester? What if you waste 45 minutes a day?

Gaining or losing weight comes down to calories burned vs. calories consumed. Burn more calories than you take in, and you'll lose weight. Burn less than you take in, and you'll gain weight. Simple. Let's study some aspects of weight change.

2. George weighed 160 lb when he started college. If he gains just 0.25 lb each month for 4 years of college, how much will he weigh? Suppose he doesn't change his habits after graduation, and continues that modest-sounding weight gain for the next 10 years after college. How much will he weigh for his 10th college reunion?

3. A rule of thumb used by nutritionists is that to lose 1 lb of body fat, you need to burn 3,500 calories above what you take in. If you burn 450 more calories than you take in each day, how long will it take to lose 1 lb? What about 10 lb?

4. An average-sized person will burn about 350 calories in an hour of walking at a fairly brisk pace. How many calories would you burn if you walk an hour a day for 6 months? How many pounds of body fat would that correspond to?

1-3 Applications

Name ______________________________

5. The figure below is made up of some number of smaller squares, like this one: ☐

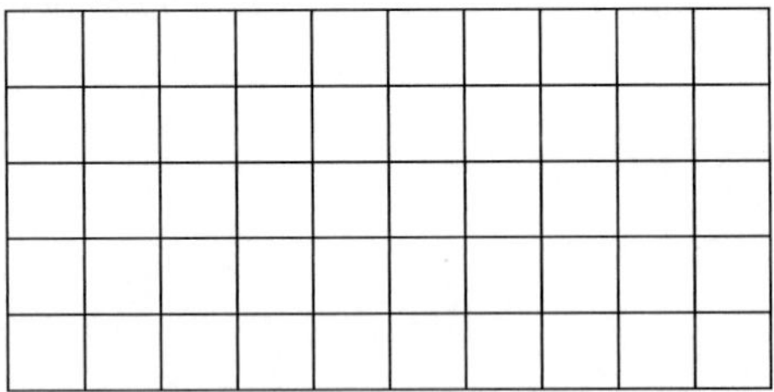

Calculate the number of small squares using first addition, then multiplication.

6. Why do you think the size of a two-dimensional object like this rectangle is measured in square units, like square inches?

7. Now let's add an extra dimension—literally! The next figure is made up of some number of little cubes, like this one:

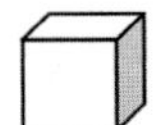

Calculate the number of small cubes using first addition, then multiplication.

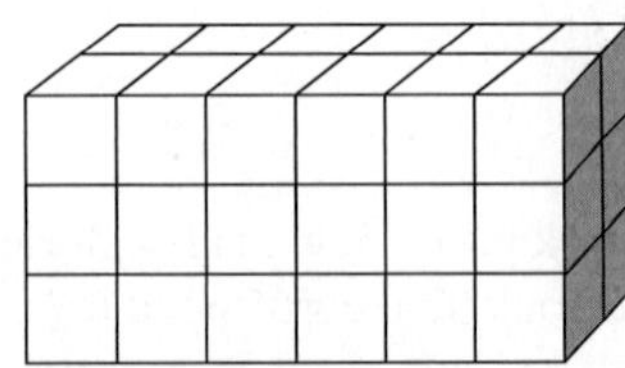

8. If each of the cubes is an inch long on every side, what units do you think would be used to measure the volume of the entire figure? Why?

Lesson 1-4 Prep Skills

SKILL 1: WORK WITH EXPONENTS

An **exponent** is a way to represent repeated multiplication. For example, 4^3 means $4 \times 4 \times 4$.

- $2^5 = 2 \cdot 2 \cdot 2 \cdot 2 \cdot 2 = 32$
- $(-4)^4 = (-4)(-4)(-4)(-4) = 256$ — The parentheses show that the negative is included in the number raised to the exponent.
- $-4^4 = -(4 \cdot 4 \cdot 4 \cdot 4) = -256$ — Only the 4 is affected by the exponent.
- $\left(\frac{1}{5}\right)^2 = \frac{1}{5} \cdot \frac{1}{5} = \frac{1}{25}$

SKILL 2: USE ORDER OF OPERATIONS

Order of operations is a group of agreed-upon rules that allow us to write calculations involving more than one operation without always having to use parentheses to indicate which operation to do first. The order is as follows, with operations higher on the list being performed before those lower on the list.

(1) Parentheses
(2) Exponents
(3) Multiplication or division
(4) Addition or Subtraction

Perform operations in parentheses first. If there are multiple parentheses, work from innermost outward.
For operations that live at the same level, work from left to right.

- $12 + 3 \cdot 4 = 12 + 12 = 24$ — Multiplication before addition
- $4(3)^2 = 4(9) = 36$ — Exponent before multiplication
- $81 \div (12 - 3) = 81 \div 9 = 9$ — Perform the subtraction first because it's in parentheses.
- $$\begin{aligned} 52 - 8 \div 4 \cdot 6 &= 52 - 2 \cdot 6 \\ &= 52 - 12 \\ &= 40 \end{aligned}$$ Division before multiplication because it's to the left
- $$\begin{aligned} (-2(3+1))^2 - 4 &= (-2 \cdot 4)^2 - 4 \\ &= (-8)^2 - 4 \\ &= 64 - 4 \\ &= 60 \end{aligned}$$ Addition in innermost parentheses first, then multiplication inside outer parentheses. Exponent before subtraction
- $\frac{7-9}{3+5} = \frac{-2}{8} = -\frac{1}{4}$ — In a fraction, the fraction bar acts like grouping symbols: interpret the fraction as $(7 - 9) \div (3 + 5)$.

SKILL 3: USE THE DISTRIBUTIVE PROPERTY

The **distributive property** is a method for multiplying some factor by a sum or difference without performing the addition or subtraction first. The factor is multiplied by each of the terms inside the parentheses.
In symbols:

$$a(b + c) = ab + ac \qquad a(b - c) = ab - ac$$

- $8(3 - 5) = 24 - 40 = -16$ — Matches what you get if you use order of operations
- $4(y + 6) = 4y + 24$ — In this case, you CAN'T perform the sum first.
- $-4(2x - 8) = -8x + 32$ — Notice that the distributed factor is negative and BOTH signs inside the parentheses changed.

PREP SKILLS QUESTIONS

Perform each operation.

1. a. 3^5 b. $(-2)^6$ c. $\left(\frac{1}{2}\right)^4$ d. -5^2
2. a. $8 - 10 \times 2$ b. $4 \cdot (-3) + 12$ c. $19 + -12 \div 3$
3. a. $4 + 3(9 - 1)^2$ b. $-5^2 \cdot 4 - 2$ c. $3 + 4 \cdot 10 \div 2$
4. a. $((8 + 1) \div 3)^4$ b. $6(2 - 5(3 - 4)^2)$ c. $(8 - 3 \cdot 2) - (6 \div 2)^2$ d. $\frac{2(3 + 4)}{11 - 9}$
5. Use the distributive property to perform each multiplication.

   a. $10(8 - 3)$ b. $-6(2 + 4)$ c. $x(14 - x)$ d. $-12(2y - 9)$

Lesson 1-4 Avoiding Empty Pockets

LEARNING OBJECTIVES

- ☐ 1. Distinguish between simple interest and compound interest.
- ☐ 2. Distinguish between linear and exponential growth.
- ☐ 3. Interpret exponents as repeated multiplication.
- ☐ 4. Simplify numeric expressions involving exponents and the order of operations.

©Stockbyte/Punchstock RF

The most powerful force in the universe is compound interest.
—Albert Einstein

Einstein was a smart guy who knew a little something about powerful forces, so that quote is a bit of an eye-opener. In this lesson, we'll study interest, a subject that everyone in our society needs to know something about (unless they want to end up like our friend to the right). If you have a trust fund that will afford you the opportunity to pay cash for a home, a car, a yacht . . . good for you. Feel free to send us a generous donation, care of the publisher. If not, you'll need to borrow money many times during your lifetime, and the amount of interest you pay will be an important part of budgeting your money.

Then there's saving money, another important part of a financial plan. Putting money in your sock drawer is *not* a sound investment strategy: You'll want to invest in a variety of accounts that will add interest and grow your money. In either case, understanding the power of compound interest will set you on a path to financial success.

0. Explain why you think what you learn in this lesson is likely to be useful in your life.

1-4 Group

1. Discuss any experiences you've had with some of these important financial situations: being responsible for a credit card or checking account, borrowing money to buy a car or home, starting a savings or retirement account, or others.

2. Each of the situations listed in Question 1 is distinct, but all have some common traits. List any you can think of.

Interest is a fee paid for the use of someone else's money. If you borrow money to buy a car, you pay the lending institution for using their money at the time of purchase. If you put money into a savings account, the bank pays you for having access to your money, which they in turn can use to lend to other people.

Simple interest is calculated as a percentage of the original amount of money. For example, if you deposit \$100 in an account that pays 4% interest per year, at the end of 1 year, you would have 4% of \$100 extra in the account. Since 4% of \$100 is $0.04 \times \$100 = \$4$, you would have earned \$4 in simple interest, and the account would now be worth \$104. In the second year, you'd earn another \$4 in simple interest, making the account worth \$108.

Compound interest, on the other hand (the force that Einstein was so enamored with), is interest that is paid not just on the original amount deposited, but on interest previously earned as well. To help you to understand the difference between simple and compound interest, let's look at two different accounts:

Account 1: You deposit $1,000 into an account that pays 5% simple interest on that $1,000 each year.

Account 2: You deposit $1,000 into an account that pays 5% of the account balance at the beginning of the current year in interest each year.

3. Answer the following questions about Account 1. Getting an answer is nice, but make sure that you think carefully about how the amount is growing and look for patterns.

How much interest will you earn in the . . .		How much will be in the account at the end of the . . .	
. . . first year?		. . . first year?	
. . . second year?		. . . second year?	
. . . third year?		. . . third year?	

4. Repeat Question 3 for Account 2.

How much interest will you earn in the . . .		How much will be in the account at the end of the . . .	
. . . first year?		. . . first year?	
. . . second year?		. . . second year?	
. . . third year?		. . . third year?	

Computing Simple and Compound Interest

To find the simple interest on an amount of money, you multiply the original amount by the interest rate (written as a decimal) and the amount of time. The units for the time (often years) should match the units on interest rate (often percent per year).

To find the compound interest on an amount of money, first calculate the simple interest for the first year and add it to the original amount to get the new balance after 1 year. For subsequent years, compute the interest as a percentage of the new balance, rather than the original amount.

Did You Get It

Try this problem to see if you understand the concepts we just studied. The answer can be found at the end of the Portfolio section.

1. A small business invests $30,000 in a fund that pays 7% interest per year. Find the amount of interest that would be earned in 2 years if the interest is (a) simple and (b) compound.

5. Fill in the next table by finding the amount in the account at the end of each number of years. You've already found four of the amounts in the table. Hopefully.

	Account 1	Account 2
Start	\$1,000.00	\$1,000.00
After 1 year	\$1,050.00	\$1,050.00
After 2 years		
After 3 years		
After 4 years		
After 5 years		

6. Suppose you wanted to find the value of each account after 20 years. Why would that be a much harder calculation for Account 2 than Account 1?

7. This is the calculation used to find the amount after the first year in either account:

Amount = \$1,000 + 0.05(\$1,000)

Previous amount plus 5% of previous amount

Can you think of a way to simplify the calculation so that all you have to do is one multiplication?

8. Why does the value of the compound interest account grow faster?

9. Here's a way to describe the difference in growth of the two accounts: Account 1 increases in value by __________ every year, while Account 2 has its value multiplied by __________ every year. So the value of Account 1 at any time looks like \$1,000 + __________ × (the number of years passed), and the value of Account 2 at any time looks like \$1,000(__________)(__________) . . . , where the multiplication occurs the same number of times as the number of years that have passed.

10. In each of the accounts, the value is growing, but in different ways. Fill in the blank lines in the following box.

The Difference Between Linear and Exponential Growth

In Account 1 (the simple interest account), the values grow by ______________ the same constant number.

This type of growth is called **linear growth**.

In Account 2 (the compound interest account), the values grow by ____________________ by the same constant number.

This type of growth is called **exponential growth**.

Did You Get It

2. Suppose that one savings account is opened with a \$500 balance and earns 4% simple interest. To find how the value grows each year, what would you add to the previous year's amount?
3. Suppose that another savings account is opened with a \$700 balance and earns 4% compound interest. To find how the value grows each year, what would you multiply the previous year's balance by?

1-4 Class

1. Here's a key observation about the repeated multiplication that occurs when finding the value of Account 2 in the group activity: Repeated multiplication can be expressed more concisely using an ______________________.

Let's continue working with the two accounts from the Group portion of this lesson. With an initial investment of \$1,000 and 5% compound interest, the value of that account after 7 years would be given by

$$1{,}000(1.05)^7$$

2. Without using a calculator, write that expression without any exponents. Don't perform the calculation: Just write out what it would look like.

3. What is the value of the account after 7 years?

4. In Account 1, where the interest is simple, the account value goes up by 0.05(\$1,000) = \$50 each year. Write a calculation that would find the value after 7 years, then find the value.

Let's take a stab at summarizing what we've learned about linear and exponential growth. Use the calculations for simple and compound interest on this page as models.

5. With linear growth, the value after a certain time is calculated using:

Starting value __________________ a constant __________________ a certain number of time periods.

6. With exponential growth, the value after a certain time is calculated using:

Starting value __________________ a growth factor ________________________________.

7. Without trying to give away the answer to Question 6 too much, part of it involves a power. What is the significance of the exponent?

Calculating Values for Linear and Exponential Growth

When a quantity increases based on linear growth, the value after a certain number of time periods can be calculated using this pattern:

New value = Original value + amount of growth per time period × number of time periods

When a quantity increases based on exponential growth, the value after a certain number of time periods can be calculated using this pattern:

$$\text{New value} = \text{Original value} \times (1 + \text{the growth rate as a percentage})^{\text{number of time periods}}$$

8. Use what you've learned about linear and exponential growth to calculate the value after 20 years in each of the accounts we've been studying. Make sure you write out the calculation that you did.

Did You Get It

4. An investment of $20,000 earns 7.5% interest. Find the value of the investment after 15 years if the interest is (a) simple and (b) compound.

9. If you borrow \$500 at 3% simple interest, the amount you'd owe after 4 years is given by ____________ + 500(____________)(4).

10. To correctly evaluate that expression, you'll need to remember the order of operations, which says that ____________________ comes before ____________________.

Math Note

If you need some review on order of operations, make sure you check out the prep skills resources for this lesson.

11. Evaluate the expression to find how much you'd owe.

12. Perform the calculation below, then compare your result to Question 11. Describe what you did in terms of order of operations. $500(1 + 0.03 \cdot 4)$

13. The reason you got the same answer for Questions 11 and 12 is the ____________________ property of multiplication over addition. (You DID get the same answer, right?)

14. Describe the property you referred to in Question 13 in words, then make up an example of a calculation that uses that property.

1-4 Portfolio

Name ____

Check each box when you've completed the task. Remember that your instructor will want you to turn in the portfolio pages you create.

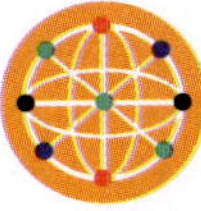

Technology

1. ☐ Create a spreadsheet that uses formulas to recreate the table describing the two bank accounts on page 49. Then extend the table to show the value of each account for the next 30 years. A template to help you get started can be found in the online resources for this lesson.
2. ☐ Do a Web search for different ways to remember the order of operations and find one you like. Warning: You may come across some that are not exactly family friendly.

Online Practice

1. ☐ Include any written work from the online assignment along with any notes or questions about this lesson's content.

Applications

1. ☐ Complete the Applications problems.

Reflections

Type a short answer to each question.

1. ☐ Describe your understanding of the difference between simple and compound interest. How does each relate to the terms "linear growth" and "exponential growth"?
2. ☐ Why do you think Einstein felt so strongly about compound interest?
3. ☐ What's the point of having an order of operations in math?
4. ☐ Name one thing you learned or discovered in this lesson that you found particularly interesting.
5. ☐ What questions do you have about this lesson?

Looking Ahead

1. ☐ Complete the Prep Skills for Lesson 1-5.
2. ☐ Read the opening paragraph in Lesson 1-5 carefully and answer Question 0 in preparation for that lesson.

Answers to "Did You Get It?"

1. a. \$4,200 **b.** \$4,347 **2.** \$20 **3.** 1.04 **4. a.** \$42,500 **b.** \$59,177.55

Answers to "Prep Skills"

1. a. 243 **b.** 64 **c.** $\frac{1}{16}$ **d.** -25

2. a. -12 **b.** 0 **c.** 15

3. a. 196 **b.** -102 **c.** 23

4. a. 81 **b.** -18 **c.** -7 **d.** 7

5. a. $10 \cdot 8 - 10 \cdot 3 = 80 - 30 = 50$

b. $-6 \cdot 2 + (-6) \cdot 4 = -12 + (-24) = -36$

c. $14x - x^2$

d. $-24y + 108$

1-4 Applications

Name ________________________________

Due to a variety of factors including fuel prices and adverse weather patterns, the cost of many groceries is expected to increase by as much as 5% per year. Let's illustrate the effects on consumers using the example of a frozen pizza that currently costs $5.60.

1. If the price of the pizza does go up by 5% next year, find the amount of increase and the new cost.

2. Complete the table below by calculating the new price of several other items for the same 5% increase. Notice the formula in cell C4 which reminds you of the most efficient way to do this calculation.

C4 fx =1.05*B4

	A	B	C
1	Item	Old price	New price
2	Small potato chips	$0.75	$0.79
3	Half-gallon of milk	$1.79	
4	Frozen pizza	$5.60	$5.88
5	Pound of ground chuck	$3.49	
6	Cooking sherry	$9.49	
7	Bag of chicken breast	$12.29	

3. If the pizza goes up by 5% again the following year, fill in each blank to complete the calculation you'd use to find the price after two 5% increases.

$5.60(1.\square\square)^{\square}$

4. How much would the pizza cost after two 5% increases?

5. Fill in the table below, which shows the cost of the $5.60 pizza after each number of 5% increases. Thinking about your answer to Question 3 should help make the calculations go quicker.

# of increases	New price
1	
2	
3	
4	
10	

6. How does the price of the pizza after 10 years of 5% increases compare to what the price would be if it went up by just 5% of the original cost each year?

1-4 Applications

Name ________________________________

Identify each scenario as illustrating either linear growth, exponential growth, or neither. Show enough work to justify your answer, and if you can, find the next value on the list. A review of the colored box on page 51 will be a BIG help.

7.

	A	B
1		**Taxi Fare**
2	Start	$3.30
3	After 1 mile	$5.70
4	After 2 miles	$8.10
5	After 3 miles	$10.50
6	After 4 miles	$12.90
7	After 5 miles	$15.30

8.

	A	B
1		**Account value**
2	Start	$12,000.00
3	After 1 year	$13,200.00
4	After 2 years	$14,520.00
5	After 3 years	$15,972.00
6	After 4 years	$17,569.20
7	After 5 years	$19,326.12

9.

	A	B
1		**Population**
2	Start	10,000
3	After 1 year	10,200
4	After 2 years	10,404
5	After 3 years	10,612
6	After 4 years	10,824
7	After 5 years	11,041

10.

	A	B
1		**Salary**
2	Start	$40,000
3	After 1 year	$41,500
4	After 2 years	$43,000
5	After 3 years	$44,500
6	After 4 years	$46,000
7	After 5 years	$47,500

11.

	A	B
1		**Stock price**
2	Start	$40.00
3	After 1 month	$42.00
4	After 2 months	$44.50
5	After 3 months	$48.00
6	After 4 months	$52.00
7	After 5 months	$60.00

Lesson 1-5 Prep Skills

SKILL 1: READ INFORMATION FROM A TABLE

The key to finding information accurately from a table is to carefully read the row and column headings so that you know exactly what information is being provided.

Smoking Rates in Canada, 2014

Age group	Males	Females	Total
12–19	8.3%	7.1%	7.7%
20–34	29.6%	19.4%	24.6%
35–44	25.9%	15.5%	20.6%
45–64	22.9%	17.5%	20.2%
65 and older	10.7%	8.4%	9.4%

- Smoking is most prevalent among males in the 20–34 age group (29.6%).
- In every age group, men are more likely to smoke than women.
- The lowest rate is among females 12–19.
- The overall smoking rate is highest for the 20–34 age group.

SKILL 2: SET A TIME SCALE

When data are provided over some time period, like the smoking rate in the United States for selected years from 1965 to 2013, we typically don't use the year, day, clock time, etc. when representing that data graphically. Instead we decide on a certain time as time zero, then use years, days, hours, etc. after that starting time.

Year	1965	1970	1980	1990	2000	2002	2003	2004	2007	2010	2013
% smokers	42.4	37.4	33.2	25.5	23.3	22.5	21.6	20.9	20.8	19.3	17.8

- If we call 1965 time 0, then 1970 would be time 5 (5 years after 1965), 1980 would be time 15, and so on:

Years after 1965	0	5	15	25	35	37	38	39	42	45	48
% smokers	42.4	37.4	33.2	25.5	23.3	22.5	21.6	20.9	20.8	19.3	17.8

SKILL 3: CHOOSE AN APPROPRIATE SCALE FOR A NUMBER LINE

It's a bad idea to always mark off a number line by ones: If we're interested in large numbers, we'd need to keep marking for a long while. Instead, we choose a consistent scale and label it on the number line.

- If we want to graph a variety of numbers between 0 and 20, starting at 0 and counting each tick mark as five numbers would be a good idea:

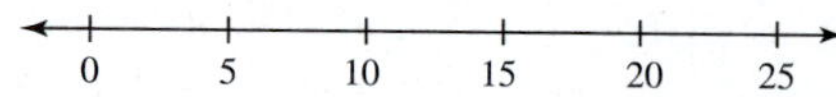

- If we want to graph numbers between −50 and 50, a different scale would be necessary:

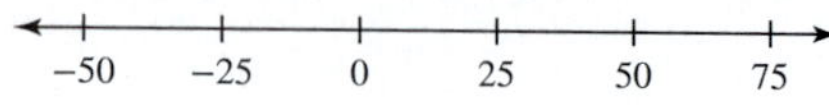

PREP SKILLS QUESTIONS

1. For the table summarizing smoking rates in Canada:
   a. What group had the lowest overall rate?
   b. Who is more likely to smoke? A female in the 20–34 range, or anyone in the 35–44 range?
   c. What happens to the total smoking rates as age goes up?

2. The first table describes the temperature on a certain day between noon and 8 P.M. In the second table, fill in a time scale with noon corresponding to time zero. Also, fill in a descriptive header at the beginning of the top row.

Time	Noon	1:00 PM	2:00	3:00	4:00	5:00	6:00	7:00	8:00
Temperature	42	45	47	48	48	49	46	40	37

Temperature	42	45	47	48	48	49	46	40	37

3. Label an appropriate scale on the number line below, then plot and label each of the numbers on the given list.
   a. −4, 0, 12, 15, −9

   b. 650, −3,000, 2,500, −500, 1,200

Lesson 1-5 A Coordinated Effort

LEARNING OBJECTIVES

- ☐ 1. Use a rectangular coordinate system.
- ☐ 2. Connect data to graphs.
- ☐ 3. Interpret graphs.

Believe you can and you're halfway there.
—Theodore Roosevelt

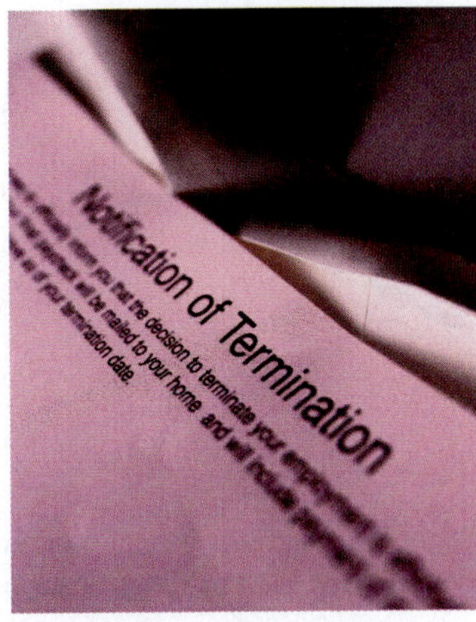

©Janis Christie/Getty Images RF

We've already used the word "graph" in this book, when referring to bar graphs. The bar graphs we looked at weren't just pretty pictures, or bars drawn at random heights for no particular reason: They were used to illustrate and understand real data. But the types of graphs we'll study in this unit might be the most misunderstood feature of math: People tend to think of them as just "plotting points and connecting the dots." Nothing could be further from the truth! Just like bar graphs, the graphs that illustrate connections between two variables are all about a visual representation of useful information. So before we talk about the mechanics and terminology involved, we'll use unemployment numbers to vividly illustrate what it's really all about.

0. How do you think the graphs we study in this lesson will be different from the bar graphs we studied earlier?

1-5 Class

Most people know that the economy in the United States went through a very rough patch starting in late 2007. How much we've recovered depends on who you ask, as there are conflicting numbers and viewpoints. One common way to measure the strength of the economy is through unemployment numbers.

These tables display the average annual unemployment rate for the years 1992 to 2015. This is an example of **time-series data**, which are quantitative data that have been collected at different points in time.

A **time-series graph** displays values on the y axis compared to equally spaced time intervals on the x axis.

Year	'92	'93	'94	'95	'96	'97	'98	'99	'00	'01	'02	'03
Rate (%)	7.5	6.9	6.1	5.6	5.4	4.9	4.5	4.2	4.0	4.7	5.8	6.0

Year	'04	'05	'06	'07	'08	'09	'10	'11	'12	'13	'14	'15
Rate (%)	5.5	5.1	4.6	4.6	5.8	9.3	9.6	8.9	8.1	7.4	6.2	5.3

1. Use the table to write a verbal description of trends in the unemployment rate over that 24-year period.

With enough effort, you were probably able to write a reasonable description. But because there's so much data in the table, spotting the trends isn't exactly a simple thing to do. Next, let's look at the same data displayed in graphical form.

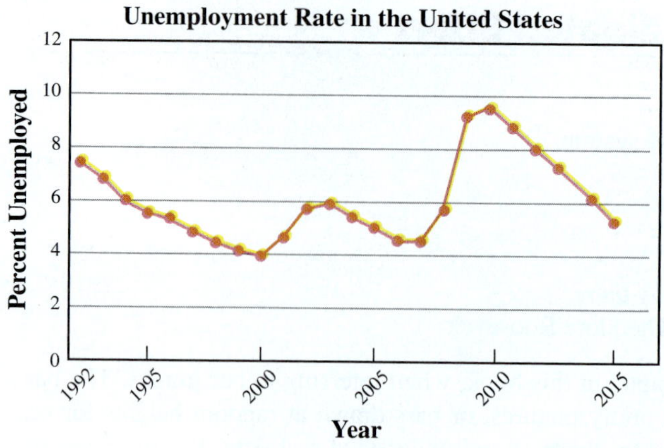

When I look at this graph, two things occur to me: It's a lot easier to see the trends than it was from looking at the table, and the graph looks kind of like a sea monster with a really big nose, which is totally irrelevant but still pretty cool.

2. Use the graph to write a verbal description of trends in the unemployment rate, then explain why the graph makes it easier than the table did.

Without the numbers running along the bottom side of the graph and down the left side, we wouldn't be able to understand any of the information the graph provides. Those numbers provide the **scale** for the graph, and they're ALWAYS crucial in drawing a graph. Each of the number lines that we write the scale on is called an **axis** (the plural of this word is **axes**).

Based on figures from January to October of 2016, the average unemployment rate for that year was expected to be about 4.9%. We can add that piece of information to the graph by finding 2016 on the horizontal axis and 4.9 on the vertical axis, then drawing imaginary lines up from 2016 and right from 4.9 until the lines meet: That's where we put the point corresponding to 2016 and 4.9%.

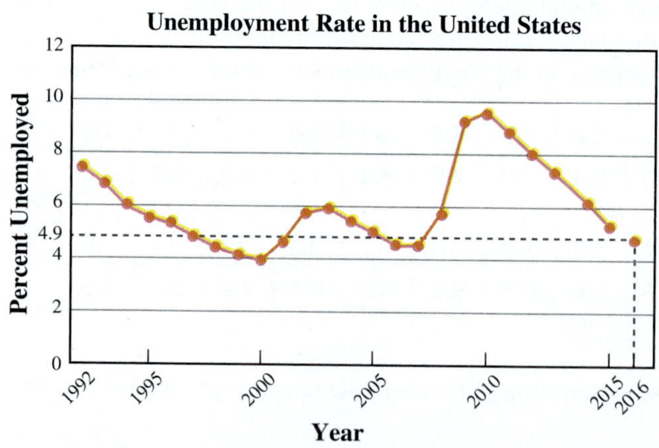

Notice that when we found the location to indicate that the unemployment rate was 4.9% in 2016, the imaginary lines we drew formed a rectangle with the two axes. That's why we call this system of graphing a **rectangular coordinate system**. Each of the numbers we used to locate that point are called **coordinates**. The horizontal axis is usually called the ***x* axis** and the vertical axis is usually called the ***y* axis**. The point where the two axes meet is called the **origin**.

Since we didn't need to worry about negative years or negative unemployment rates, the graph we drew earlier only showed positive values along each axis. But there are plenty of examples of data where negative values make perfect sense, so a rectangular coordinate system is often set up like this:

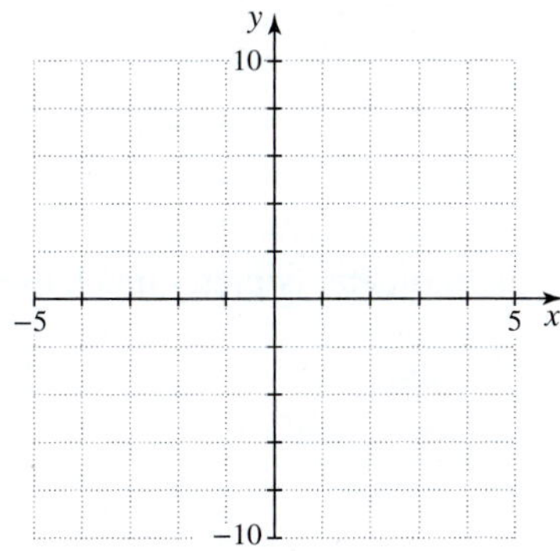

The process of locating information on a rectangular coordinate system, like the 4.9% unemployment rate in 2016, is called **plotting points**. We identify locations by writing the two coordinates together inside parentheses, like this: (2016, 4.9).

3. What are the coordinates of the origin in the rectangular coordinate system above?

4. Look carefully at the numbers on each axis. What distance does each box on the grid represent along the x axis? What about the y axis?

5. Plot each of the following points on the rectangular coordinate system above. Label the coordinates of each point.

a. (4, 0)
b. (0, −8)
c. (−3, −4)
d. (2, 7)

Math Note

To be more specific about the term "scale," the distances you found in Question 4 are usually referred to as the scale for each axis.

Did You Get It

Try this problem to see if you understand the concepts we just studied. The answer can be found at the end of the Portfolio section.

1. Draw a rectangular coordinate system, then plot and label the following points on it: (0, −30), (2, 50), (−4, −60), (−1, 10). Make sure you choose an appropriate scale for each axis, and clearly label the scale on your graph.

1-5 Group

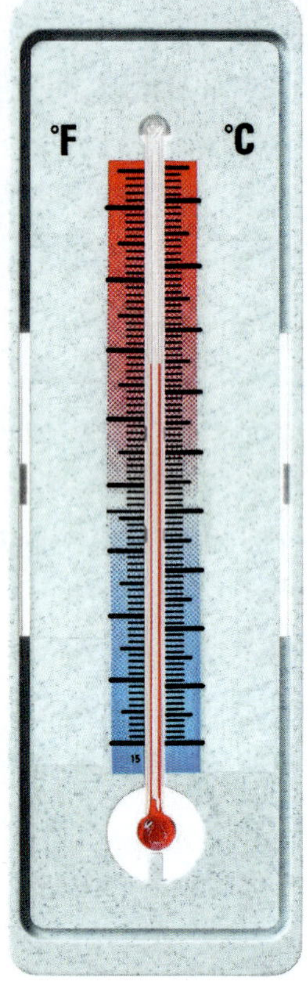

©McGraw-Hill Education/ Ken Cavanagh, photographer

1. What temperature is this thermometer displaying?

2. Explain why the thermometer is pretty much useless.

3. What are the coordinates of the point drawn on the graph?

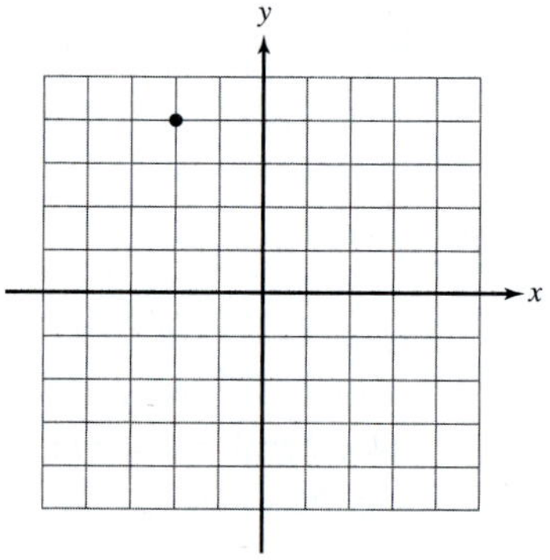

4. Explain why the graph is pretty much useless.

Important Features of a Good Graph

1. EVERY graph has to have a clearly labeled scale on each axis. A graph with no scale labeled is every bit as useless as a thermometer with no numbers on it.
2. There's no reason the scale has to be the same on both axes. If a graph contains points like (10, 4,000) and (20, 3,000), using the same scale on the x and y axes would lead to a graph that's very difficult to read (try it!).
3. If there are certain points on a graph that are important for some reason, you should label the coordinates of those points directly on the graph.

Being able to understand the connection between a graph and the information that it illustrates is by far the most important skill in graphing. If you can't interpret the meaning of a graph, it's just really bad art!

Almost everyone is interested in gas prices: They can have a very real effect on a household budget. The time-series data in the table below show the average price of a gallon of regular unleaded gas in the United States in January of each year from 2006 to 2016.

5. Write ordered pairs of the form (Years after 2006, Price per gallon) for each pair of values, then plot the points on the graph.

Year	Years after 2006	Jan. price per gallon ($)	Ordered pair (*x*, *y*)
2006	0	2.32	
2007	1	2.27	
2008	2	3.05	
2009	3	1.79	
2010	4	2.73	
2011	5	3.09	
2012	6	3.40	
2013	7	3.35	
2014	8	3.32	
2015	9	2.11	
2016	10	1.97	

6. Connect the points you plotted to draw a graph, and describe what that graph illustrates. Then add a verbal label to each axis that describes the information it represents.

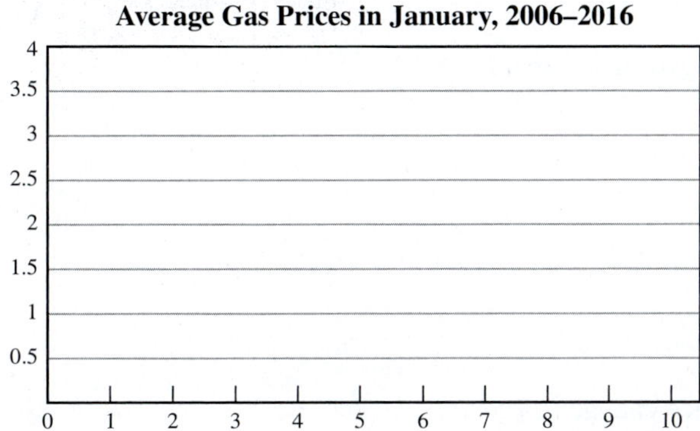

For Questions 7–13, explain how you got your answer using BOTH the table and the graph.

7. What was the highest price of January gas between 2006 and 2016? Is the answer based on the graph different? Why?

8. When was gas most expensive? (Again, you may have two slightly different answers.)

9. When did gas prices stay relatively level from one year to the next?

10. Over what time spans were gas prices rising?

11. Over what time spans was gas getting cheaper?

12. What was the largest single price change in 1 year?

13. Explain why we didn't bother to include negative values along either axis when drawing the rectangular coordinate system for the gas price graph.

Did You Get It 

2. Estimate the price of gas in January 2011 using the graph, and write a description of how you got your answer. Then write a description of how your answer compares to the price you can find in the table.

14. If I offered you one of the two accounts detailed below, which would you choose, and why?

	Account 1	Account 2
Start	$1,000.00	$1,000.00
After 1 year	$1,060.00	$1,050.00
After 2 years	$1,120.00	$1,102.50
After 3 years	$1,180.00	$1,157.63
After 4 years	$1,240.00	$1,215.51

15. When we plot points on a coordinate system that correspond to pairs of data, we call the result a **scatter diagram** or **scatter plot**. For the bank accounts in the two tables below, create a scatter diagram for each. First, you'll need to complete the table using skills we practiced earlier in the course. After writing ordered pairs, decide on an appropriate scale for each axis, then plot each point. It would probably be a good idea to use different colors or different markers for each account. (Note: We do NOT connect the points on a scatter plot!)

Time (yrs)	Account 1	Ordered pair
Start	$1,000.00	
After 1 year	$1,060.00	
After 2 years	$1,120.00	
After 3 years	$1,180.00	
After 4 years	$1,240.00	
After 5 years		
After 10 years		
After 15 years		
After 20 years		
After 25 years		

Time (yrs)	Account 2	Ordered pair
Start	$1,000.00	
After 1 year	$1,050.00	
After 2 years	$1,102.50	
After 3 years	$1,157.63	
After 4 years	$1,215.51	
After 5 years		
After 10 years		
After 15 years		
After 20 years		
After 25 years		

16. Explain why you chose the scale that you did for each axis.

17. Use the two scatter plots to write a verbal description of the differences between the growth of the two accounts. More detail is better!

18. In some cases, if the pattern of points in a scatter plot is relatively clear, we can connect the points to complete a graph. Do that now, drawing two graphs that represent the growth of the two accounts in Question 15. How would you describe the two graphs verbally?

Math Note

In Unit 3, we'll delve deeper into when it's appropriate to connect the points on a scatter plot, and how to find a graph that best fits a given plot.

Did You Get It

3. Draw a scatter plot for the data in the table, which represents the U.S. national surplus/deficit for even-numbered years for the period from 1998 to 2016. Positive numbers indicate the government taking in more money than it spent; negative numbers indicate the government spending more than it took in. All numbers are in billions of dollars. (Seriously. Billions.) Use years after 1998 on the horizontal axis, not the actual year.

Year	Surplus/Deficit
1998	70
2000	236
2002	−158
2004	−413
2006	−248
2008	−459
2010	−1,294
2012	−1,100
2014	−492
2016	−550

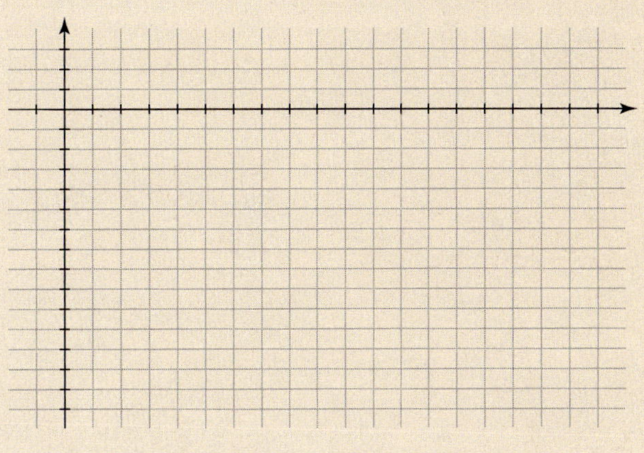

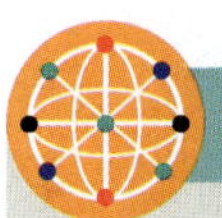

Using Technology: Creating a Scatter Plot

To create a scatter plot in Excel:

1. Type the values that will go on the x axis in one column. (This would correspond to the number of years in Question 15.)
2. Type the values that will go on the y axis in one column. (This would correspond to the value of the account in Question 15.)
3. Use the mouse to drag and select all the data in those two columns.
4. With the appropriate cells selected, click the **Insert** tab, then **Charts**, and click on Scatter. Then choose the type of scatter diagram you want. Options include plotting only the points, connecting the points with curves, and connecting the points with line segments.

You can add titles and change colors and other formatting elements by right-clicking on certain elements, or using the options on the **Charts** menu. Try some options and see what you can learn!

See the Lesson 1-5 Using Tech video in class resources for further information.

1-5 Portfolio

Name ______________________________

Check each box when you've completed the task. Remember that your instructor will want you to turn in the portfolio pages you create.

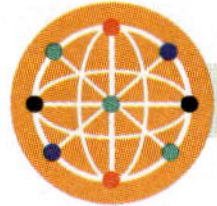

Technology

1. ☐ Use Excel to create two different scatter diagrams for the ordered pairs in the table on page 71. The first should just have the points; the second should connect the points with curves. A template to help you get started can be found in the online resources for this lesson.

Online Practice

1. ☐ Include any written work from the online assignment along with any notes or questions about this lesson's content.

Applications

1. ☐ Complete the Applications problems.

Reflections

Type a short answer to each question.

1. ☐ If someone says that the point of graphing is plotting points and connecting the dots, how would you explain to him how very, very wrong he is? It'll be tough, but try to be nice.

2. ☐ Why do you think we use the word "ordered" in "ordered pair"?

3. ☐ Explain the advantages of graphed data over data in table form.

4. ☐ Name one thing you learned or discovered in this lesson that you found particularly interesting.

5. ☐ What questions do you have about this lesson?

Looking Ahead

1. ☐ Complete the Prep Skills for Lesson 1-6.

2. ☐ Read the opening paragraph in Lesson 1-6 carefully and answer Question 0 in preparation for that lesson.

Answers to "Did You Get It?"

1.

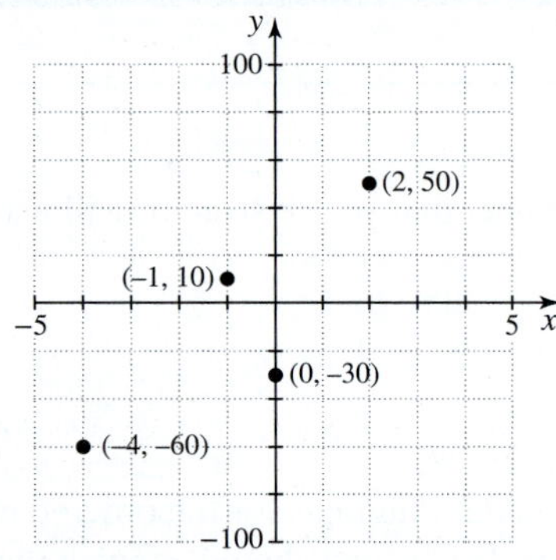

2. It looks a little above \$3, maybe \$3.10 or so. This isn't as precise as the \$3.09 we can get from the table.

3.

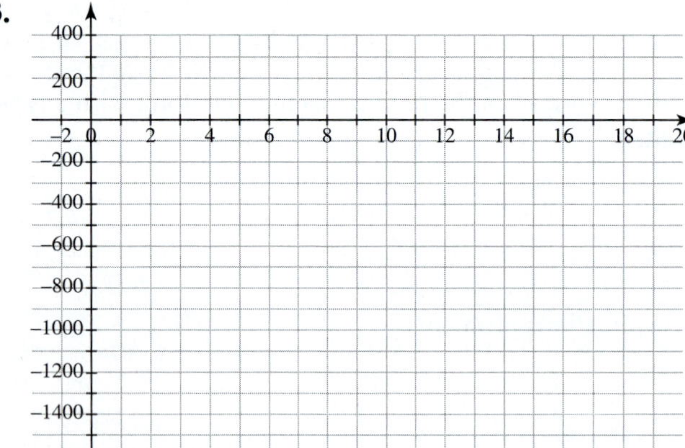

Answers to "Prep Skills"

1. **a.** 12–19 **b.** Anyone in the 35–44 range

   c. From ages 12–19 to 20–34 the rate increases, but then the rate decreases as age goes up.

2.

Hours since 12 P.M.	0	1	2	3	4	5	6	7	8

3. **a.** –10 –5 0 5 10 15

   b. –3,000 –1,500 0 1,500 3000 4500

1-5 Applications

Name ______________________________

The hourly temperatures for Champaign, Illinois, on October 29, 2016, are given in the table. Use hours after midnight (NOT the actual time) as first coordinates, and the temperature as second coordinates.

Time	Temperature	Ordered Pair
12:00 AM	61°	
1:00 AM	63°	
2:00 AM	62°	
3:00 AM	62°	
4:00 AM	62°	
5:00 AM	64°	
6:00 AM	65°	
7:00 AM	63°	
8:00 AM	63°	
9:00 AM	68°	
10:00 AM	71°	
11:00 AM	74°	
12:00 PM	79°	
1:00 PM	81°	
2:00 PM	81°	
3:00 PM	81°	
4:00 PM	79°	
5:00 PM	77°	
6:00 PM	73°	
7:00 PM	70°	
8:00 PM	69°	
9:00 PM	67°	
10:00 PM	66°	
11:00 PM	66°	
12:00 AM	63°	

1. Write an ordered pair for each time and temperature pairing. Remember, the first coordinate is hours after midnight.

2. Decide on an appropriate scale for each axis and create a scatter plot for this data. Then connect the points with a smooth curve. Make sure that the lowest height on the graph corresponds to zero degrees.

Temperatures in Champaign, IL on October 29, 2016

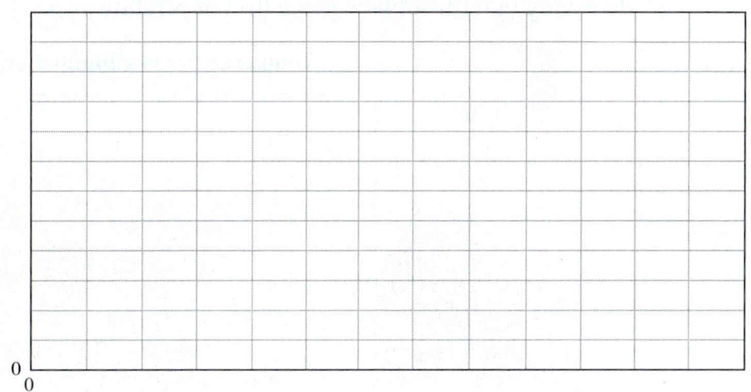

3. What do you need to find in the table to find when the temperature was highest during the day?

4. What do you need to look for on the graph to find when the high temperature was reached?

1-5 Applications

Name ______________________________

5. Use your graph to estimate time spans when the temperature was increasing.

6. Use your graph to estimate time spans when the temperature was decreasing.

7. Draw a second graph for the data: This time make the lowest height on the graph correspond to 50°. Why is this second graph deceiving in terms of how much the temperature varies?

Temperatures in Champaign, IL on October 29, 2016

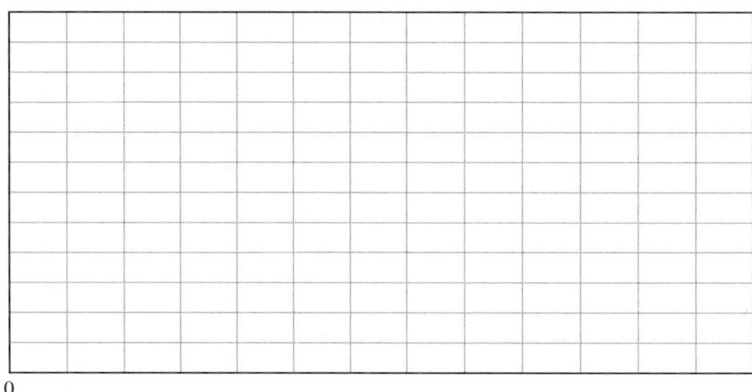

8. Draw scatter plots for the data in Questions 7 and 8 of Lesson 1-4 Applications, then connect the points with a smooth curve. Describe the shape of each graph, and describe conclusions you can draw for linear and exponential growth graphs.

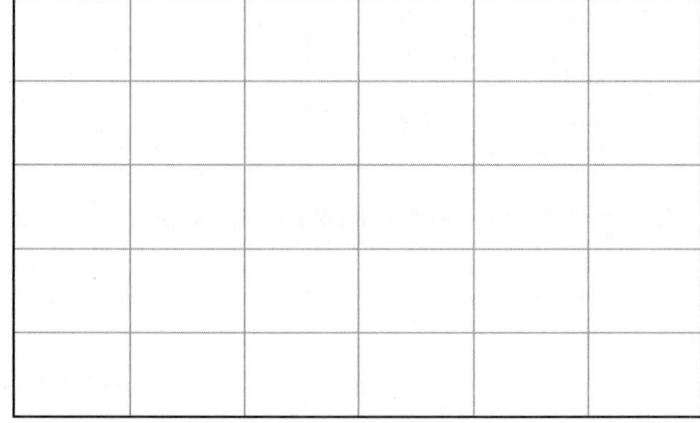

Lesson 1-6 Prep Skills

SKILL 1: REDUCE FRACTIONS

A fraction is in **lowest terms** or is **reduced** if there are no factors (other than 1) common to the numerator and denominator.

- $\frac{4}{6}$ is not in lowest terms because 2 is a factor of both 4 and 6.
- $\frac{6}{25}$ is in lowest terms because the factors of 6 are 1, 2, 3, and 6, while the factors of 25 are 1, 5, and 25.

The process of writing fractions in lowest terms is called **reducing fractions.** It's useful to do because it's easier to work with and interpret fractions when they're in lowest terms. For example, it's a lot easier to see that $\frac{5}{2}$ is two and a half than it is to see that $\frac{180}{72}$ is two and a half. To reduce a fraction, divide BOTH the numerator and denominator by any common factors.

- $\frac{4}{6} = \frac{4 \div 2}{6 \div 2} = \frac{2}{3}$ The greatest factor common to 4 and 6 is 2.
- $\frac{30}{12} = \frac{30 \div 6}{12 \div 6} = \frac{5}{2}$ The greatest factor common to 30 and 12 is 6.

SKILL 2: CONVERT BETWEEN FRACTIONS, DECIMALS, AND PERCENTS

To write a fraction in decimal form, you just need to perform the division, either by hand or with a calculator.

- $\frac{3}{5} = 0.6$

If the decimal doesn't terminate, it's usually appropriate to round. If a certain number of decimal places isn't required, use your judgment based on the context: How accurate would you want your answer to be in a given situation? Note that since you're rounding, the decimal is just an approximation of the exact fractional value.

- $\frac{3}{7} \approx 0.43$

To write a decimal in percent form, move the decimal two places to the right.

- $0.435 = 43.5\%$

To write a percent in decimal form, move the decimal two places to the left.

- $8.2\% = 0.082$

To write a percent in fraction form, write the fraction as $\frac{\text{Percent}}{100}$ and reduce. This makes sense because "percent" literally means "per hundred."

- $70\% = \frac{70}{100} = \frac{7}{10}$

PREP SKILLS QUESTIONS

1. Write each fraction in lowest terms.

   a. $\frac{12}{16}$ b. $\frac{65}{80}$

2. Write each percent in decimal form.

   a. 60% b. 12.5%

3. Write each fraction in decimal and percent forms.

   a. $\frac{3}{5}$ b. $\frac{12}{16}$

4. Write each decimal in percent form.

   a. 0.13 b. 0.058

5. Write each percent in fractional form and reduce.

   a. 75% b. 15%

Lesson 1-6 What Are the Chances?

LEARNING OBJECTIVES

- ☐ 1. Compute and interpret basic probabilities.
- ☐ 2. Translate a probability to a percent chance.
- ☐ 3. Recognize the difference between theoretical and empirical probability.

The 50–50–90 rule: Anytime you have a 50–50 chance of getting something right, there's a 90% probability you'll get it wrong.

—Andy Rooney

I'm planning on playing golf tomorrow, and my trusty Weather Puppy app tells me that there's a 10% chance of rain. So what exactly does that mean? Will it rain for 10% of the day tomorrow? Actually, a forecast like that is really just an educated guess: The forecaster is saying that there's about a 1 in 10 chance that it will rain at some point tomorrow, which means I'll probably stay dry on the golf course. For our purposes, "probably" is the key word in the last sentence. That word indicates a certain likelihood that something will occur. In this lesson, we'll study *probability,* which is a way to assign a number or percentage to the likelihood of something occurring, like rain in Fairfield, Ohio, tomorrow.

Courtesy of Weather Puppy. Background photo: ©federico stevanin/Shutterstock RF

0. Write about a time you've heard the word "probability" used outside of school.

1-6 Class

The **probability** of an event occurring is a description of how likely it is that the event will actually happen. Probability can be described using a number ranging from 0 to 1, a percentage between 0% and 100%, or using words like *impossible, unlikely, even chance, likely,* or *certain.*

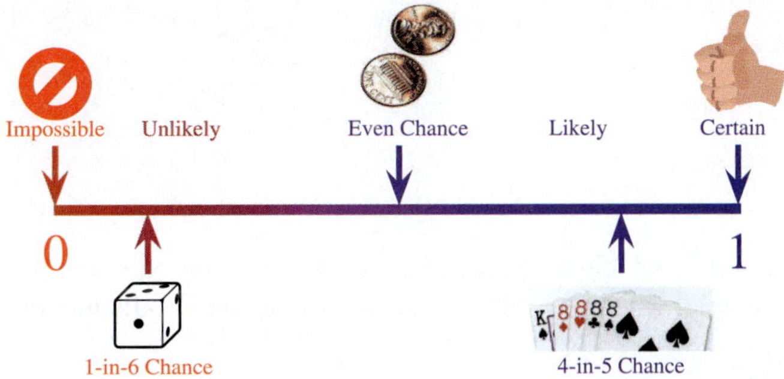

1. The probability of rolling a 4 with one standard six-sided die is $\frac{1}{6}$ because there is one side with 4 dots, and six sides total that could be showing when the die stops. Write this probability in equivalent forms as a decimal (rounded to three places) and as a percentage (rounded to one decimal place).

2. Write a sentence describing the percent chance of rolling a 4 with one standard die.

3. If you rolled a die 300 times, how many times would you expect it to land on 4?

4. How sure are you that you'd get the number of 4s you'd expect in 300 rolls? Explain.

When we're thinking about a situation in which we'd like to compute some probabilities, we use the word **outcome** to refer to different possible results. For example, when you roll a single die, there are six distinct outcomes: the whole numbers from 1 through 6.

5. When a standard coin is flipped, how many different outcomes are possible in terms of which side lands face up?

6. Based on your answer to Question 5, what is the percent chance that a single coin lands on heads when flipped?

7. Write the probability from your answer to Question 6 as a fraction between zero and one, and describe the likelihood verbally.

Computing a Probability

The probability that a certain outcome occurs is the number of ways that outcome can occur divided by the total number of outcomes. To represent the probability of an outcome, we use a capital *P*, and briefly describe the outcome in parentheses following the *P*. So a formula for probability is

$$P(\text{outcome}) = \frac{\text{Number of ways that outcome can occur}}{\text{Total number of outcomes}}$$

For example, the probability of rolling a 4 with one die is

$$P(\text{rolling } 4) = \frac{1}{6} \text{ or just } P(4) = \frac{1}{6}$$

This would be read as "*P* of 4 is one sixth."

Did You Get It

Try this problem to see if you understand the concepts we just studied. The answer can be found at the end of the Portfolio section.

1. Suppose that you decide to take 1 day off per week from your phone, and just turn it off and leave it at home all day. If you choose the day of the week randomly, what is the probability that you'll go without your phone on Wednesday? What about on a day that ends with the letter *y*? Use probability notation to write your answers.

8. If you flipped a penny 80 times, about how many times would you expect it to land heads up?

9. Would you definitely get the number of heads you found in Question 8? Discuss.

10. The probability diagram on page 75 contains a picture of a good hand in poker: four 8s. If you were to randomly pick a card from that hand, what's the probability that it would be an 8? Write as a fraction between zero and one and as a percent. Use probability notation.

11. Write a sentence describing the percent chance of drawing the king from that hand of cards.

Each of the probabilities that we've calculated so far can be found without doing any actual experiments: We know for sure that when we roll a die, there are six possible results, and only one of them is a 4. This type of probability is known as **theoretical probability**. Our calculations are based on observing how many outcomes are possible, how many fit a certain criteria, and assuming that each of the outcomes is as likely as any other to occur. If the die were weighted strangely to make 4 more likely to come up than other numbers, our probability calculation wouldn't be accurate anymore.

In that case, we might turn our attention to **empirical probability**. We could roll the die a bunch of times—let's say 100—and record how many times 4 comes up. If we then divide the number of times we got 4 by the total number of rolls, that would give us an approximate percent chance of rolling 4.

1-6 Group

Empirical Probability Lab

Supplies needed: A standard playing card, a dime, and scotch tape.

The point: If you flip a standard playing card and let it fall to the floor, the theoretical probability of it landing face up is $1/2$. We can change the probability by making one side heavier; then it would make sense to compute an empirical probability.

The procedure: Tape the dime to the back side of the playing card. Then flip the card from eye level or higher, letting it fall to the floor without hitting anything. Note whether or not the front of the card lands face up. Repeat as many times as you think you need to in order to get a reasonable probability; any less than 20 would be a pretty bad idea. Record your results in the table using tally marks (|):

Number of flips	Times landing face up

1. Based on the results from your table, what is the percent chance that the card will land face up?

2. Write the percent chance as a probability between zero and one, then write a sentence describing how likely you think it is that the card will land face up.

3. Did you find the results of this experiment surprising based on how the card was altered? Explain.

4. Why was empirical probability a better idea than theoretical probability for this experiment?

> **Math Note**
>
> Experts have estimated that the probability of being struck by lightning is three times greater than the probability of winning $1 million or more in a lottery.

Did You Get It

2. The 2016 Cleveland Indians won 94 games and lost 67 during the regular season. What was their percent chance of winning any given game? What was the probability?

5. What's wrong with the logic displayed by Rachel in the following conversation? Explain in depth.

Rachel: I have a 50-50 chance of passing this exam.
Ross: Why do you say that? That doesn't sound very optimistic.
Rachel: Well, it's just like flipping a coin—there are two possibilities. I'll either pass or I won't. So I have a 50% chance of passing.

6. Make a list of some things you can do to change the probability of passing an exam.

7. There are 45 applicants for two really great jobs, and you're one of them. If every applicant has an equal chance, what's the probability that you'll get the job? Write your answer using probability notation.

8. Explain why theoretical probability is probably not a very realistic way of deciding your chances of landing the job in Question 7.

9. The promotional materials for a certain program at a community college boast that out of their 480 most recent graduates, 450 are currently working in the field. If you graduate in this program, describe what you think your chances are of getting a job. Include as many aspects as you can think of.

In many cases, you can use the results of existing surveys to compute empirical probabilities, eliminating the need to do an entire experiment on your own.

In 2016, *USA Today* reported on the results of a study where 1,005 adults were asked how long it took them to know if they could stay in a new job long term. The number of people giving each response is shown in the table.

Response	Number of people
Less than a week	231
Within a month	402
Half a year	241
Within a year	60
Over a year	71

10. What's the probability that a randomly selected worker knew within a week if she could stay in her job long term?

11. What's the percent chance that a randomly chosen worker took more than a month to decide if she could stay in her job long term?

12. Write a sentence or two explaining how likely you think it is that your professor decided if he or she could stay in a job long term in a half year or less.

Did You Get It

3. What's the probability that a random worker feels like it took him just about a month to decide if he could stay in his job long term?
4. What's the percent chance that a random worker took a month or less to decide if he could stay in his job long term?

13. As we pointed out, 1,005 people were surveyed. The survey was done online. How accurate do you think the probabilities computed using this survey are when used to describe the feelings of all American workers? Discuss.

1-6 Portfolio

Name ______________________________

Check each box when you've completed the task. Remember that your instructor will want you to turn in the portfolio pages you create.

Online Practice

1. ☐ Include any written work from the online assignment along with any notes or questions about this lesson's content.

Applications

1. ☐ Complete the Applications problems.

Reflections

Type a short answer to each question.

1. ☐ If someone asked you "What is probability, and why is it called that?", how would you answer?
2. ☐ Describe the relationship between probability and percent chance.
3. ☐ Describe the difference between theoretical and empirical probability. Which do you think is more reliable?
4. ☐ Name one thing you learned or discovered in this lesson that you found particularly interesting.
5. ☐ What questions do you have about this lesson?

Looking Ahead

1. ☐ Complete the Prep Skills for Lesson 1-7.
2. ☐ Read the opening paragraph in Lesson 1-7 carefully and answer Question 0 in preparation for that lesson.

Answers to "Did You Get It?"

1. $P(\text{Wednesday}) = \frac{1}{7}$; $P(\text{Ends in y}) = 1$
2. 58.4%; $P(\text{Winning}) = 0.584$
3. $P(\text{Within a month}) = 0.4$
4. About 63%

Answers to "Prep Skills"

1. **a.** $\frac{3}{4}$ **b.** $\frac{13}{16}$
2. **a.** 0.6 **b.** 0.125
3. **a.** 0.6; 60% **b.** 0.75; 75%
4. **a.** 13% **b.** 5.8%
5. **a.** $\frac{3}{4}$ **b.** $\frac{3}{20}$

1-6 Applications

Name ______________________________

1. According to the United States Census Bureau, just about 51.5% of the adults in America are women. What would you say is the probability that the next adult you run into is female? What's the probability of that person being male?

2. There are 535 members of the United States Congress. Based on Question 1, how many would you expect to be female?

3. In reality, there are 104 women in Congress. What do you think that statistic says about politics in this country?

4. On one game show, contestants get to roll one die one time, and if they roll a number greater than 4, they win an all-expenses paid trip to Bora Bora (which, as it turns out, is an actual place. I used to think it was made up.) What is the probability of winning? Describe how good you'd feel about your chances if you got to play that game.

©Tero Hakala/123RF

5. In one math class, there are 30 students, 12 of whom are "nontraditional" in terms of age. If you randomly pick one person from that class, what's the probability that he or she will be of nontraditional age?

6. An online simulator helps students to understand their chances of passing a course based on study habits, attitude, attendance, work hours, sleep hours, and amount of homework completed. One student runs the simulation 10 times, with results shown in the table. Based on these results, discuss how likely you think it is that the student will pass. More detail is better.

Simulation	Result
1	Fail
2	Pass
3	Pass
4	Pass
5	Pass
6	Fail
7	Pass
8	Pass
9	Fail
10	Pass

7. What do you think is the probability that YOU will pass the math course you're in? Is this a set value, or can it change? If so, how?

Lesson 1-7 Prep Skills

SKILL 1: EVALUATE POWERS OF TEN

When 10 is raised to a positive power, the result is always 1 followed by a number of zeros. That number of zeros matches the exponent.

- $10^7 = 10{,}000{,}000$ Exponent is seven: seven zeros

When 10 is raised to a negative power, the result is always a number less than one. The decimal form begins with a number of zeros and ends with 1. The number of zeros is one less than the size of the exponent.

- $10^{-4} = 0.0001$ Exponent is negative four: three zeros

SKILL 2: WRITE REPEATED MULTIPLICATION AS AN EXPONENT

Positive integer exponents are used to represent repeated multiplication. If a number, like 5, is multiplied by itself three times, we write it as 5^3.

- $12 \cdot 12 \cdot 12 \cdot 12 \cdot 12 = 12^5$

Negative integer exponents also represent repeated multiplication, but the negative power indicates that the base should be put into the denominator of a fraction.

- $\frac{1}{3 \cdot 3 \cdot 3 \cdot 3} = 3^{-4}$

SKILL 3: MULTIPLY EXPONENTIAL EXPRESSIONS

To multiply two expressions involving exponents that have the same base, you simply keep the base as is, and add the exponents.

- $12^3 \cdot 12^6 = 12^9$ Bases are the same: add the exponents

SKILL 4: DIVIDE EXPONENTIAL EXPRESSIONS

To divide two expressions involving exponents that have the same base, you simply keep the base as is, and subtract the exponents.

- $\frac{7^8}{7^2} = 7^6$ Bases are the same: subtract the exponents

PREP SKILLS QUESTIONS

1. Evaluate 10^5
2. Evaluate 10^{-3}
3. Write the expression in exponent form: $8 \cdot 8 \cdot 8 \cdot 8 \cdot 8 \cdot 8 \cdot 8$
4. Write the expression in exponent form, with no fraction:

   $$\frac{1}{10 \cdot 10 \cdot 10 \cdot 10 \cdot 10}$$

5. Perform each multiplication, writing your answer with a single exponent:

   $10^5 \cdot 10^{12}$ $\quad$ $8^{-3} \cdot 8^2$

6. Perform each division, writing your answer with a single exponent:

   $\frac{5^7}{5^3}$ $\quad$ $\frac{10^4}{10^{11}}$

Lesson 1-7 Debt: Bad. Chocolate: Good

LEARNING OBJECTIVES

- ☐ 1. Convert numbers between decimal and scientific notation.
- ☐ 2. Describe the significance of writing numbers in scientific notation.

©iStockphoto/Getty Images RF

Worrying is like paying on a debt that may never come due.
—Will Rogers

According to an article posted to the *Motley Fool* website in May 2016, the average household in the United States has a little over $93,000 in debt. This is the amount of money that they need to pay back at some point. That's not a good thing, and sounds like an awful lot. Until you realize that as I write this, the federal government's total debt is—get this—$19,804,594,378,489. If I were a betting man, which I'm not on account of being too cheap to risk losing money, I'd be willing to bet that over three quarters of the population can't even pronounce that number, let alone comprehend how ridiculously large it is. (For the record, it's 19 trillion, 804 billion, 594 million, 378 thousand, 489 dollars.) If you decided to pitch in and help out, sending $10 to the government to help defray some of the debt, your contribution would amount to 0.0000000000505% of the debt. Thanks for the effort, big spender!

In certain fields, like governmental finance and the sciences, it's common to work with incredibly large or incredibly small numbers. Using regular decimal notation, this can be pretty cumbersome: do 0.0000004 and 0.000000004 look very different to you? One is a *hundred times as big* as the other! In this lesson, we'll use federal debt and melted chocolate to study a special way of writing very small or very large numbers in a more readable, convenient way.

0. What do you think are some of the reasons we'll be interested in an efficient way to work with really large and really small numbers in this lesson?

1-7 Class

If you track the national debt over the last 90 years or so, it can be modeled fairly accurately by the type of exponential growth we studied in Lesson 1-4. On average, the debt has grown by about 7.9% per year. If that trend continues, by 2050, it would be $276,205,000,000,000, and by 2100 it would be $12,300,000,000,000,000.

If you're like most people (including me, so don't feel bad) you have NO idea what those numbers actually mean. They're just too big to get any perspective on them other than "wow, those are really big." To make it easier to understand and work with numbers of ludicrous size, we'll use a clever method for writing them using exponents.

Consider the number 300, which can also be written as 3×10^2, which is of course 3×100. If you're thinking "that's silly," hang in there . . . there is a point. Take a look at the first handful of powers of 10:

$10^1 = 10 \qquad 10^2 = 100 \qquad 10^3 = 1{,}000 \qquad 10^4 = 10{,}000 \qquad 10^5 = 100{,}000$

Do you see the pattern? Each extra power of 10 puts another zero on the decimal equivalent. In fact, the power indicates exactly the number of zeros included. Here's why this is helpful: Five billion is a tremendously large number: 5,000,000,000. Notice that there are nine zeros, so we can write five billion as 5×10^9, which is more concise and easier to read. If that makes sense to you, then you understand the idea behind *scientific notation.*

Now let's take a look at some debt numbers. In 1930, the national debt was about $16 billion. By 1985, it had grown to about $1.823 trillion. Whatever, right? Both are huge numbers. But let's look at them in scientific notation:

$1930: \$1.6 \times 10^{10} \qquad 1985: \$1.823 \times 10^{12}$

Because the first part of a number in proper scientific notation is ALWAYS a number between 0 and 10, you can easily compare the sizes of numbers by looking at the size of the exponent. The second exponent (12) is 2 bigger than the first (10). That tells us that the second debt has two more digits to the left of the decimal point in its decimal expansion, which makes it 100 times as big. Simple!

1. Earth is about 2.57×10^7 miles from Venus, and about 4.67×10^9 miles from Pluto. About how many times further away is Pluto? (Careful . . . there's more to think about than the size of the exponent.)

Writing Large Numbers in Scientific Notation

1. Write a decimal point at the end of the number, then move it left so that there is just one digit before it.

2. Multiply the result by 10 to some power. The power will be the number of places you moved the decimal.

Example: 3,968,000

3,968,000. (decimal moved six places)

3.968 (the extra zeros are unnecessary)

3.968×10^6

For Questions 2 and 3, write each number in scientific notation.

2. The number of people in the United States in November 2015: 322,146,000

3. The distance in miles from Earth to the Moon: 238,900

We can also use scientific notation to write very small numbers, using negative powers of 10:

$10^{-1} = \frac{1}{10} = 0.1$ $10^{-2} = \frac{1}{10^2} = 0.01$ $10^{-3} = \frac{1}{10^3} = 0.001$ $10^{-4} = \frac{1}{10^4} = 0.0001$

This time, the pattern is that when the negative exponent gets larger by one (meaning more negative), the decimal point slides one spot to the left. So, for example, 0.03 is the same as 3×10^{-2}.

Writing Small Numbers in Scientific Notation

1. Move the decimal point right so that there is just one nonzero digit before it. Drop all zeros that come before that digit.
2. Multiply the result by 10 to some power. The power will be the negative of the number of places you moved the decimal.

Example: 0.000437
0.000437 (decimal moved four places)
4.37
4.37×10^{-4}

In Questions 4 and 5, write each number in scientific notation.

4. A typical flu virus measures between 0.000 000 08 and 0.000 000 12 meters in length.

5. *E. coli* is a common bacterium that can cause intestinal issues. A typical specimen is about 0.000 007 meters long.

6. Based on the size of the exponent in scientific notation, which is bigger: a small flu virus, or a typical *E. coli* sample? How many times bigger?

Did You Get It

Try this problem to see if you understand the concepts we just studied. The answer can be found at the end of the Portfolio section.

1. Convert each number to scientific notation.
   a. 0.0098 b. 1,478 c. 54,302,000 d. 0.000 0058

Now that we're good at converting from decimal to scientific notation, it would seem appropriate, if not expected, to work on the reverse: converting back from scientific to decimal notation. So that's just what we'll do.

Converting from Scientific to Decimal Notation

1. If the power of 10 is positive, move the decimal point to the right the same number of places as the exponent. You might need to put in zeros as placeholders when you run out of digits. Putting in commas will make it easier to read a large number.

Example: 4.01×10^8

4.01 (decimal moved eight places)

401,000,000 (fill in extra zeros)

2. If the power of 10 is negative, move the decimal point to the left the same number of places as the exponent, again putting in zeros as needed.

Example: 3.2×10^{-5}

3.2 (decimal moved five places)

0.000032

7. Write each number in decimal notation.

a. The number of red blood cells per microliter of blood in an average healthy man: 5.4×10^6

b. The mass of an average human ovum: 3.6×10^{-9} kg

Did You Get It

2. Convert each number to decimal notation.

a. 6.3×10^7 b. 2.973×10^{-3} c. 4.999×10^5 d. 9.103×10^{-6}

Now let's see if you really understand scientific notation.

8. When a number written in scientific notation has a positive exponent on 10, what can we say for sure about the number?

9. If a number between 0 and 1 is written in scientific notation, what can we say about the exponent on 10?

10. Fill in the blanks: If a number in scientific notation has a negative exponent on 10, to convert to decimal form, move the decimal place ____________ because the number is ______________.

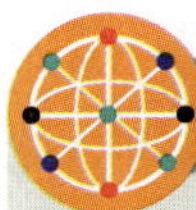

Using Technology: Scientific Notation in Calculators and Excel

Spreadsheets and graphing calculators are both programmed to provide answers in scientific notation only if the size of the result is especially large or small.

TI-84 Plus Calculator

```
99*125
                12375
990000*125000
            1.2375E11
(1.5E12)(3.2E-8)

                48000
```

Source: Texas Instruments

Notice that whether the numbers are entered in decimal or scientific notation, the result is in scientific notation only if it's very large. The calculator displays 1.2375×10^{11} as 1.2375E11.

To input 1.5×10^{12} in scientific notation:

Press 1.5 2nd , 12. This is listed as "EE" on the calculator.

If you want to force the results of calculations to be in scientific notation, put the calculator in scientific mode:

Press MODE ▷ ENTER to set the calculator to scientific mode. The result:

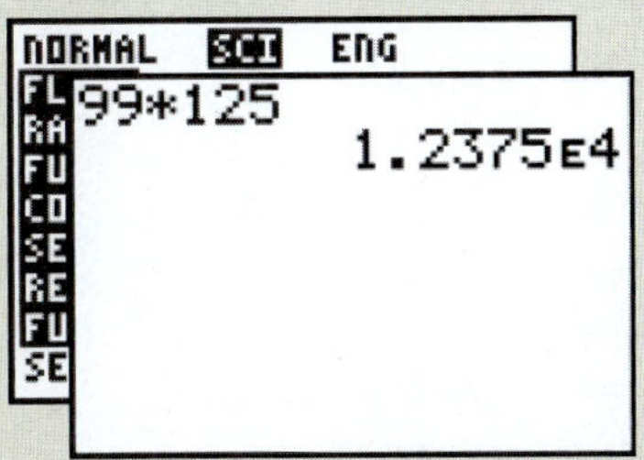

Excel

C1 ▾ f_x =A1*B1

	A	B	C
1	99	125	12375
2	990000	125000	1.24E+11
3	1.50E+12	3.20E-08	48000
4			
5	99	125	1.24E+04

Notice that in rows 1–3, Excel decides whether or not to put the result in scientific notation depending on its size.

To input 1.5×10^{12} in scientific notation, as in cell A3, type 1.5E12. Excel will interpret this as scientific notation.

If you compare rows 1 and 5, you can see that the product is the same in each case: 99 * 125. The difference in the format of the output is based on the formatting option chosen for the cell. In cell C1, the **General** format was chosen from the **Number** menu, in which case Excel decides on the format as mentioned. In cell C5, **Scientific** was chosen from the **Number** menu, so the result is shown in scientific notation regardless of its size.

See the Lesson 1-7 Using Tech video in class resources for further instruction.

11. Write the result of the calculation on the calculator screen in scientific and decimal notation.

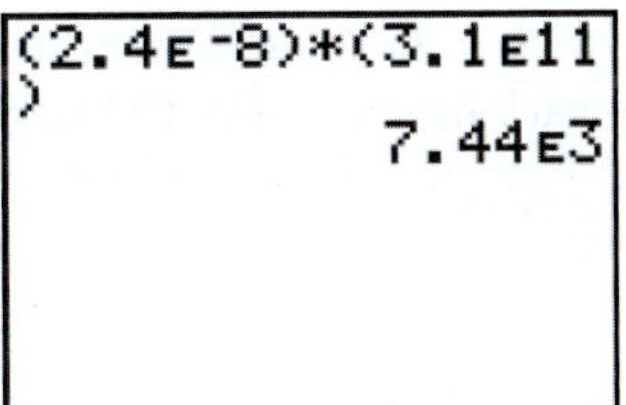

In the opening paragraphs for this lesson, we pointed out that when a large number is written in decimal form, it can be pretty difficult to even know how to pronounce the number. In scientific notation, however, it's easier, provided that you know the powers of ten that correspond to various numbers. Some common examples are provided in the following box.

Number	Corresponding power of 10
Hundred	10^2
Thousand	10^3
Million	10^6
Billion	10^9
Trillion	10^{12}

12. The size of the national debt in dollars for selected years from 1930 to 2015 is shown in the table and the graph. In the third column of the table, write how the debt would be read aloud.

Year	Debt in dollars	Pronunciation
1930	1.60E + 10	$16 billion
1940	5.10E + 10	
1950	2.57E + 11	
1960	2.86E + 11	
1970	3.71E + 11	
1980	9.08E + 11	$908 billion
1990	3.233E + 12	
2000	5.674E + 12	
2010	1.3562E + 13	
2015	1.8151E + 13	

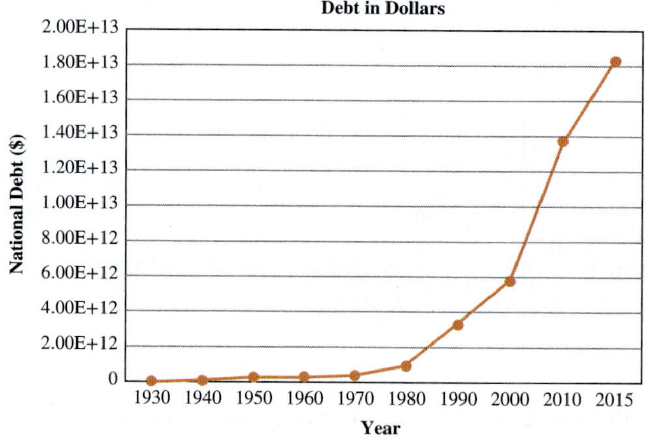

13. The population of the United States was roughly 1.23×10^8 in 1930, and 3.2×10^8 in 2015. If the national debt were to be paid off by dividing the total amount evenly among all citizens, how much would each have owed in 1930? What about in 2015?

Did You Get It

3. The United States government spent about $\$3.7 \times 10^{12}$ in 2015. How much did they spend on average for each person in the country?

1-7 Group

A fun fact: You can use a microwave oven and a bar of chocolate to measure the speed of light. Really! Microwaves are electromagnetic waves, just like visible light. The difference is the wavelength of these waves: For visible light, the wavelengths are in the neighborhood of 10^{-6} m, while microwaves are close to 10^{-2} m. Basically, microwaves emitted in an oven make molecules in food vibrate, which causes heat. The first places to heat up are separated by half of a wavelength, so by using something that starts to melt visibly in certain spots, we can calculate the wavelength. We can then use that to calculate the speed of light, which I think we can all agree is pretty darn cool.

If your teacher is unusually adventurous, he or she might bring in a microwave and chocolate bars, in which case you can perform the experiment on your own. Start with around 20 seconds, and make sure you take out the carousel: The chocolate has to stay stationary. (And now you know why a lot of microwaves have a carousel: Food heats more thoroughly at the spots corresponding to half of the wavelength.)

In case no microwave and chocolate are available, I did the experiment at home and took a picture for you, which is reproduced below. It's a life-size scale, so you can measure the distance between the melted spots with a ruler.

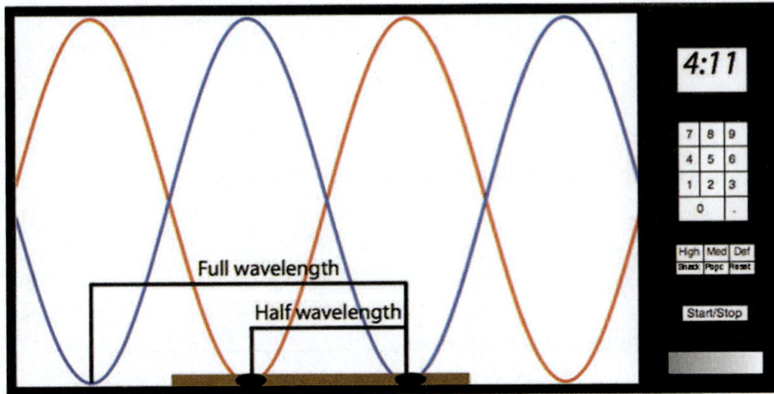

Courtesy of Dave Sobecki

1. Measure the distance between melted spots on your chocolate, trying to measure from center to center. Record your distance below, remembering that this is half the wavelength of the microwaves.

   Distance: ___________ cm

2. Convert the half-wavelength to meters. (Think about what the prefix "centi" means!)

   Distance: _____________ m

3. Write the full wavelength: _____________ m

4. If you have any, eat the chocolate.

5. For electromagnetic waves, the speed of light is the product of the wavelength and the frequency, which is the number of cycles per second (Hz). The frequency of most microwaves is 2.45 GHz: one GHz represents 10^9 cycles per second. Write the microwave frequency in scientific notation.

6. Use your answers to Questions 3 and 5 to calculate the speed of light in meters per second.

7. Use your phone or computer to look up the speed of light and write it here: _______________________

8. Find the error in your speed of light calculation. This is the difference between your calculation and the actual speed of light. What percent is this difference of the actual speed of light? (This is known as the percent error.) If your answer is WAY off, you might want to check the microwave to see if the frequency is different from 2.45 GHz.

Math Note

There's a link to a really cool video about how microwave ovens work in the online resources for this lesson.

9. This one requires some thought, but it will help you in the homework if you're successful. When a number is written in scientific notation with a negative exponent, it's almost always going to begin with a decimal point and a string of zeros in decimal notation. Looking over the conversion process, develop a rule for the number of zeros to start with.

10. Describe a couple of quantities that your group finds interesting that would be efficiently written in scientific notation. Describe why you think scientific notation would be reasonable for the quantities.

1-7 Portfolio

Name ____

Check each box when you've completed the task. Remember that your instructor will want you to turn in the portfolio pages you create.

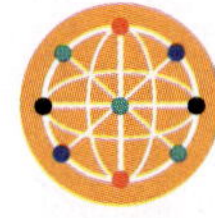

Technology

1. ☐ Watch the video on scale linked in the online resources for Lesson 1-7. Then tell everyone you know how cool it is. Even people you don't like.

Online Practice

1. ☐ Include any written work from the online assignment along with any notes or questions about this lesson's content.

Applications

1. ☐ Complete the Applications problems.

Reflections

Type a short answer to each question.

1. ☐ What is the point of scientific notation?

2. ☐ When a number is in scientific notation, how can you tell if it's really large or really small? Explain.

3. ☐ Name one thing you learned or discovered in this lesson that you found particularly interesting.

4. ☐ What questions do you have about this lesson?

Looking Ahead

1. ☐ Complete the Prep Skills for Lesson 1-8.

2. ☐ Read the opening paragraph in Lesson 1-8 carefully and answer Question 0 in preparation for that lesson.

Answers to "Did You Get It?"

1. **a.** 9.8×10^{-3} **b.** 1.478×10^{3} **c.** 5.4302×10^{7} **d.** 5.8×10^{-6}
2. **a.** 63,000,000 **b.** 0.002973 **c.** 499,900 **d.** 0.000009103
3. $11,562.50

Answers to "Prep Skills"

1. 100,000
2. 0.001
3. 8^7
4. 10^{-5}
5. 10^{17}; 8^{-1}
6. 5^4; 10^{-7}

1-7 Applications

Name ______________________________

1. According to the *Motley Fool,* in 2014 there were 148.6 million personal federal tax returns filed, and the average amount of tax paid per filer was \$9,118. Write each of these numbers in scientific notation.

©danielfela/Shutterstock RF

2. How much money total did the government take in from these taxpayers? Perform the calculation in scientific notation, and write how your answer would be pronounced.

3. An American dollar bill is 4.3×10^{-3} inches thick. If all of the tax revenue from Question 2 was paid in dollar bills, how many inches high would it be if they were all put in one stack? Write your answer in decimal notation.

4. To convert a number of inches into a number of miles, you divide by $12 \times 5{,}280$, which is the number of inches in a mile. How many miles high is our stack of tax proceeds?

5. The earth is 9.3×10^{7} miles from the sun. What percentage of the way to the sun would that big old stack of dollar bills reach? Write in both scientific and decimal notation.

1-7 Applications

Name ______________________________

6. Okay, enough silliness: back to more realistic calculations. The total amount of money taken in by the federal government in 2014 was about 3.02 trillion dollars. What percentage of this revenue came from personal income taxes? (Refer back to Question 2.)

7. Among those earning 10 million dollars or more, the average amount of taxes paid was $\$7.88 \times 10^6$. That's an awful lot of money. But what percentage is that of the government's revenue? Write in both decimal and scientific notation.

8. How did the amount of money taken in by the government in 2014 compare to the amount spent, which was $\$3.51 \times 10^{12}$? What can you conclude? (A good answer will say something about the national debt.)

Lesson 1-8 Prep Skills

SKILL 1: READ NUMBERS FROM A DIAGRAM

In Lesson 1-8 it will be really useful to identify information found in different parts of a diagram. The key skill is to clearly identify which part of a diagram is relevant in a given setting.

- The diagram below shows the number of students in the third-floor classrooms of a campus building during the 11:30–12:25 time block on my campus.

- There were 24 students in room 302.
- There were 4 more students in room 305 than there were in room 304.
- If there were 40 students combined in rooms 324 and 327, there must have been 40 − 33 = 7 students in room 327.

PREP SKILLS QUESTIONS

1. How many students were in room 309?
2. How did the number of students in room 305 compare to the combined number in rooms 301 and 302?
3. If there were 235 students total on that floor, how many were in room 327?

Lesson 1-8 What's Your Type?

©Stockbyte/Punchstock RF

LEARNING OBJECTIVES

- ☐ 1. Analyze how your personality type affects how you interact with others.
- ☐ 2. Create and interpret Venn diagrams.
- ☐ 3. Describe sets using appropriate terminology.

The meeting of two personalities is like the contact of two chemical substances; if there is any reaction, both are transformed.

—Carl Gustav Jung

In almost every walk of life, from social, to academic, to professional, working with and interacting with other people is important. But how many people actually take the time to think and learn about what may be the single most important aspect of their lives? In my view, one of the best things about higher education is being exposed to people with different backgrounds and ideas. When you learn to embrace the value of diversity, a new world of learning and social opportunities opens before you. In this lesson, we'll learn about different personality types and think about how your type affects the way you interact with others in a learning environment.

In preparation for this class, you should have taken an online personality assessment. The result is a four-letter code that describes your specific personality type. The code is made up of a combination of 8 letters in four pairs, so we'll begin by learning what each letter represents.

First pair: E (Extrovert) or I (Introvert)

Extroverts draw energy from action, preferring to act, then reflect, then adjust actions based on that reflection. Introverts expend energy from action, and prefer to think first, then act upon the thoughtful planning that they've done.

Second pair: S (Sensing) or N (Intuition)

Those with a preference for sensing prefer to base their thoughts and actions on concrete things that they can observe with their five senses; they distrust hunches and "gut feelings." The intuition people tend to trust information that is more abstract or theoretical, and information gathered from previous thoughts and experiences shared by others.

Third pair: T (Thinking) or F (Feeling)

Thinking and feeling are known as the decision-making functions; both are used to make conscious, rational decisions based on the data received from either sensing or intuition. Those with a thinking preference tend to make decisions based on what seems most logical and reasonable with a bit less regard for emotional impact. The feeling preference indicates a tendency toward decision making from a more emotional and empathetic viewpoint, trying to achieve balance and consensus among all people affected by a decision.

Fourth pair: J (Judging) or P (Perceiving)

These types are about the style you prefer when interacting with the outside world. A person with the judging preference is most likely to approach actions with a plan in place, preferring organization, preparation, and staying on schedule. The perceiving style folks are perfectly fine with "playing it by ear," making up the plan as they go. They're flexible and adaptive, and like new opportunities and challenges to pop up.

0. If you took the online personality inventory, write your four-letter code here. If you're unable to take the online inventory, choose one letter from each pairing that you feel best fits your personality, then write the code here.

1-8 Class

Now that we know a little bit about personality characteristics, we'll use a new type of graph known as a **Venn diagram** to see how your personality type compares to others, and later you'll reflect on how this might affect working with others in groups. Venn diagrams are used to represent sets of objects (in this case the letters from the Myers Briggs personality test) and visualize what they have in common, and how they differ.

1. Pair up with someone else in the class, and write your partner's code here: _____
   (It would be a good idea to pick someone who doesn't have the exact same type as you.)

2. Label the two circles in the Venn diagram with one of your names on each line.

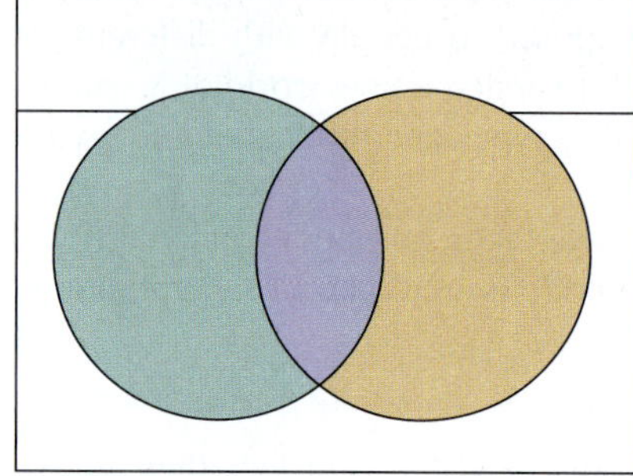

3. Write the letters your personality type has in common with your partner's in the overlapping portion of the two circles.

4. Write any letters in your personality type but NOT your partner's inside the circle labeled with your name, but outside the overlapping portion.

5. Write any letters in your partner's personality type but NOT yours inside the circle labeled with your partner's name, but outside the overlapping portion.

6. Write the remaining letters that don't appear in either code outside the circles, but inside the rectangle.

7. This Venn diagram shows which personality traits you have in common with your partner, and which you each have separately. Write a description of what the diagram says to you.

Venn diagrams are used all the time in studying *sets*. A **set** is any collection of objects, like the letters that make up your personality type. We'll prefer to work with sets that are **well-defined**: A set is well-defined if you can objectively determine (based on facts, not opinion) whether an object does or does not belong in a set.

8. Which of the following sets is well-defined? Why?
a. The set of all good shows available to binge watch on Netflix
b. The set of shows with all seasons available on Netflix

Each object in a set is called an **element** or a **member** of a set. One method of describing a set is called the **roster method**, in which elements are listed between braces, with commas between the elements. The order in which we list elements isn't important: $\{2, 5, 7\}$ and $\{5, 2, 7\}$ are the same set.

Often, we will name sets by using a capital letter. We might name the set of objects in the left circle in the diagram in Question 2 using the letter *A*.

9. Use the roster method to list the elements of the left circle: $A = \{$_____, _____, _____, _____$\}$

The **universal set** is the set of all objects under consideration in a given situation (not all the objects in the universe, which would be silly).

10. In the case of the letters from the personality test, the universal set could be written as: $U =$ ____________________

The **complement** of a set is the set of all objects from the universal set that are not in that set. The complement of a set *A* is denoted A'.

11. For the set *A* that you wrote above, what is A'?

Notice that there are four distinct regions in a Venn diagram illustrating two sets A and B. We'll want to number the regions for reference; we use Roman numerals so that we don't confuse the number of the region with elements in the set.

12. Write the appropriate description next to each region. Pick your answer from the following list:

the elements in set B that are not in set A.
the elements in both sets A and B.
the elements in the universal set that are in neither set A nor set B.
the elements in set A that are not in set B.

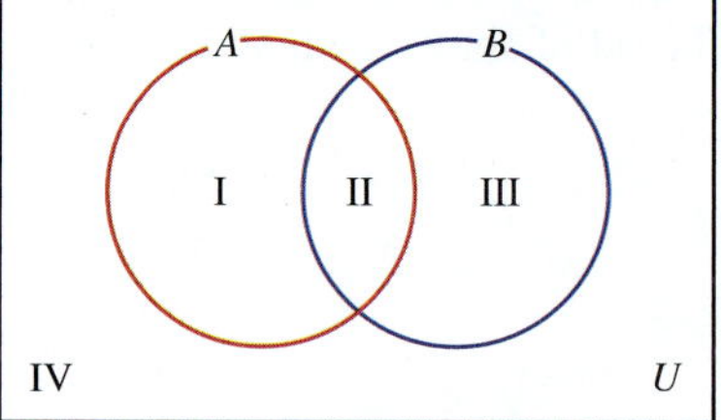

Region I represents

Region II represents

Region III represents

Region IV represents

13. The set of elements that are in both sets A and B is called the **intersection of A and B**, and is symbolized $A \cap B$. Write the region or regions in the Venn diagram that correspond to $A \cap B$.

14. The set of elements that are in at least one of sets A and B is called the **union of A and B**, and is symbolized $A \cup B$. Write the region or regions in the Venn diagram that correspond to $A \cup B$.

Did You Get It

Try these problems to see if you understand the concepts we just studied. The answers can be found at the end of the Portfolio section.

1. If the universal set is the days of the week, write the set D of days beginning with the letter T or S using the roster method. Then do the same for the complement of set D.
2. Which region(s) in the Venn diagram correspond(s) to the set A'?

1-8 Group

We can also draw Venn diagrams with three circles to study interactions between three different sets. In this activity, you'll compare characteristics with two other people in your group.

1. Complete the following Venn diagram by placing the letter next to each statement in the appropriate location. For example, if the two people corresponding to the top two circles like to study late at night and the third doesn't, you should put A in the portion marked with the arrow.

A. Like to study late at night
B. Like to study early in the morning
C. Like to get work done early
D. Like to put work off to the last minute
E. Will miss very few classes
F. Usually miss a lot of classes
G. Working a job for 20 or more hours per week
H. Working a job for less than 20 hours per week
I. Taking 15 hours or more of classes
J. Taking less than 15 hours of classes
K. Have children
L. Do not have children
M. Feel confident in most math courses
N. Do not feel confident in most math courses

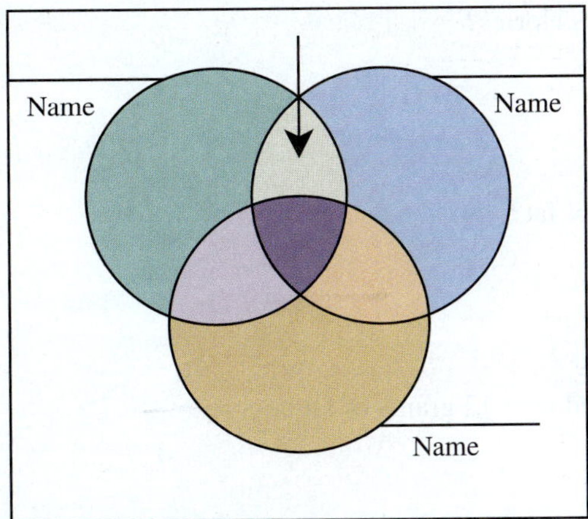

2. Write any observations you find interesting from the completed diagram.

Math Note

It's possible to draw Venn diagrams for more than three sets, but it gets pretty complicated, making them kind of difficult to interpret.

We've seen Venn diagrams where the stuff written inside the circles represents actual elements of the set represented by that circle. Often, when using Venn diagrams to organize information, we'll instead label portions of the diagram with the number of elements that live in that particular set.

The next Venn diagram illustrates the result of a study done on 400 entrees at 75 campus cafeterias. Use the results summarized in the Venn diagram to answer Questions 3–6.

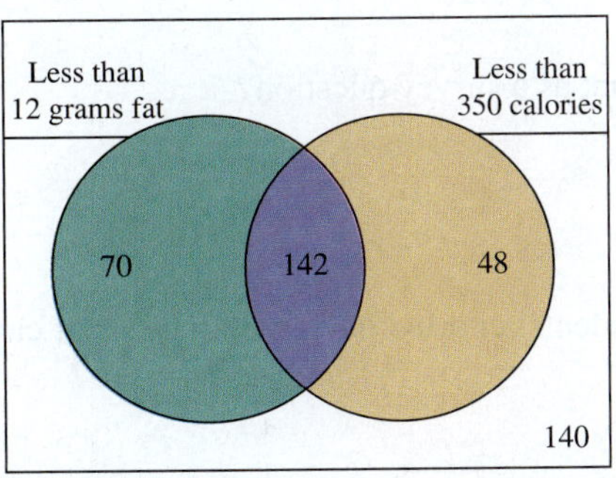

3. How many meals had less than 12 grams of fat and less than 350 calories?

4. How many meals had less than 350 calories?

5. How many had less than 12 grams of fat?

6. How many had over 350 calories and over 12 grams of fat?

Arranging the results of surveys is a really common application of Venn diagrams. The way that questions are asked in a survey can have a big effect on the results. Here are two different ways that students could be surveyed on usage of credit and debit cards:

What type of card do you carry? Choose one	
Credit	☐
Debit	☐
Neither	☐

What type of card do you carry? Choose one	
Credit	☐
Debit	☐
Both	☐
Neither	☐

7. According to Sallie Mae's 2013 "How America Pays for College" report, 800 undergraduate students were asked about their credit and debit card habits: 616 reported carrying a debit card and 240 a credit card. Which of the two questions above was asked? How can you tell?

8. Why does the first question kind of stink as a survey question?

9. The survey reported that 216 of the students reported having both types of cards. Which region in the Venn diagram should be labeled with 216? Label it!

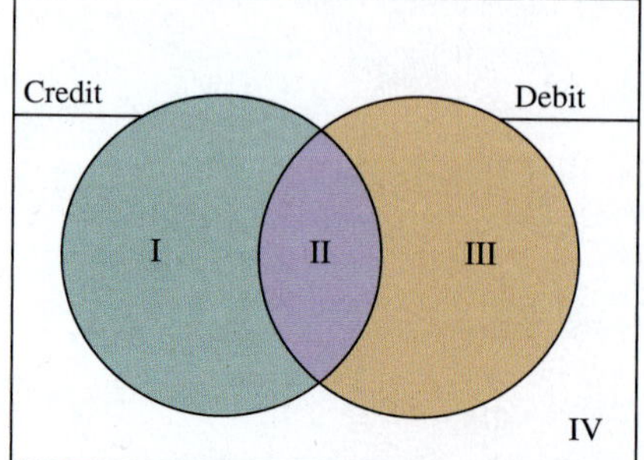

10. Does the fact that the report said that 616 students reported carrying a debit card mean that all 616 carried ONLY a debit card? Explain.

In Questions 11–13, label appropriate sections of the Venn diagram as you find the requested information.

11. How many students carried only a debit card? Think carefully about Questions 9 and 10.

12. How many carried only a credit card?

13. How many carried neither card?

Did You Get It

3. In a 2013 survey reported by Public Policy Polling, 500 Americans were asked about their soft drink preferences. There were 240 who reported drinking regular soda, 185 who drink diet soda, and 15 who drink both. Use a Venn diagram to answer each question.

   a. How many drink only diet soda?
   b. How many drink neither?

According to a survey conducted by the National Pizza Foundation that I just now made up, out of 113 customers surveyed, 47 prefer pizza with pepperoni, 56 prefer sausage, and 30 prefer onion. These results are displayed in the next Venn diagram.

14. According to the information given in the previous paragraph, 47 customers prefer pizza with pepperoni. But notice that 47 doesn't appear in the diagram anywhere. Why not?

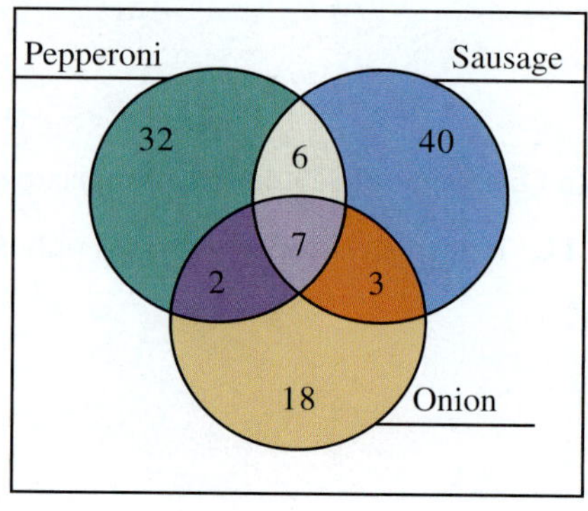

15. How many customers prefer pizza with only sausage (and none of the other two ingredients)?

16. How many customers prefer pizza with all three ingredients?

17. How many prefer pizza with sausage and onion?

18. I hate to call anyone boring, but . . . how many go the boring route (none of these items)?

1-8 Portfolio

Name ______________________________

Check each box when you've completed the task. Remember that your instructor will want you to turn in the portfolio pages you create.

Technology

1. ☐ Find a Venn diagram online based on a topic that interests you. Copy and paste the image into your portfolio, then explain why you found it interesting, and what you learned from it.

Online Practice

1. ☐ Include any written work from the online assignment along with any notes or questions about this lesson's content.

Applications

1. ☐ Complete the Applications problems.

Reflections

Type a short answer to each question.

1. ☐ Think about what you learned about yourself from studying your personality type. How can you use this information in your academic and social lives?
2. ☐ Now consider your personality type as well as the Venn diagram at the top of page 109. How can you use what you learned to help you in this class, and other classes where group work is important?
3. ☐ Name one thing that you learned or discovered in this lesson that you found particularly interesting.
4. ☐ What questions do you have about this lesson?

Looking Ahead

1. ☐ Complete the Prep Skills for Lesson 1-9.
2. ☐ Read the opening paragraphs in Lesson 1-9 carefully and answer Question 0 in preparation for that lesson.

Answers to "Did You Get It?"

1. $D = \{$Tuesday, Thursday, Saturday, Sunday); $D' = \{$Monday, Wednesday, Friday$\}$
2. Regions III and IV
3. 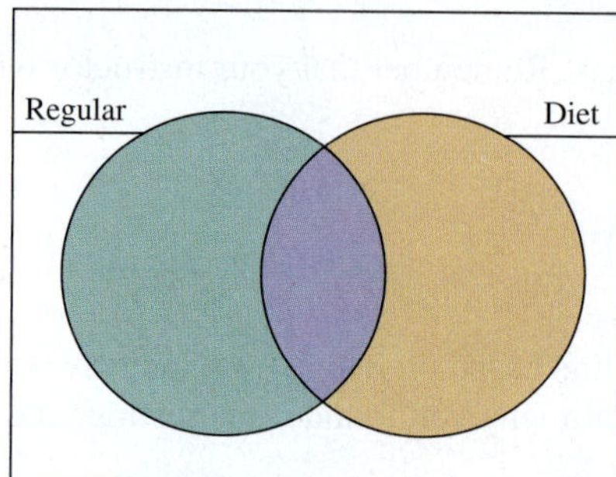

   a. Diet only: 170; **b.** neither: 90

Answers to "Prep Skills"

1. 19 students
2. There are 12 fewer students in room 305 than in rooms 301 and 302 combined.
3. 24 students

1-8 Applications

Name ______________________________

According to this diagram from Boston Children's Hospital, there are a lot of similarities between the symptoms you experience when you have a cold and symptoms caused by allergies.

1. List the set of cold symptoms using the roster method.

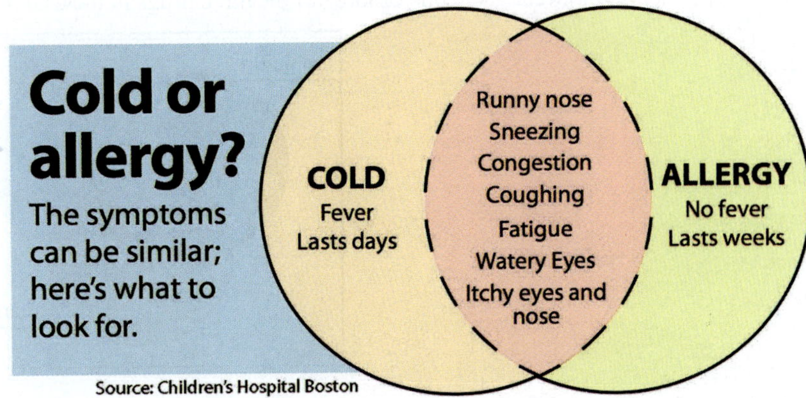

2. If C is the set of cold symptoms and A is the set of allergy symptoms, write the set $C \cap A$

3. Describe the symptoms that help you distinguish between a cold and allergies.

4. If C is the set of cold symptoms and A is the set of allergy symptoms, how would we represent the set below?

   {Fever, lasts days, runny nose, sneezing, congestion, coughing, fatigue, watery eyes, itchy eyes and nose, no fever, lasts for weeks}

In a survey of 130 college students, 108 use Instagram, 37 use Facebook, and 21 use both. Use this information to fill in the Venn diagram to the right, then use it to answer the questions.

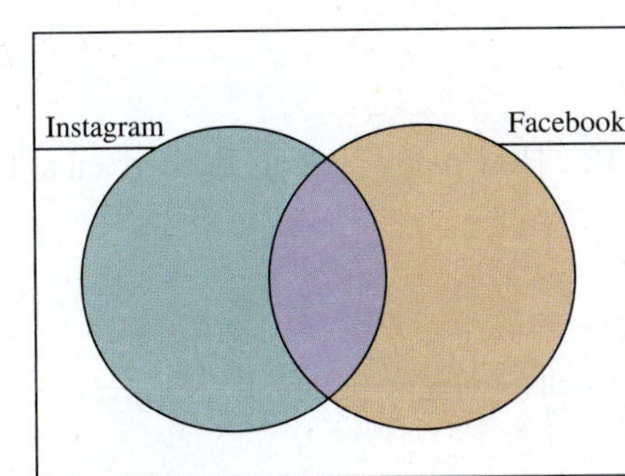

5. How many use Instagram only?

6. How many use Facebook only?

7. How many use neither?

1-8 Applications

Name ______________________________

As the different types of bacteria that cause illness evolve into drug-resistant strains, it's becoming more common for doctors to treat infections with multiple antibiotics. Tuberculosis is one of the diseases where multiple medications are often used. A group of 200 patients with active TB infections enrolled in a clinical trial to test the effects of various antibiotics. The Venn diagram shows the number of patients who were given one or more of the three drugs in the trial. Use the diagram to answer Questions 8–13.

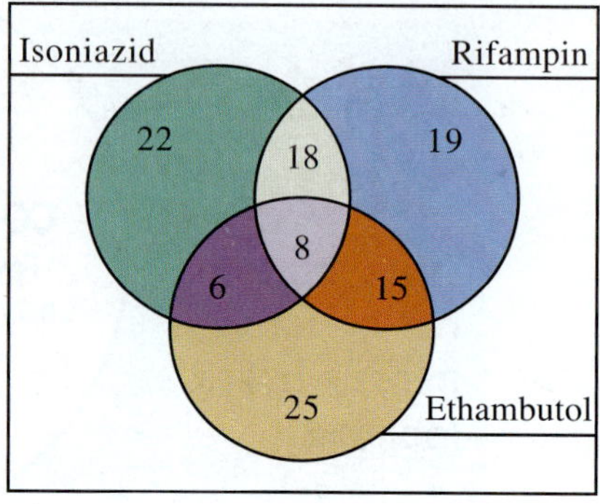

8. How many patients were given all three drugs?

9. How many patients were given just one drug?

10. How many patients were given Rifampin and Ethambutol but not Isoniazid?

11. How many patients were given exactly two of the three drugs?

12. How many patients were given at least Isoniazid and Ethambutol?

13. How many patients were given a placebo containing none of those three drugs?

Lesson 1-9 Prep Skills

SKILL 1: READ DATA FROM A LIST

Unorganized data can be difficult to read, especially if there are a lot of values, or many identical or similar values. In that case, rewriting the list in order is helpful.

The scores for 20 students on a 30-point quiz are listed here.

18	30	27	26	18	29	29	30	30	24
27	30	12	15	22	28	27	27	24	25

If we order the list, we get this:

12	15	18	18	22	24	24	25	26	27
27	27	27	28	29	29	30	30	30	30

- The most common scores are 27 and 30.
- The lowest score was 12 and the highest was 30.

SKILL 2: DRAW A BAR GRAPH

A bar graph is used to compare the sizes of different data values when those values come from distinct categories. On the list of quiz scores above, we see that there were 2 scores less than 18, 2 from 18–20, 1 from 21–23, 4 from 24–26, and 11 from 27–30. We could represent these numbers with a bar graph:

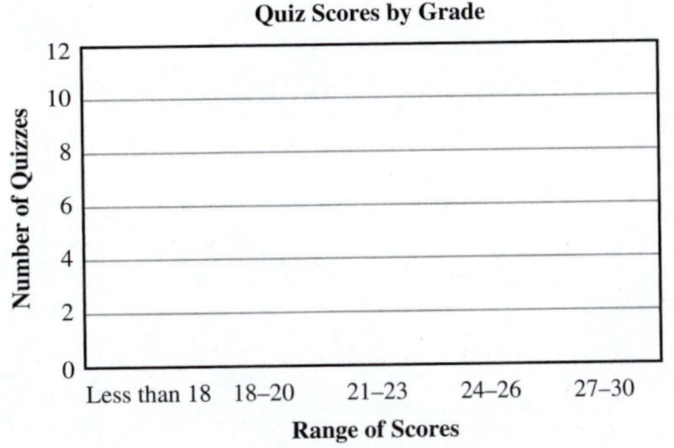

PREP SKILLS QUESTIONS

The data below are the number of classes missed in one semester by all students in a five-day-a-week math class.

2	4	0	1	1	10	6	2	3	8	0	1
3	0	18	2	0	3	3	0	1	5	4	2

1. Arrange the list in order from smallest to largest.
2. What number of days missed was most common?
3. What were the least and most days missed by anyone in the class?
4. Draw a bar graph that illustrates how many students missed 0 or 1 day, 2–4 days, 5–8 days, and more than 8 days.

Lesson 1-9 News in the Data Age

LEARNING OBJECTIVES

- ☐ 1. Explain the difference between a population and a sample.
- ☐ 2. Organize data with frequency distributions and histograms.
- ☐ 3. Analyze data with stem and leaf plots.

Errors using inadequate data are much less than those using no data at all.

—Charles Babbage

©scanrail/Getty Images RF

With the rise of online news sites and social media, many people have declared the death of traditional news sources in our country. But are newspapers really dead? Are they even sick? Statements like "Everyone just gets news online now" or "Who even reads a newspaper anymore?" aren't evidence—they're opinions. In order to decide if there really has been a significant decline in traditional news sources, we need to do better than listen to anecdotes or read opinions in 140 characters or less.

Instead, we would want to gather some **data**, which are measurements or observations that are gathered for an event under study. The term "data age" is coming into vogue to describe this era because of the vast amount of data that is available to anyone with Internet access and a web browser. But the expense of having access to so much data is that knowing how to gather data appropriately, and how to organize it into a useful form, is more important than it's ever been. So that's what we'll study in this lesson.

0. What types of data do you think it would be useful to look for in studying whether traditional newspapers are dying?

1-9 Class

There are a number of different approaches we could take in trying to evaluate the condition of traditional newspapers, but all of them would begin with gathering relevant data. We could study things like circulation numbers, advertising revenue, how many people read print newspapers compared to online news, and many others.

The table below shows the average daily circulation for newspapers in America for certain years from 2001 to 2015.

Year	2001	2003	2005	2007	2009	2011	2013	2015
Avg. circulation (thousands)	55.6	55.2	53.3	50.7	45.7	43.4	40.7	37.7

Source: Pew Research Center

1. What preliminary conclusions can you draw from the data in the table?

2. Draw a scatter plot of the data. Does this reinforce the conclusions you drew in Question 1, or would you adjust your conclusions now?

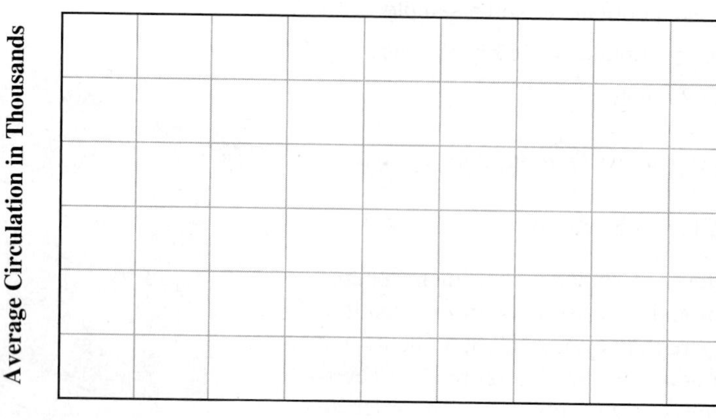

The data that we've seen so far seem to indicate a clear negative trend for newspaper circulation. But where did that data come from? If this takes into account ALL print newspapers in the country, that would certainly be more compelling than if the numbers came from just a handful. This brings up our first key idea in gathering data.

A **population** consists of all subjects under study. A **sample** is a representative subgroup or subset of a population.

3. In the case of the newspaper circulation numbers, the population would be ______________________________.

4. A sample would be ____.

If we're going to use a sample to draw conclusions on a population, it's important that the sample is **representative**. A sample is considered representative if many different types of members of a population are included.

For Questions 5–8, which of the following do you think would be a representative sample when studying newspaper circulation?

5. Daily newspapers in the 10 biggest cities in the United States

6. One newspaper randomly chosen from each of the 50 states

7. All newspapers affiliated with a national media conglomerate

8. All newspapers in the country are listed alphabetically, and every 10th one is chosen.

> **Math Note**
>
> The word "data" is plural, so we say "data are" not "data is." The singular version is "datum." You don't hear that word very often because it's awfully hard to draw a meaningful conclusion from one tiny piece of information.

The circulation numbers that we've seen so far (as near as I can tell) are population data: While it's not 100% clear from the source, it appears that these numbers are an average of all print newspapers in America. The next data we study are not.

The next table shows the results of a 2015 survey in which adults who read newspapers were asked, "In what format do you read newspapers?"

Format	Percentage
Print only	51%
Print/Desktop computer	11%
Print/Desktop/Mobile device	14%
Desktop only	5%
Desktop/mobile	7%
Print/mobile	7%
Mobile only	5%

9. Which would be most effective in illustrating this data: scatter plot, pie chart, or bar graph? Why?

10. How can you tell that this data came from a sample, rather than a population?

11. It's a relatively common belief that print newspapers are becoming a thing of the past. Do you think the data above supports that idea? Explain.

Did You Get It

Try this problem to see if you understand the concepts we just studied. The answers can be found at the end of the Portfolio section.

1. Suppose that we're interested in the sources that students at your campus use to get news. Classify each of the groups below as population, representative sample, or not representative sample.
   a. A researcher stands next to a box where students can pick up a daily newspaper and asks where they typically get news.
   b. Every student on campus is required to answer a survey on news sources.
   c. Students are divided by class rank, and each student in each class with a student ID number that ends in 4 is surveyed.

1-9 Group

Frequency Distributions

The data collected for a statistical study are called **raw data**. Raw data can be really difficult to interpret; in order to describe situations and draw conclusions, we need to organize the data in a meaningful way. Two methods that we will use are *frequency distributions* and *stem and leaf plots*. First up, we'll study **categorical frequency distributions**, which are a way to organize data that are divided into distinct categories, like gender, your class standing in school, or conferences for college football teams.

A categorical frequency distribution is a table in which you list different categories, then count up how many individuals in a sample or population fall into each category and list those frequencies. In Questions 1 and 2, you'll create a categorical frequency distribution.

1. Twenty-eight college students were surveyed about what source they are most likely to get news from. Their responses are listed below. What should the categories be for this data? Why?

Social media	Print media	TV	Social media
Internet	Internet	Print media	Radio
TV	Radio	Internet	Internet
Print media	Internet	Internet	TV
Radio	Social Media	Internet	Radio
TV	TV	Social media	Radio
Internet	Internet	TV	Print media

2. Build a categorical frequency distribution by writing in your categories, then putting tally marks in the second column as you count the number of individuals in each category.

Categories	Tally marks	Frequency

3. Use your distribution to write some conclusions on the sources typical college students use to keep up on recent news. Do you think that there's enough data for your conclusions to be reliable?

As we saw in Lesson 1-1, we can use a bar graph to illustrate categorical data.

4. Below are two sets of axes that can be used to draw a bar graph illustrating the data in your frequency distribution. Draw a bar graph on each set of axes.

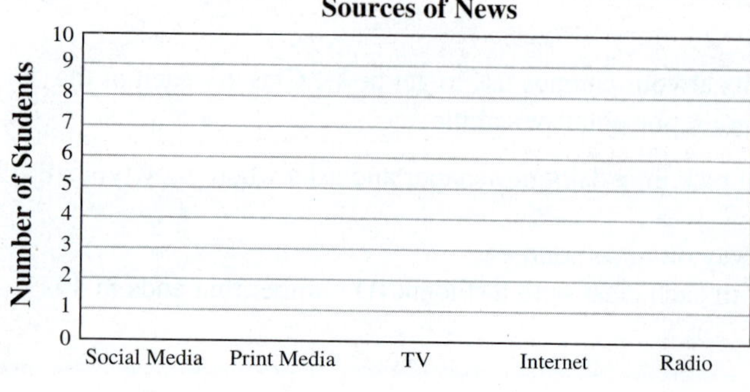

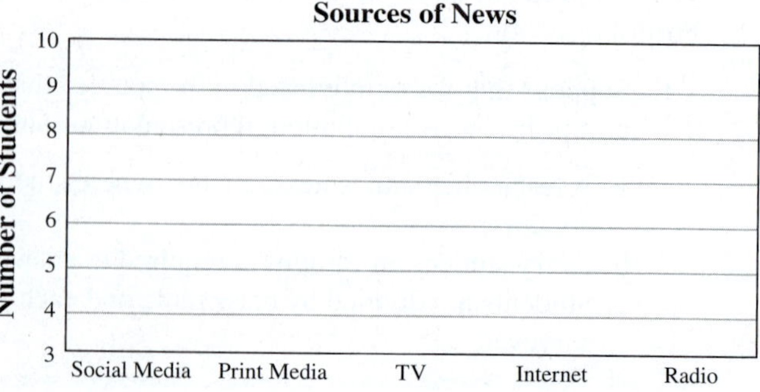

5. Which bar graph would you use if you wanted to exaggerate the difference between the number of students who indicated print media compared to radio? Explain why that graph would be misleading.

Did You Get It

2. A health-food store recorded the type of vitamin pills 35 customers purchased during a 1-day sale. Construct a categorical frequency distribution for the data.

C C C A D E C E E A B D C E C E C C

C D A B B C C A A E E E E A B C B

3. Draw a bar graph based on the results of your frequency distribution.

Another type of frequency distribution that can be constructed uses numerical data and is called a **grouped frequency distribution**. In a grouped frequency distribution, the numerical data are divided into **classes**. For example, if you gathered data on the weights of people in your class, there's a decent chance that no two people have the exact same weight. In that case, everyone would be in a separate category, and each category would have one weight in it. This, of course, would make the frequency distribution no more useful than the original list of raw data!

So it would be reasonable to group people into weight ranges, like 100–119 pounds, 120–139 pounds, and so forth. In the 100–119 pound class, we call 100 the **lower limit** and 119 the **upper limit**.

We'll illustrate by organizing the following data. While the data we've seen indicate that print newspapers have been declining in circulation for many years, most news sources now have a significant online presence. The table lists the number of unique visitors (in thousands) to a number of different online news sources for the month of January 2015.

Site	Unique visitors (in thousands)	Site	Unique visitors (in thousands)
AZCentral.com	6.6	NYDailyNews.com	25.9
BostonGlobe.com	9.8	NYPost.com	22.9
ChicagoTribune.com	12.0	NYTimes.com	54.0
Chron.com	14.4	OregonLive.com	6.3
Cleveland.com	6.5	OrlandoSentinel.com	5.6
DailyMail.co.uk	51.1	SeattleTimes.com	6.1
DallasNews.com	7.0	SFGate.com	19.0
Freep.com	10.6	Telegraph.co.uk	16.8
Independent.co.uk	11.5	TheGuardian.com	28.2
LATimes.com	25.2	USAToday.com	54.5
Mirror.co.uk	12.0	WashingtonPost.com	47.8
Newsday.com	6.0	WashingtonTimes.com	7.0
Nola.com	6.0		

6. Why would a categorical frequency distribution be a bad idea for this data? (Hint: What would the categories be?)

Rather than using each individual data value as a separate category, we'll divide the entire range of data values into evenly spaced portions.

7. Note the classes we chose for dividing the data, and complete a frequency distribution by tallying up the number of sites that fall into each class.

> **Math Note**
>
> In a statistics class, you'll learn a lot about how to make good choices of classes when organizing data with a grouped frequency distribution.

Class	Tally marks	Frequency
5–14.9		
15–24.9		
25–34.9		
35–44.9		
45–54.9		

When data are organized in a grouped frequency distribution, we draw a special kind of bar graph to illustrate. This type of graph is called a **histogram**, and it differs from a bar graph in two main ways. First, the horizontal axis isn't labeled with category names, like Radio or 5–14.9. Because the data values are numerical, we'll label the horizontal axis in number line form, like the scatter plots and time-series graphs we've studied.

As a consequence of that labeling, we get a second difference: The bars won't be physically separated by blank space; instead, they'll be touching each other.

8. Using the axes shown, draw a histogram based on your frequency distribution.

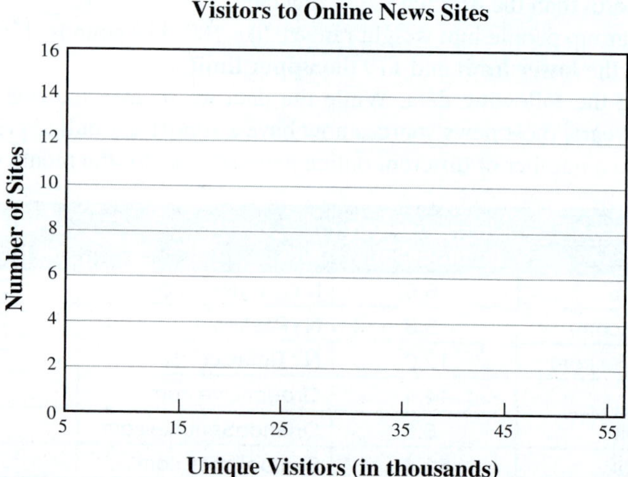

Comparison of Bar Graphs and Histograms

	Bar graph	Histogram
Use	Illustrating categorical data	Illustrating numerical data
Title	Included at top of graph	Included at top of graph
Bars	Include spaces between bars	Edges of bars touch: no spaces
Labels	Vertical scale is frequency or % Horizontal labels are categories	Vertical scale is frequency or % Horizontal labeled with a numeric scale

9. Write down at least three observations you can make based on your frequency distribution and histogram.

10. Look back at the raw data on visitors to news websites. Do any of the data values live exactly on one of the borders of the bars in your histogram?

11. Suppose that a different news site, let's just call it Wuzzup.com, had exactly 25.0 thousand visitors in January 2015. Looking *strictly at your histogram,* which bar would that data point raise the height of? In other words, which group of values does it go into, the ones from 15–25, or the ones from 25–35?

12. Now look at your *frequency distribution.* Which class would Wuzzup.com go into?

Did You Get It

4. The online homework system I use in my calculus classes allows me to track how much time students spend working on homework. In my two Calc 2 sections last semester, here are the average number of hours spent per week for each student. Construct a grouped frequency distribution and histogram for the data. Your classes should be 1–3, 4–6, 7–9, 10–12, 13–15, and 16–18.

1	2	6	7	12	13	2	6	9	5
18	7	3	15	15	4	17	1	14	5
4	16	4	5	8	6	5	18	5	2
9	11	12	1	9	2	10	11	4	10
9	18	8	8	4	14	7	3	2	6

Stem and Leaf Plots

Another way to organize data is to use a stem and leaf plot (sometimes called a stem plot). Each data value or number is separated into two parts. For a two-digit number such as 53, the tens digit, 5, is called the **stem,** and the ones digit, 3, is called its **leaf.** For the number 72, the stem is 7, and the leaf is 2. For a three-digit number, say 138, the first two digits, 13, are used as the stem, and the third digit, 8, is used as the leaf. For values rounded to the tenths place, like 8.4, you can use the value to the left of the decimal place as stem and the tenths place as leaf. In any case, the very last digit is used as the leaf, and what comes before is the stem. Note that we'll include a key with our plot that clarifies what the stems and leaves represent.

Market penetration has been a traditional print newspaper metric. It measures the number of newspapers sold in each market as a percentage of households and includes online readers as well. The top 25 papers in America in terms of market penetration are listed below, along with the percentage of households in their metropolitan area that they reach.

Paper	Market penetration	Paper	Market penetration
Arizona Daily Star	56	*Memphis Commerical Appeal*	52
Arkansas Democrat-Gazette	62	*Milwaukee Journal-Sentinel*	59
Austin American-Statesman	51	*Minneapolis Star Tribune*	52
Buffalo News	64	*New Orleans Times-Picayune*	64
Chattanooga Times Free Press	53	*Oklahoman*	49
Cincinnati/Kentucky Enquirer	54	*Portland Oregonian*	51
Columbus Dispatch	53	*Richmond Times-Dispatch*	58
Des Moines Register	68	*Rochester Democrat and Chronicle*	75
El Paso Times	52	*San Antonio Express-News*	53
Gannett Wisconsin	68	*San Diego Union-Tribune*	51
Honolulu Star-Advertiser	59	*Syracuse Post-Standard*	63
Kansas City Star	56	*Washington Post*	57
Louisville Courier-Journal	58		

13. If you were to build a grouped frequency distribution, what classes would you choose?

14. What will you choose to be the stems for this data? What about the leaves?

15. Do you think the data in this table come from a representative sample of all American newspapers in terms of market penetration? Why or why not?

16. List all of the first digits that appear in percentages in the table. (You don't need to repeat them if they appear more than once.)

17. Write your answers from Question 16 in order under "Stems" in the table below. Write the lowest value first, then larger values as you work your way downward.

18. Go through the data values in the table, and one by one write the second digit in each value under "Leaves," next to the corresponding stem. For example, since the first data value is 56, you'd write a 6 in the leaves column next to 5 in the stems column. Separate leaves that correspond to the same stem with some space, but not with commas. Put in ALL second digits, including repeats.

Stems	**Leaves**

Key: 5|6 means 56%

19. When drawing a stem and leaf plot, it's traditional to write the leaves in order from least to greatest. This makes it a lot easier to see what the spread of values looks like within one stem. Finish your stem and leaf plot by ordering the leaves.

Stems	**Leaves**

Key: 5|6 means 56%

Constructing a Stem and Leaf Plot

Step 1: Identify stems and leaves. The leaves will typically be the last digit in numeric data; the stems will be the digits that come before.

Step 2: Write "Stems" and "Leaves" with a horizontal line beneath them, and a vertical line separating them, extending downward.

Step 3: Write the stems you chose in order under "Stems," in order from least to greatest.

Step 4: Working through the data values one at a time, write each leaf to the right of the vertical line next to the corresponding stem. Leave spaces between leaves that are in the same row.

Step 5: Redraw your plot, putting all leaves within a given row in order from least to greatest.

Step 6: Write a key under your plot that describes what the data values mean. For example, if one of the stems is 51 and a corresponding leaf is 6, and this represents a data value of 51.6, write "51|6 means 51.6."

20. Write at least three observations you can make about the market penetration data from looking at your stem and leaf plot.

Did You Get It

5. It's no secret that gas prices have fluctuated wildly in the last few years, but from 1980–1999, they were surprisingly stable. According to the Energy Information Administration, the following data represent the average price (in cents) per gallon of regular unleaded gas for those years. Draw a stem and leaf plot of the data.

119	131	122	113	112	86	90	90	100	115
114	113	111	111	115	123	123	123	106	117

21. Compare grouped frequency distributions, histograms, and stem and leaf plots. How are they similar? Which do you think makes it easier to analyze a set of data?

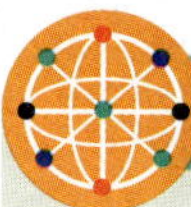

Using Technology: Using the Sort Feature on a Calculator and a Spreadsheet

To sort data in Excel:

1. Enter the data.
2. Select the cells containing the data.
3. Click on the **Data** tab.
4. Click on **Sort**.
5. Select options and click OK.

To sort data on a TI-84 Calculator:

1. Enter the data in list **L1**. (Access L1 by first pressing STAT and ENTER.)

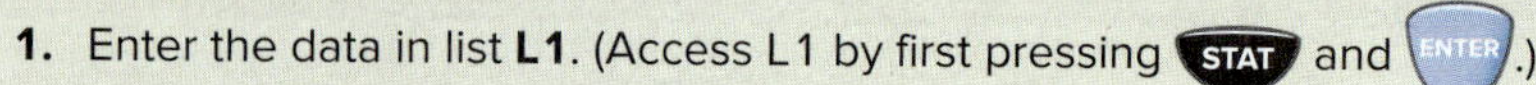

2. Press 2nd, STAT, ›, ENTER to access the **SortA(** feature.
3. Press 2nd, STAT, ENTER to input L1 after **SortA(**.
4. Press the) key and then press ENTER.
5. Go back to L1 and the data you entered should be sorted.

See the Lesson 1-9 Using Technology Video in class resources for further instruction.

1-9 Portfolio

Name ____

Check each box when you've completed the task. Remember that your instructor will want you to turn in the portfolio pages you create.

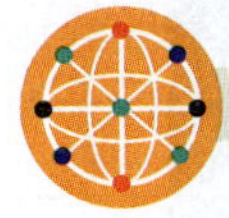

Technology

1. ☐ Excel can be used to do two tasks from this lesson that are both useful and tedious to do by hand: sorting data and building a grouped frequency distribution. In the online resources for this lesson, there's a spreadsheet that has two of the data sets from this lesson entered: student responses to the media survey, and usage numbers for 25 selected online media outlets. Your first job is to sort the data based on the instructions that appear on the spreadsheet. Note that there's a different worksheet for each data set, and you can change from one to another using the tabs along the bottom. After sorting, there will be further instructions on the template for building the frequency distribution.

Online Practice

1. ☐ Include any written work from the online assignment along with any notes or questions about this lesson's content.

Applications

1. ☐ Complete the Applications problems.

Reflections

Type a short answer to each question.

1. ☐ What is the point of sampling in statistics? What's the difference between a sample and a population?
2. ☐ How do frequency distributions, histograms, and stem and leaf plots help to organize data?
3. ☐ Why is it bad to have too few or too many categories in a grouped frequency distribution?
4. ☐ Name one thing you learned or discovered in this lesson that you found particularly interesting or useful.
5. ☐ What questions do you have about this lesson?

Looking Ahead

1. ☐ Complete the Prep Skills for Lesson 2-1.
2. ☐ Read the opening paragraph in Lesson 2-1 carefully and answer Question 0 in preparation for that lesson.

Answers to "Did You Get It?"

1. a. Not representative sample **b.** Population **c.** Representative sample

2.

Type	Tally marks	Frequency
A	𝍸 I	6
B	𝍸	5
C	𝍸 𝍸 II	12
D	III	3
E	𝍸 IIII	9

3.

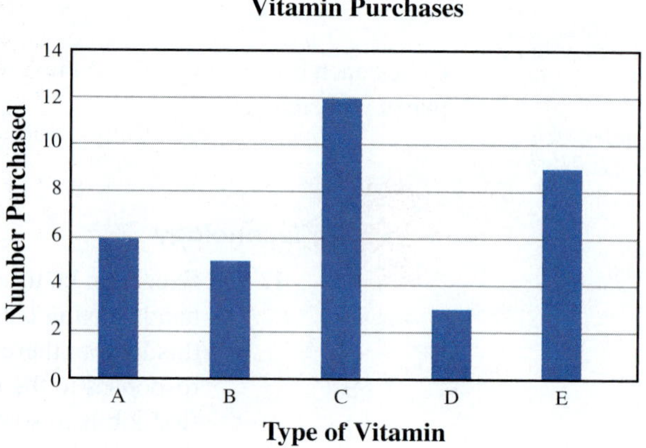

4.

Class	Tally marks	Frequency
1–3	𝍸 𝍸	10
4–6	𝍸 𝍸 IIII	14
7–9	𝍸 𝍸	10
10–12	𝍸 I	6
13–15	𝍸	5
16–18	𝍸	5

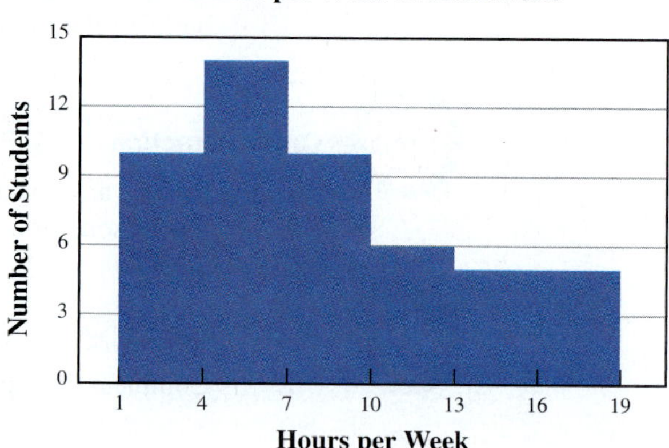

5.

Stems	Leaves
8	6
9	0 0
10	0 6
11	1 1 2 3 3 4 5 5 7 9
12	2 3 3 3
13	1

Key: 13|1 means 131

Answers to "Prep Skills"

1.

0	0	0	0	0	1	1	1	1	2	2	2
2	3	3	3	3	4	4	5	6	8	10	18

2. 0 days **3.** 0 days; 18 days

4.

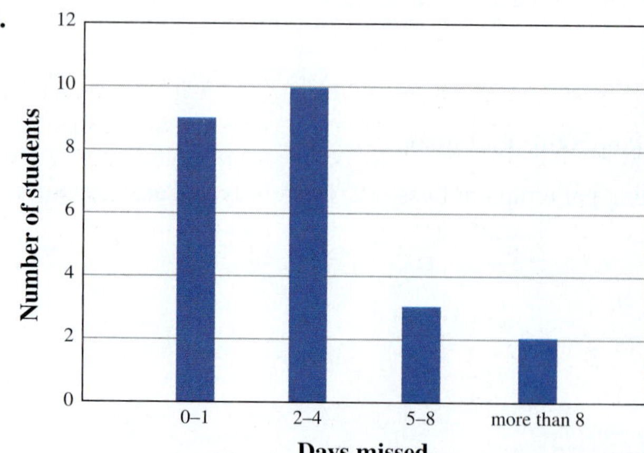

1-9 Applications

Name ______________________________

Thirty random people at a subway station in New York City were asked what their most common source of news is: *The New York Times* (T), the *Post* (P), the *Daily News* (D), or some other source (O).

1. Do you think this would be a representative sample for all New Yorkers? Why or why not?

2. Responses are summarized below. Construct a frequency distribution for the data.

T	O	O	D	P	P	O	P	T	D
T	P	D	D	O	O	O	D	P	O
D	D	O	O	T	P	P	P	O	O

3. What single source do you think was the most common source of news? Can you be certain?

It can be difficult to compare monetary amounts over a large time span because of inflation and changes in the cost of living. For example, the median new home price in the United States in 1980 was about \$65,000; in 2015 that number was about \$296,000. In short, \$65,000 in 1980 doesn't compare to \$65,000 in 2015 because it's worth a LOT less.

To combat this, dollar comparisons are often done in inflation-adjusted dollars, which allows more accurate comparison. So if something had actually cost \$100 in 1980, the inflation-adjusted cost might be more like \$350 in 2015 dollars. The next data set we'll study lists amounts of money in inflation-adjusted 2015 dollars.

1-9 Applications

Name ______________________________

This table shows overall advertising revenue for the American newspaper industry (in billions of inflation-adjusted 2015 dollars) for the years 1980 to 2015.

Year	Ad revenue	Year	Ad revenue	Year	Ad revenue	Year	Ad revenue
1980	42.9	1989	61.8	1998	62.4	2007	51.7
1981	42.8	1990	58.1	1999	64.9	2008	41.6
1982	42.3	1991	52	2000	66.3	2009	30
1983	48.3	1992	51	2001	58.1	2010	27.6
1984	52.5	1993	51.6	2002	56.5	2011	25
1985	55.1	1994	53.8	2003	58.4	2012	22.3
1986	56.9	1995	55.5	2004	60	2013	20.7
1987	61	1996	57	2005	59.8	2014	19.4
1988	62.1	1997	59.9	2006	57.3	2015	17.9

Source: Columbia Journalism Review

4. Build a grouped frequency distribution for the data. Your first class should be 10–19.9, and all others should be the same size.

5. Which class contains the most number of values? Write a sentence explaining what this means using units on any numbers used.

6. Draw a histogram for the ad revenue data on the axes at right. Don't forget to fill in a scale on both axes! You should also include labels on the vertical and horizontal axis describing what those scales represent.

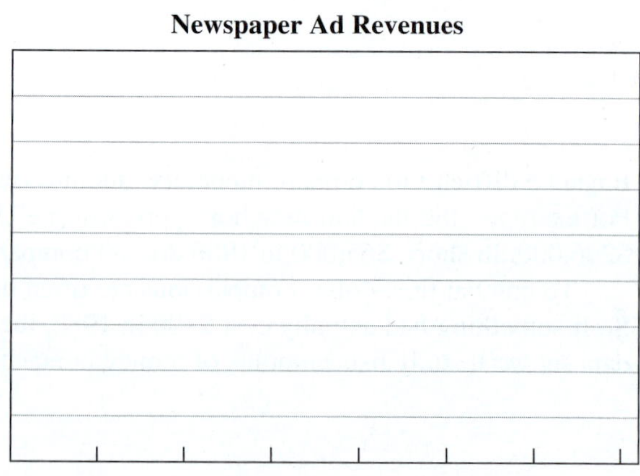

1-9 Applications

Name ______________________________

The stem and leaf plot below represents the number of daily newspapers in each of the 50 states plus the District of Columbia as of November 2016. Use it to answer Questions 7–10.

Stems	Leaves
0	3 5 7 7 8 8 9 9
1	0 0 0 3 5 5 5 6 9 9
2	0 1 3 5 5 5 7 8 9 9
3	0 2 7 8 8
4	5 9
5	0 0 6 8 9
6	5 6
7	6 6 9
8	9
9	0
11	8
20	0
22	0
23	0

Key: 4|5 means 45 newspapers

7. How many states had 656 newspapers?

8. How many states had 25 newspapers?

9. How many states had less than 30 newspapers? More than 100?

10. What's the largest number of newspapers in any state? Smallest?

1-9 Applications

Name ______________________________

The data below are the average weekday individually paid print circulations (in thousands) of select newspapers in the United States as of September 2015. For example, the first row shows that the *Arizona Republic* sells 164,000 copies of their print edition on an average weekday.

Paper	Daily circulation	Paper	Daily circulation
Arizona Republic	164	*Minneapolis Star-Tribune*	184
Atlanta Journal-Constitution	92	*New York Daily News*	228
Boston Globe	140	*New York Post*	245
Chicago Sun-Times	118	*New York Times*	528
Chicago Tribune	266	*Newark Star-Ledger*	114
Cleveland Plain Dealer	153	*Newsday*	217
Dallas Morning News	140	*Orange County Register*	110
Denver Post	156	*Philadelphia Inquirer*	138
Honolulu Star-Advertiser	94	*San Diego Union-Tribune*	117
Houston Chronicle	169	*Tampa Bay Times*	141
Las Vegas Review-Journal	90	*USA Today*	299
Los Angeles Times	328	*Washington Post*	330

11. Build a stem and leaf plot to organize the data.

Stems	Leaves

Key: 13 | 8 means 138,000 copies sold

1-9 Applications

Name ____

12. How useful do you think the stem and leaf plot is in this case? Explain your answer.

13. Based on EVERYTHING you've seen in this lesson, how would you assess the claim that the American newspaper industry is dying?

14. Of the three methods we used to represent data in this lesson (frequency distribution, histogram, stem and leaf plot), which in your opinion is easiest to interpret? Why do you feel that way?

Unit 1 Language and Symbolism Review

Carefully read through the list of terminology we've used in this Unit. Consider circling the terms you aren't familiar with and looking them up. Then test your understanding by using the list to fill in the appropriate blank in each sentence. Hint: One word is used twice.

$A \cap B$
$A \cup B$
axis
bar graph
categorical frequency distribution
classes
commutative
complement
compound interest
coordinates
data
degrees
element
empirical probability
exponential growth
grouped frequency distribution
histogram
interest
intersection
like quantities
linear growth
lower limit
origin
perimeter
pie chart
plotting points
population
probability
raw data
rectangular coordinate system
representative sample
roster method
sample
scale
scatter diagram or scatter plot
scientific notation
set
simple interest
stem and leaf plot
theoretical probability
time-series data
time-series graph
union
universal set
upper limit
Venn diagram
well-defined
x axis
y axis

1. A ______________ is a diagram used to compare the relative sizes of different parts of a whole.
2. A circle can be divided into 360 equal units of measure, which we call ______________.
3. A ______________ is a visual way to compare the sizes of different values.
4. Polling is a really common and useful way to gather some ______________, which are measurements or observations that are gathered for an event under study.
5. If we wanted to learn about characteristics of all the students at your college, the ideal approach would be to poll every single one of them. But in most cases, that's not particularly realistic. So instead, we'd likely choose a ______________ of students from the larger ______________ of all students at your school.
6. Two quantities can only be added when they are ______________.
7. Addition is ______________: The order in which you add doesn't matter.
8. The ______________ of a figure is found by taking the sum of the lengths of each side of that figure.
9. Multiplication is ______________: The order in which you multiply doesn't matter.
10. ______________ is a fee paid for the use of someone else's money.
11. Interest that is calculated as a percentage of the original amount of money is called ______________.
12. ______________ is interest that is paid not just on the original amount deposited, but on interest previously earned as well.
13. With ______________ the values grow by adding the same constant number.
14. With ______________ the values grow by multiplying by the same constant number.
15. ______________ are quantitative data that have been collected at different points in time. A ______________ displays values on the y axis compared to equally spaced time intervals on the x axis.
16. The ______________ for a graph is used to describe the distance between the marks on an ______________ for a graph.

17. The numbers we use to locate a point on a graph are called ______________.

18. The system we use for plotting points on a graph is called the ______________.

19. The horizontal axis is usually called the ______________ and the vertical axis is usually called the ______________.

20. The point where the two axes meet is called the ______________.

21. The process of locating information on a rectangular coordinate system is called ______________.

22. When we plot points on a coordinate system that correspond to pairs of data, we call the result a ______________.

23. The ______________ of an event occurring is a description of how likely it is that the event will actually happen.

24. If we assume that both sides of a coin are equally likely to land up, we are using ______________.

25. If we performed an experiment and observed 1,000 flips of a coin to determine whether one side was more likely than the other to land up, we would be using ______________.

26. ______________ is often used to express really large or really small numbers using powers of 10.

27. ______________ are used to represent sets of objects and visualize what they have in common, and how they differ.

28. A ______________ is any collection of objects.

29. A set is ______________ if you can objectively determine (based on facts, not opinion) whether an object does or does not belong in a set.

30. Each object in a set is called an ______________ or a member of a set.

31. One method of describing a set is called the ______________, in which elements are listed between braces, with commas between the elements.

32. The ______________ is the set of all objects under consideration in a given situation.

33. The ______________ of a set is the set of all objects from the universal set that are not in that set.

34. The set of elements that are in both sets *A* and *B* is called the ______________ of *A* and *B*, and is symbolized ______________.

35. The set of elements that are in at least one of sets *A* and *B* is called the ______________ of *A* and *B*, and is symbolized ______________.

36. A ______________ is a subset of a statistical population that accurately reflects the members of the entire population.

37. The data collected for a statistical study are called ______________.

38. A ______________ is a way to organize data that are divided into distinct categories, like gender, your class standing in school, or conferences for college football teams.

39. A type of frequency distribution that can be constructed using numerical data is called a ______________.

40. In a grouped frequency distribution, the numerical data are divided into ______________.

41. If one of the classes in a grouped frequency distribution was 100–119 pounds, we would call 100 the ______________ and 119 the ______________.

42. When data are organized in a grouped frequency distribution, we draw a special kind of bar graph to illustrate. This type of graph is called a ______________.

43. A way of organizing data, where each data value or number is separated into two parts, is called a ______________.

Unit 1 Technology Review

This is a short review of the technology skills we've used in this unit. In each case, rate your confidence level by checking one of the boxes, If you feel like you're struggling with these skills, consult the online resources for extra practice.

1. Make a pie chart with Excel. (Lesson 1-1)
2. Make a bar graph with Excel. (Lesson 1-1)
3. Use Excel to perform addition. (Lesson 1-2)
4. Use the auto-fill feature in Excel. (Lesson 1-3)
5. Create a scatter plot with Excel. (Lesson 1-5)
6. Use scientific notation. (Lesson 1-7)
7. Sorting a data set. (Lesson 1-9)

Answer each question based on the technology skills from this unit.

1. Make a pie chart from the data in the table.

	A	B
1	Grades on an Exam	Number of Students
2	A	4
3	B	7
4	C	5
5	D	1
6	F	3
7		

2. Make a bar graph from the data in the table.

	A	B
1	Day of the Week	Total Number of Absences by Day of the Week
2	Monday	12
3	Tuesday	4
4	Wednesday	6
5	Thursday	8
6	Friday	16
7		

3. Write a formula to add all of the values in column B and display the result in cell B19.

	A	B
1	Date	Charge
2	2-Dec	$24.51
3	4-Dec	$45.21
4	4-Dec	$11.89
5	5-Dec	$19.29
6	8-Dec	$16.00
7	9-Dec	$188.22
8	11-Dec	$36.78
9	12-Dec	$44.31
10	15-Dec	$31.97
11	16-Dec	$228.75
12	20-Dec	$22.01
13	21-Dec	$3.75
14	22-Dec	$19.56
15	23-Dec	$9.85
16	24-Dec	$55.08
17	24-Dec	$103.47
18		
19	Total	

4. Use the fill-down feature to complete a spreadsheet like this one, down to row 52.

	A
1	Value
2	$1,000
3	$1,020
4	$1,040
5	$1,060
6	$1,080
7	
8	

5. Use the data in the table to create a scatter plot.

	A	B
1	Age (years)	Height (inches)
2	0	17
3	1	23
4	2	30
5	3	32
6	4	39
7		

6. The values given came from either a calculator or a spreadsheet. Write each value in scientific notation:

a. 3.45E+7

b. 1.72E-04

Use a calculator or a spreadsheet to calculate each value and write the result in scientific notation.

c. $(1.2 \times 10^{12})(8 \times 10^{3})$

d. $\dfrac{1}{(389)^{10}}$

7. Sort the list of quiz scores in ascending order.

	A
1	Scores
2	18
3	15
4	17
5	12
6	13
7	19
8	20
9	15
10	16
11	14
12	13
13	18
14	20
15	10
16	18
17	17
18	

Unit 1 Learning Objective Review

This is a short review of the learning objectives we've covered in Unit 1. In each case, rate your confidence level by checking one of the boxes, and then answer the question. If you feel like you're struggling with these skills, consult the lesson referenced next to the objective and see the online resources for extra practice.

1. Analyze personal time management for a week of activities. (Lesson 1-1)
2. Solve problems involving percentages. (Lesson 1-1)
3. Create and interpret pie charts. (Lesson 1-1)
4. Create and interpret bar graphs. (Lesson 1-1)
5. Identify circumstances where addition or subtraction is possible. (Lesson 1-2)
6. Add or subtract quantities. (Lesson 1-2)
7. Interpret multiplication as repeated addition. (Lesson 1-3)
8. Multiply or divide quantities. (Lesson 1-3)
9. Distinguish between simple interest and compound interest. (Lesson 1-4)
10. Distinguish between linear and exponential growth. (Lesson 1-4)
11. Interpret exponents as repeated multiplication. (Lesson 1-4)
12. Simplify numeric expressions involving exponents and the order of operations. (Lesson 1-4)
13. Use a rectangular coordinate system. (Lesson 1-5)
14. Connect data to graphs. (Lesson 1-5)
15. Interpret graphs. (Lesson 1-5)
16. Compute and interpret basic probabilities. (Lesson 1-6)
17. Translate a probability to a percent chance. (Lesson 1-6)
18. Recognize the difference between theoretical and empirical probability. (Lesson 1-6)
19. Convert numbers between decimal and scientific notation. (Lesson 1-7)
20. Describe the significance of writing numbers in scientific notation. (Lesson 1-7)
21. Analyze how your personality type affects how you interact with others. (Lesson 1-8)
22. Create and interpret Venn diagrams. (Lesson 1-8)
23. Describe sets using appropriate terminology. (Lesson 1-8)
24. Explain the difference between a population and a sample. (Lesson 1-9)
25. Organize data with frequency distributions and histograms. (Lesson 1-9)
26. Analyze data with stem and leaf plots. (Lesson 1-9)

1. Use the information about where I spend my 168 hours in a week to complete the table and to create a pie chart. Round to the nearest tenth of a percent and then to the nearest whole degree.

Activity	Number	Percent	Degrees
Teaching Class	15		
Grading & Office Hours	20		
Preparing for Class	10		
Sleep	56		
Other	67		

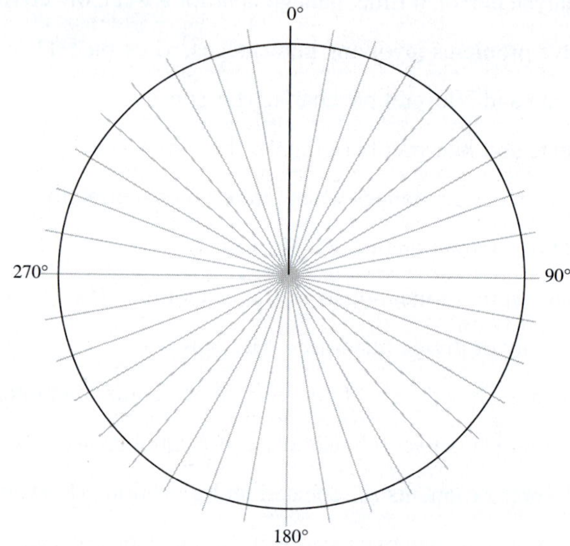

Starting in 2011, the tax rate in Illinois changed to 5% of personal income. Previously, the tax rate was 3% of personal income. These tax rates are represented on the bar graphs provided. The data is the same, but the two graphs use different scales.

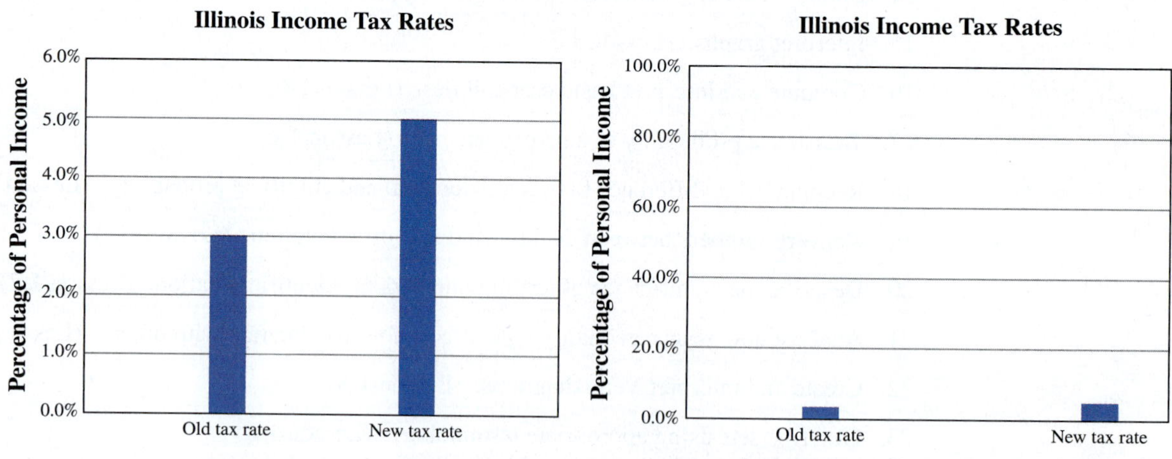

2. Using the old tax rate, how much income tax would an individual pay who makes $32,000 in personal income?

3. Using the new tax rate, how much income tax would an individual pay who makes \$32,000 in personal income?

4. Which graph do you think would be used by someone wanting to argue that the tax increase was a huge increase, and which would be used by someone wanting to show that the increase wasn't a big deal? Explain.

Perform each addition or subtraction, if possible. If not, be sure to explain why.

5. $\frac{3}{8} + \frac{1}{6}$

6. $12x - 5x$

7. 20 miles + 30 hours

Perform each multiplication or division.

8. $\frac{3}{5} \times \frac{7}{12}$

9. $-12 \div \frac{4}{3}$

10. $-2(7)(5)$

Simplify each expression.

11. $10 + 5 \times 3$

12. $(10 + 5)3$

13. $\frac{-3^2 + 12}{6(2)}$

14. Rewrite the following expression so that it no longer contains addition. Do not calculate the value!

$50 + 50 + 50 + 50 + 50 + 50 + 50 + 50 + 50 + 50$

15. Rewrite the following expression so that it no longer contains an exponent. Do not calculate the value!

$(1.05)^6$

16. A small aircraft has taken off from an airport. The elevation of the plane has been recorded at various times throughout the flight, shown in the table here. Record each pair of values as an ordered pair and then plot the points on the graph and connect them with lines.

Flight time *x* (min)	Altitude *y* (ft)	Ordered Pair (*x*, *y*)
0	0	
10	4,000	
20	3,000	
30	2,000	
40	2,000	
50	1,000	
60	0	

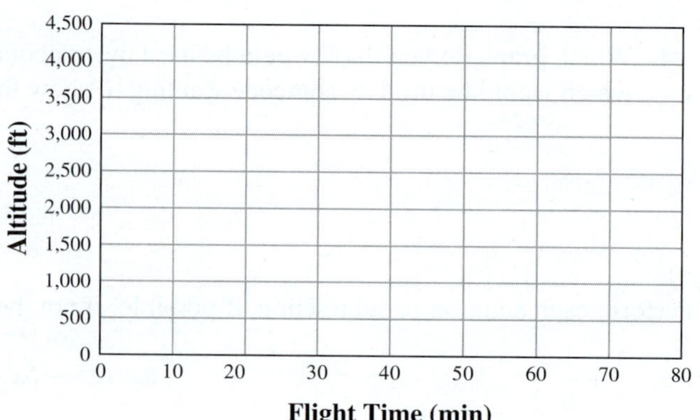

As you answer each question, consider both the table and the graph when you explain your answer.

17. What was the highest altitude reached by the plane?

18. How long after the flight began did the plane reach this highest altitude?

19. When was the plane flying level?

20. How long was the flight?

21. When was the altitude of the plane increasing?

22. When was the altitude of the plane decreasing?

©Ekaterina Minaeva/Stockimo/Alamy

23. The holiday mint M&M's contain 3 colors: red, white, and green. If you reached your hand in the bag and grabbed one M&M, it might be reasonable to calculate the probability of selecting a red one to be 1/3 or about 0.33. Would making this assumption be using empirical or theoretical probability? Explain.

Suppose you dumped out all of the M&M's from the bag and counted all of them (without eating any) and found that there were 78 red, 96 green, and 82 white. Now if you randomly selected one M&M, determine each probability as a fraction and as a decimal rounded to two decimal places.

24. P(red) =

25. P(green) =

26. P(white) =

27. Are each of the three probabilities you just calculated examples of empirical or theoretical probabilities? Explain.

28. Write a sentence describing the percent chance of drawing a red M&M from the bag based on your answer to question 25.

Write each number in standard decimal notation.

29. 5.62×10^{-4}

30. 3.902×10^{7}

Write each number in scientific notation.

31. 43,450,000,000,000

32. 0.000000728

33. A number often thrown out is that 400 million M&M's per day are produced in the United States. Assuming that number is produced every day of the year, how many M&M's would be produced in a year in the United States? Write the answer in scientific notation.

Consider the two accounts shown.

	A	B	C	D	E
1		Account #1			Account #2
2	Start	$10,000.00		Start	$10,000.00
3	After 1 year	$10,600.00		After 1 year	$10,600.00
4	After 2 years	$11,200.00		After 2 years	$11,236.00
5	After 3 years	$11,800.00		After 3 years	$11,910.16
6	After 4 years	$12,400.00		After 4 years	$12,624.77
7	After 5 years			After 5 years	

34. Account ____ is an example of linear growth because we add ________ to each value to obtain the next value.

35. Account ____ is an example of exponential growth because we multiply each value by ________ to obtain the next value.

36. Which account illustrates simple interest? Explain why.

37. Which account illustrates compound interest? Explain why.

38. What could be typed in B7 to calculate the value in Account #1 after 5 years?

39. What could be typed in E7 to calculate the value in Account #2 after 5 years?

40. Find the value that should appear in cell B7 (it should be the amount in Account #1 after 5 years).

41. Find the value that should appear in cell E7 (it should be the amount in Account #2 after 5 years).

42. Assuming the trend continues for each account, write a numerical calculation that will provide the amount in Account #1 after 60 years.

43. Assuming the trend continues for each account, write a numerical calculation that will provide the amount in Account #2 after 60 years.

44. Find the amount in Account #1 after 60 years.

45. Find the amount in Account #2 after 60 years.

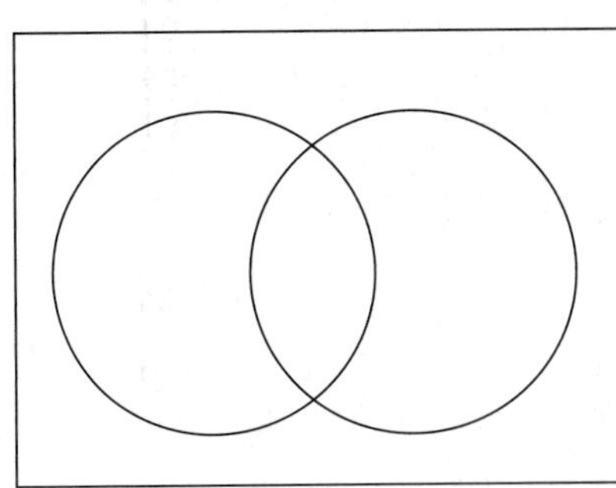

In order to report on which presidential candidate the students at an Illinois community college were supporting, a writer for the school paper surveyed 60 students at that school. They were asked which presidential candidate they thought would do a good job. Of those surveyed, 19 thought Clinton would do a good job, 15 thought Trump would do a good job, and 3 thought both would do a good job.

46. Describe the sample and the population in the previous scenario.

47. Which is more likely to be a representative sample of the student population of a particular school? Explain why.

a. Asking all students who attend a political rally for one of the candidates.

b. Asking all students who went through the cafeteria over a two-hour period.

48. Complete the Venn diagram with the appropriate values.

49. How many students thought Clinton would do a good job?

50. How many students thought only Clinton would do a good job?

51. How many students thought Trump would do a good job?

52. How many students thought only Trump would do a good job?

53. How many students thought neither would do a good job?

54. How many students thought both would do a good job?

Consider the data from the table showing the heights of 15 people at a family gathering.

Name	Height (inches)
Bob	72
Bill	65
Lisa	60
Manuel	42
Hannah	38
Braxton	55
Ava	51
Jake	64
Charlotte	62
Nate	68
Emmett	57
John	54
Mario	60
Ben	61
Luigi	47

55. Create a stem and leaf plot using the first digit as the stem and the second digit as the leaf for each number.

Stems	Leaves

56. How many people had a height under 5 feet?

57. If we let A represent the set of names of all people at this gathering whose height is under 5 feet, write that set using the roster method.

58. Give a verbal description of what would be meant by A' in this situation.

59. Write the set A' using the roster method.

60. Complete the frequency distribution, and then use the frequency distribution to make a histogram for the height data.

Class	Tally Marks	Frequency
30–39		
40–49		
50–59		
60–69		
70–79		

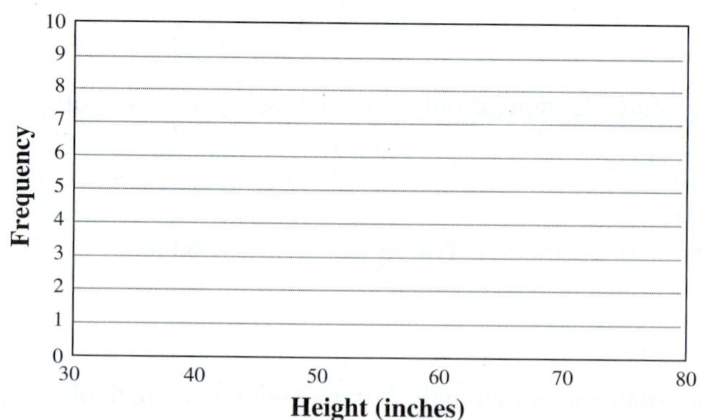

61. Which class contains the largest number of people?

Unit 2
Making Sense of It All

©Colin Anderson/Blend Images RF

Outline

Lesson 2-1 Prep Skills

SKILL 1: CALCULATE A PERCENTAGE

To compute what percentage of a whole a certain amount is, you perform two steps.

Step 1: Divide the portion by the whole. This gives you the percentage in decimal form.

Step 2: Convert that decimal to percent form by moving the decimal two places to the right. This corresponds to multiplying the decimal form by 100.

- $20/50 = 0.4$, which is 40%. So 20 is 40% of 50.
- $110/150 \approx 0.733$, which is 73.3%. So 110 is about 73.3% of 150.
- $8/120 \approx 0.067$, which is 6.7%, so 8 is about 6.7% of 120.
- $200/150 \approx 1.33$, which is 133%, so 200 is about 133% of 150.

SKILL 2: PERFORM CALCULATIONS INVOLVING ORDER OF OPERATIONS IN FRACTIONS

When we have additions or subtractions above or below a fraction bar, order of operations would usually tell us to do division first. But to avoid having to always put parentheses around the entire numerator and denominator, we instead treat the fraction bar as if it implies grouping symbols around the numerator, and around the denominator. That means we perform any calculations in the numerator first, then the denominator, then finally we divide or write the fraction in reduced form, depending on the context.

- $\dfrac{8+12+22+15+9}{5} = \dfrac{66}{5} = 13.2$
- $\dfrac{12}{7+3-1-5} = \dfrac{12}{4} = 3$
- $\dfrac{100+250-50+300}{10-5+20+15} = \dfrac{600}{40} = 15$

SKILL 3: ORDER A LIST OF NUMBERS

Here's an efficient method for putting a list of numbers in order:

Step 1: Find the lowest number on the list and write it down, then cross it off the original list.

Step 2: Repeat until there are no more numbers on the original list.

Step 3: Count to make sure you have the same number of entries on your ordered list as there were on the original list.

PREP SKILLS QUESTIONS

1. Fill in each blank.

   a. 120 is ______% of 500.

   b. 11 is ______% of 20.

   c. 7,000 is ______% of 12,500.

   d. $8.32 is ______% of $41.20.

2. Perform each calculation. Convert fractions to decimals rounded to the nearest tenth.

   a. $\frac{9 + 12 + 27 + 11 + 16 + 4}{6}$

   b. $\frac{100}{12-3-14-5}$

   c. $\frac{-8 + 4 + 3 + 16}{10 - 5 - 2 + 8}$

3. Put this list in order from least to greatest.

   20.3 19.1 15.0 15.7 9.2 12.3 19.3 16.8 12.3 10.6 17.7 15.6 15.2

Lesson 2-1 Did You Pass the Test?

LEARNING OBJECTIVES

- ☐ 1. Consider strategies for preparing for and taking math tests.
- ☐ 2. Understand the impact of a single question, or a single exam.
- ☐ 3. Calculate, interpret, and compare measures of average.

©Blend Images - Hill Street Studios/Getty Images RF

Tell me and I'll forget; show me and I may remember; involve me and I'll understand.

—Chinese proverb

When's the last time you were tested? If you think of that only in an academic sense, you may be consulting a calendar. If you're more of a metaphorical person, you may be thinking "I get tested every day." Life tests us in many different ways, and how we deal with tests goes a long way to determining our character. In fact, that's one of the reasons that an education is so important. The tests that you take in college classes are a metaphor for some of the most important events in life: With the pressure on, can you perform your best? What have you learned, and can you apply that knowledge when you need to?

0. Discuss some ways an instructor might decide how well or how poorly a class did on a test.

2-1 Class

In this lesson, we're going to look at exams in a math course. How do you prepare? How is your grade for the exam determined? What impact does this score have on your grade in the course? How does an instructor determine how well the class as a whole understood the material?

Answer each question in your groups and then prepare to share your answers with the class.

1. What are some strategies you use to prepare for an exam?

2. What are some habits you have or have seen that are not helpful when preparing for an exam?

3. What are some strategies you use during an exam?

4. When a test has been graded and returned, students (and instructors) are usually interested in knowing what the "class average" was. Explain the meaning of the word "average" in your own words. Don't think in terms of formulas, but what it tells you.

	A	B	C
1	Student	Exam 1 (%)	Exam 2 (%)
2	Michael	80	89
3	Andy	77	93
4	Pam	68	84
5	Jim	81	88
6	Dwight	96	91
7	Stanley	54	75
8	Phyllis	75	54
9	Kevin	81	86
10	Creed	71	0
11	Darryl	89	83
12	Gabe	56	64
13	Toby	81	65
14	Holly	92	73

In math and stats, the term "average" is sort of a generic term, with several interpretations. Loosely defined, the average of a list of numbers is the most typical value. But what exactly does that mean? The one that appears most often? The one right in the middle? Or some mathematical combination?

There are three different measures that we'll study in this lesson, and all of them fall under the umbrella of "average." And in some cases, they can take on completely different values. This means that interpretation becomes far more important than being able to do calculations.

The **mean** of a set of numbers is what you probably think of as the "average." You find it by adding all of the numbers, then dividing by how many numbers there are on the list.

5. Find the mean test score on Exam 1.

The **median** of a list of numbers is the value that lives right in the middle of the set if it's arranged in order. To find the median:

Step 1: List the numbers in order from either largest to smallest or smallest to largest (you can pick).

Step 2: If there is an odd number of values, the median is the value that has the same number of values above and below it on the list.

Step 3: If there is an even number of values, the median is the mean of the two values right in the middle of the list.

6. Find the median test score on Exam 1.

The **mode** of a list of numbers is the value that appears most often. If all values appear only once, there is no mode.

7. Find the mode of the Exam 1 scores.

Math Note

A data set can have more than one mode if there are multiple values that appear the same number of times.

Did You Get It

Try this problem to see if you understand the concepts we just studied. The answers can be found at the end of the Portfolio section.

1. Here are the number of points that Creed earned on each question on Exam 1. Find all three measures of average for these scores.

Question	1	2	3	4	5	6	7	8
Points	5	12	10	10	6	12	9	7

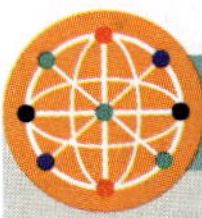

Computing Measures of Average Using Excel

To compute the three measures of average in Excel, use these commands:

- Mean: =AVERAGE(cell range)
- Median: =MEDIAN(cell range)
- Mode: =MODE(cell range)

	80
	77
	68
	81
	96
Mean	80.4
Median	80
Mode	#N/A

80
77
68
81
96
=AVERAGE(B1:B5)
=MEDIAN(B1:B5)
=MODE(B1:B5)

The samples shown here calculate measures of average for the first five scores from Test A earlier. The first spreadsheet shows the results of the calculations. The second displays the formulas that were used to get those measures of average. The "#N/A" indicates that there is no mode for the data.

See the Lesson 2-1 Using Tech video in class resources for further instruction.

2-1 Group

If it doesn't make you uncomfortable, exchange the following information with the classmates in your Unit 2 group. This will be your small group for the second unit. It would be a good idea to schedule a time for the group to meet to go over homework, ask/answer questions, or prepare for exams. You can use this table to help schedule a mutually agreeable time.

Name	Phone Number	Email	Available times

The spreadsheet summarizes the results for one student on a 15-question math test with partial credit awarded.

1. Complete the totals at the bottom of each column using addition.
2. What percent of the total possible points did this student earn? (Round all percentages in this lesson to one decimal place.)

	A	B	C
1	Problem Number	Points Possible	Points Earned
2	1	2	2
3	2	2	2
4	3	2	0
5	4	5	4
6	5	5	2
7	6	8	4
8	7	8	3
9	8	9	9
10	9	4	4
11	10	5	5
12	11	5	5
13	12	6	6
14	13	6	6
15	14	10	6
16	15	10	10
17	Totals		
18			
19			

3. What letter grade did this student get? (The grading scale is 90% = A, 80% = B, 70% = C, 60% = D.)
4. Problem 15 on the test was an application (what some folks call "story" or "word" problems). What letter grade would this student have earned if she'd wimped out and skipped that question?
5. What's the highest grade you could get on this test if you skip problem 15 because it's a scary word problem?
6. Does Question 5 sound familiar to you? Have you ever been in that position? Explain.
7. What do these questions make you think about in terms of taking tests, and the importance of trying every question?

Did You Get It

Suppose that the instructor who wrote the test in Questions 1–5 had added two more questions, one worth 8 points and another worth 5.

2. What percentage would this student have earned if she didn't even notice those questions were part of the test?
3. How much would her percentage have improved if she got both of those questions completely correct?

So here's the big question: Why three measures of average? Let's look into it. The first list below shows the percentage of points earned by 25 students on a quiz. The mean score is 69.4%. The second shows the number of hours spent by the students preparing for the quiz in their online homework system. The mean number of hours is 3.5.

Scores

100 100 100 100 100 100 100 100 100 100 100 100 88 85 82 78
71 70 60 0 0 0 0 0 0

Study Time

28 6 5.3 4.3 4 3.7 3.5 3.5 3.4 3.3 3.2 2.2 2.2 2.1 1.9 1.8 1.7 1.7 1.7
1.7 1.4 1.1 0 0 0

8. After returning the quiz, the professor scolds the class mercilessly, making four students cry and sending six others to RateMyProfessor.com to say really mean things. His justification: The average grade was a D! Ugh. Do you think this is an accurate measure of how the class performed on the quiz? Explain.

9. Compute the other two measures of average, then decide which of the three measures of average provide the most accurate description of how the class did.

10. The professor is really puzzled by that D average because students spent an average of 3.5 hours studying, which isn't bad for a quiz (but would probably be a little low for a test). Assess his confusion. Is the mean a good measure of average for the time spent studying? Why or why not?

11. Find the other two measures of average, and discuss which of the three you think provides the most accurate representation of the time spent by students.

12. Do you think the scores came from a math class? Why or why not?

At the beginning of the course, you should have been provided some information on how exactly you're going to be graded. Most instructors have their own standards, so you should make an effort to understand how your grade will be calculated. Most likely it was on a syllabus provided by your instructor.

13. How well do you feel you understand the grading standards for your course? If you don't know them, consult the syllabus or ask your instructor.

Let's look at a fairly basic points system for grading. We'll say that your course grade comes from 4 exams worth 100 points each, a homework score worth 200 points, and a final exam worth 200 points.

14. How many total points can be earned in the course?

15. What percentage of the total score is accounted for by the first exam?

16. If you don't show up for the first exam and take a 0%, then average 82% for everything else the rest of the course, what would your final percentage be?

	A	B	C	D	E	F	G	H	I
1		Exam 1	Exam 2	Exam 3	Exam 4	Homework	Final Exam	Total Points	Overall %
2	Points possible	100	100	100	100	200	200		
3	Your scores								

17. What would your final percentage be if you score 100% on Exam 1, then average 82% for the rest of the course?

	A	B	C	D	E	F	G	H	I
1		Exam 1	Exam 2	Exam 3	Exam 4	Homework	Final Exam	Total Points	Overall %
2	Points possible	100	100	100	100	200	200		
3	Your scores								

18. What does the difference between your answers to Questions 16 and 17 tell you?

2-1 Portfolio

Name ______________________________

Check each box when you've completed the task. Remember that your instructor will want you to turn in the portfolio pages you create.

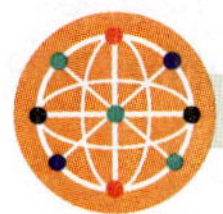

Technology

1. ☐ Create a spreadsheet that calculates final grades for the grading system described on page 158. You can use the samples there as a template. Your sheet should use formulas to add up the number of points possible, your total score when you enter individual scores for exams and homework, and your final percentage. Experiment with your spreadsheet to find the effects of one unusually bad test score. A template to help you get started can be found in the online resources for this lesson.

2. ☐ The four scores below show where one student stands going into a 200-point final. Enter the current scores, then experiment with values on the final to see the range of grades this student could get. Follow further instructions on the template.
   Exam 1: 73 Exam 2: 78 Exam 3: 84 Exam 4: 82 Homework: 164

Online Practice

1. ☐ Include any written work from the online assignment along with any notes or questions about this lesson's content.

Applications

1. ☐ Complete the Applications problems.

Reflections

Type a short answer to each question.

1. ☐ Write down a list of things that you think are most likely to affect how well you do on tests in this course. Then describe what you can do to make that information work for you.

2. ☐ There are many websites with test-prep and test-taking hints. Find one that you think is helpful, and describe some tips that you can use.

3. ☐ Describe the three measures of average we studied.

4. ☐ When a data set has one or two values that are much higher than all of the others, which measure of average is most likely to be deceiving? Why?

5. ☐ Name one thing you learned or discovered in this lesson that you found especially interesting.

6. ☐ What questions do you have about this lesson?

Looking Ahead

1. ☐ Complete the Prep Skills for Lesson 2-2.

2. ☐ Read the opening paragraph in Lesson 2-2 carefully then answer Question 0 in preparation for that lesson.

Answers to "Did You Get It?"

1. Mean: 8.9; median: 9.5; mode: 10 and 12
2. 68%
3. She could have improved by about 2.8%.

Answers to "Prep Skills"

1. a. 24 b. 55 c. 56 d. about 20.2
2. a. $\frac{79}{6} \approx 13.2$ b. -10 c. $\frac{15}{11} \approx 1.4$
3. 9.2 10.6 12.3 12.3 15.0 15.2 15.6 15.7 16.8 17.7 19.1 19.3 20.3

2-1 Applications

Name ____________________

Here's another look at a group of exam scores:

	A	B	C
1	**Student**	**Exam 1 (%)**	**Exam 2 (%)**
2	Michael	80	89
3	Andy	77	93
4	Pam	68	84
5	Jim	81	88
6	Dwight	96	91
7	Stanley	54	75
8	Phyllis	75	54
9	Kevin	81	86
10	Creed	71	0
11	Darryl	89	83
12	Gabe	56	64
13	Toby	81	64
14	Holly	92	74

1. Find the mean for all Exam 2 scores.

2. Find the median for all Exam 2 scores.

3. Find the mode for all Exam 2 scores.

4. Find the mean, median, and mode for the scores on Exam 2 if you throw out Creed's zero (dude didn't even show up . . . Come on, man!).

5. You've found six measures of average for the Exam 2 scores so far on this page: mean, median, and mode for all the scores, then for the scores without the zero. Of the six, which single one do you think is the best representation of how students performed on the test? Needless to say, you should explain why you chose that one.

6. After the Exam 2 scores were listed on the class online site (without names, of course, which would violate several federal laws), one of the students said, "Wow, we did great on this test," while the instructor said, "I disagree. The average was a low C!" Who's right? Justify your answer, including information about measures of average.

Lesson 2-2 Prep Skills

SKILL 1: USE ORDER OF OPERATIONS

Order of operations is a group of agreed-upon rules that allow us to write calculations involving more than one operation without always having to use parentheses or other grouping symbols to indicate which operation to do first. The order is as follows, with operations higher on the list being performed before those lower on the list. If there are two operations tied (like two multiplications), they are performed left to right. Grouping symbols can indicate a different order.

Parentheses

Exponents

Multiplication or division

Addition or subtraction

- $12 + 3 \cdot 4 = 12 + 12 = 24$ — Multiply $(3 \cdot 4)$ first.
- $4(3)^2 = 4(9) = 36$ — Exponent first
- $81 \div (12 - 3) = 81 \div 9 = 9$ — Calculation in parentheses $(12 - 3)$ first
- $52 - 8 \div 4 \cdot 6 = 52 - 2 \cdot 6 = 52 - 12 = 40$ — Divide $(8 \div 4)$ first, then multiply $(2 \cdot 6)$. For two operations at the same level, perform the leftmost one first.
- $\dfrac{7-9}{3+5} = \dfrac{-2}{8} = \dfrac{-1}{4}$ — Fraction bar acts like parentheses around the numerator $(7 - 9)$ and denominator $(3 + 5)$.

When a calculation contains multiple parentheses, the innermost parentheses are done first:

- $(-2\,(3 + 1))^2 - 4 = (-2 \cdot 4)^2 - 4$ — Inner parentheses $(3 + 1)$ first, then outer $(-2 \cdot 4)$

  $= (-8)^2 - 4$ — Exponent before subtraction

  $= 64 - 4$

  $= 60$

PREP SKILLS QUESTIONS

Perform each operation.

1. a. 3^5 b. $(-2)^6$ c. $\left(\frac{1}{2}\right)^4$ d. -5^2
2. a. $12 + 20 \div 2$ b. $(-4)(-6) - 7$ c. $7 \times 4 - (-60) \div 12$
3. a. $20 + 2(9 - 4)^2$ b. $-3^3 \cdot 2 - 42$ c. $12 + 2 \cdot 30 \div 5$
4. a. $((8 - 1) \times 3)^2$ b. $2{,}000(1 + 0.05)^{12}$ c. $10{,}000(1.045)^2 - (10{,}000 + 10{,}000(0.045) \cdot 2)$
   d. $(7 - (3 + 4)^2)^2$

Lesson 2-2 Ins and Outs

©Gyro Photography/amanaimagesRF/Getty Images RF

LEARNING OBJECTIVES

- ☐ 1. Distinguish between inputs (independent variables) and outputs (dependent variables).
- ☐ 2. Evaluate expressions and formulas.
- ☐ 3. Write and interpret expressions.

Experience is a hard teacher because she gives the test first, the lesson afterward.
—Vernon Law

The holy grail of energy research: a process that outputs far more energy than you put in. Develop a safe, clean process that does that, and you'll be rich beyond your wildest dreams. There's a worthwhile metaphor here: In many instances, the output (result you get) is determined by what you input. Don't practice and stay out all night before a big game, your output will probably stink. Put in a ton of work and effort, and you're likely to do your best. In this lesson, we'll study input and output from a mathematical sense, and hopefully give you a MUCH better idea of just what in the world a "variable" really is.

0. Provide a couple more examples of where an input leads to an output.

2-2 Class

First, let's talk about variables. Answer this question honestly and to the best of your ability without any outside help:

1. What is a variable?

If you're like most people, your answer was probably something like this: "A variable is a letter instead of a number." I am very sorry to inform you that, and I say this in the most caring possible way, your answer stinks. Think about what the word "variable" should mean in plain English: able to vary. And THAT is the key to understanding variables:

A **variable** is a quantity that is able to change, or vary.

Contrast this with a **constant**, which can't vary. The number 12 is a constant, because no matter what, it still has the same value. The number of hours you spend studying, on the other hand IS a variable, because it can vary according to how much effort you choose to expend.

But what about the whole "letter" thing? Isn't x a variable? Technically, the answer is no. We use letters to REPRESENT variables, since that distinguishes them from numbers, which never change. But a variable is NOT a letter: It's a *quantity* that can vary, which is usually represented by a letter. To save time, we often refer to the letter itself as a variable, but try to keep in mind it's actually the quantity represented by that letter that is really a variable.

2. Which sentence makes more sense? Explain your reasoning.
   a. The amount of time you spend studying depends on the grade you earn in a course.
   b. The grade you earn in a course depends on the amount of time you spend studying.

3. List other factors that are likely to impact your grade in this course.

Because these factors can cause a change in your course grade, and not the other way around, the course grade (which can vary) is referred to as a **dependent variable** in this situation. Other factors, like the amount of time you spend studying, are called **independent variables**. Independent variables cause changes in dependent variables. Think of it as "Your grade *depends* on the factors you identified."

4. Consider the following relationships, where one quantity or event causes another to change. Identify the independent and dependent variable in each case, and don't forget to think about WHY it makes sense that these things are variables.

   a. The age of a tree and the height of a tree

   b. The number of practice sessions and the quality of a musical performance

   c. Your score on a placement test and the math courses you've taken previously

   d. Your blood pressure and the amount of time you spend exercising each week

   e. The value of a share of Apple stock and what year it is

   f. The number of songs a band sells on iTunes and the amount of money spent on marketing

   g. The cost of a cab ride and the number of miles you're driven

h. The number of customers that want to buy a certain product and the price of that product

Did You Get It

Try this problem to see if you understand the concepts we just studied. The answers can be found at the end of the Portfolio section.

1. For each situation, describe which is the best choice for the independent and dependent variable, and justify your answer.
   a. The amount of money spent by a candidate for the presidency, and the percentage of the population that votes for that candidate.
   b. The taste test rating of a certain soft drink, and the number of teaspoons of sugar in each 12 ounces of that drink.

2-2 Group

1. A few key terms used in this section are listed below. Discuss each term with your group and note how the definitions are similar, and how they're different.

Expression – A combination of variables and constants using mathematical operations and grouping symbols

Examples: $4x + 2y$, $\dfrac{11}{t-2}$, $\sqrt{x^2 - 20}$, $(4n - 6)(2 + n)$

Equation – A statement that two quantities are equal that is built using expressions and an equal sign (=)

Examples: $3x + 2 = 7$, $y = z^2 - z - 6$, $100e^{0.02t} = 500$

Formula – An equation with multiple variables that is used to calculate some quantity that we're interested in

Examples: $A = \pi r^2$, $P = 2l + 2w$, $A = P + Prt$

In many cases, relationships like the ones we thought about in the Class portion of this lesson can be described mathematically using a formula. For example, it might cost \$2 for a cab ride plus \$2.50 for each mile; in that case we could write the formula

$$C = 2 + 2.50m$$

where C represents the variable cost of the cab ride, and m represents the variable number of miles. It's the number of miles that affects the cost, so m is the independent variable and C is the dependent variable. We might say that the number of miles is the **input**, while the cost is the **output**. Think of it like a machine: You input a number of miles, and the machine gives you back the cost.

2. If we wanted to know the cost of a 7-mile cab ride, we could replace the variable m, which represents miles, with the number 7; the result would be $C = 2 + 2.50(7)$. Finding the value of the calculation to the right of the equal sign will tell us the cost. We call this **evaluating** the expression (or formula) for cost. Find the cost of the 7-mile cab ride.

> **Math Note**
>
> There's no reason you HAVE to use a letter to represent a variable quantity: You could use a smiley face, a zodiac sign, a picture of your mom, whatever. But we'll use letters because we don't know what your mom looks like.

3. Evaluate the cost formula for $m = 4$, then attach units to your answer and write a sentence describing what it tells us. Include information about each variable.

Did You Get It

2. If a streaming service charges a monthly fee of \$3, then adds \$0.75 for each movie or TV show watched, then the formula $C = 3 + 0.75n$ describes the monthly cost C in terms of variable n, which represents the number of movies or shows watched. Use this formula to find the cost of watching 13 movies or shows.

In Lesson 1-4, we studied simple and compound interest. One of the things we discovered is that to compute the value in an account that earns simple interest, we can use this procedure:

New value = Original value + interest earned

which is

Original value + Original value × interest rate × number of time periods

Whether you realized it or not at the time, you were doing algebra! Each of the quantities in that verbal description can change, so they are variables that can be represented by symbols. Instead of writing the long verbal description above, it's a whole lot quicker and more concise to write it as a formula using letters to represent the variable quantities:

Computing Future Value Using Simple Interest

The new value of an account A (known as the **future value**) is given by

$$A = P + Prt$$

where P represents the original amount (or **principal**), r represents the interest rate in decimal form, and t represents the number of time periods.

If it makes perfect sense to you that we've written the expression describing future value twice, and the second one is simpler, then congratulations: You get algebra.

4. When using the formula $A = P + Prt$ in a given situation, you'll usually know the principal amount and the interest rate, so P and r will be constants. That leaves A and t as the only variables. Which of those variables would you say is the input? Which would you say is the output? Why?

5. Now forget about formulas for a second, and pretend we're back in Lesson 1-4, computing interest using what we know about percents. If you start with \$2,000 in an account and earn 3% simple interest each year, how much interest would you earn each year? Make sure you show your calculation.

6. Now substitute the values provided in Question 5 into the formula $A = P + Prt$ and simplify just a bit. Does your answer to Question 5 make sense based on the resulting formula?

7. Complete the table of values for your shiny new formula.

t	$A = 2{,}000 + 60t$
0	2,000
1	
2	
3	
4	
5	
6	

8. What's the value of the account after 2 years?

9. How long does it take for the account to reach a value of \$2,240?

Did You Get It

3. If \$42,000 is invested at 4.5% simple interest, write a formula that calculates the value of the account after t years. Then use your formula to find the value of the account 12 years after it was opened.

When working with formulas, especially complicated ones, calculators and spreadsheets really come in handy for doing repeated calculations. So let's have a look at some cool features that will save you a lot of work in this class—and any other class that uses formulas.

Evaluating a Formula with a Calculator or Spreadsheet

To make a table of values for an expression on a TI-84 Plus:

1. Press [Y=] to get to the equation editor screen.
2. Enter the right side of the formula you are finding values for; use [X,T,θ,n] for the independent variable.
3. Press [2nd] [WINDOW] to get to the table setup screen. Enter the first input value you want an output for next to TblStart, and the distance between input values next to ΔTbl. (In Question 1 of the group activity, these would be 0 and 50, respectively.) The **Indpnt** and **Depend** settings should be **Auto** so the calculator does all of the values automatically.
4. Press [2nd] [GRAPH] to display the table of values.

Plot1 Plot2 Plot3
\Y1=2000+60X
\Y2=
\Y3=
\Y4=
\Y5=
\Y6=
\Y7=

X	Y1
0	2000
1	2060
2	2120
3	2180
4	2240
5	2300
6	2360

X=0

See the Lesson 2-2-1 video in class resources for further information.

To make a table of values for an expression in Excel:

1. Enter the inputs you want in one column. If they're evenly spaced, you can enter the first, then use a formula to add a certain amount to the first input, and copy that formula down or use the fill-down feature. See samples.
2. Enter the formula for the expression in the output column next to the first input. Use the cell that the first input value lives in to replace the variable in the expression.
3. Copy the formula down, or use the fill-down feature to complete the table.

	A	B
1	Input	2000+60t
2	0	2000
3	1	2060
4	2	2120
5	3	2180
6	4	2240
7	5	2300
8	6	2360

	A	B
1	Input	2000+60t
2	0	=2000+60*A2
3	=A2+1	=2000+60*A3
4	=A3+1	=2000+60*A4
5	=A4+1	=2000+60*A5
6	=A5+1	=2000+60*A6
7	=A6+1	=2000+60*A7
8	=A7+1	=2000+60*A8

See the Lesson 2-2-2 Using Tech video in online resources for further information.

We also studied compound interest in Lesson 1-4, and found that when a quantity increases based on exponential growth, the value after a certain number of time periods can be calculated using this pattern:

New value = Original value × (1 + the growth rate in decimal form)$^{\text{number of time periods}}$

Once again, we can write an algebraic formula for this pattern that makes it more concise. In most cases, the time period is years, so we'll be using that in our algebraic formula.

Computing Future Value Using Compound Interest

$$A = P(1 + r)^t$$

where A is the new value, P is the principal, r is the interest rate in decimal form, and t is the number of years.

10. Evaluate this formula for $P = 10{,}000$, $t = 40$, and $r = 0.06$. Then write a sentence or two explaining what the resulting calculation tells us.

11. Write a formula that describes the value of an account that starts with \$5,000 and earns 4% compound interest.

12. Use a calculator or spreadsheet to complete this table. Round to the nearest cent.

t	*A*
0	
10	
20	
30	
40	
50	
60	

13. What's the value of this account after 20 years?

14. To the nearest year, how long does it take the account to reach a value of \$24,000? Did it reach \$24,000 before or after this time? Explain.

2-2 Portfolio

Name ______________________________

Check each box when you've completed the task. Remember that your instructor will want you to turn in the portfolio pages you create.

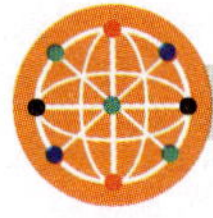

Technology

1. ☐ One of the big advantages to using Excel when doing calculations involving formulas is that Excel can easily handle formulas with multiple variables. The table feature on a graphing calculator restricts you to one. A classic problem in both algebra and calculus involves the connection between area and perimeter for rectangles. Does a certain perimeter give you a fixed area? Or do the lengths of the sides change the area even if they add up to the same perimeter? In the online template for this assignment, you'll find a table that allows you to input various lengths and widths for rectangles. Your job is to use formulas to calculate the perimeter and area, then experiment with different values and write any conclusions. The geometry formulas you need are $A = lw$ and $P = 2l + 2w$.

Online Practice

1. ☐ Include any written work from the online assignment along with any notes or questions about this lesson's content.

Applications

1. ☐ Complete the Applications problems.

Reflections

Type a short answer to each question.

1. ☐ What is a variable? Why is it wrong to answer "a letter"?
2. ☐ When two variables are related, what does it mean to say that one is the independent variable and the other is the dependent variable?
3. ☐ Describe the difference between an equation and an expression.
4. ☐ How do the terms input and output apply to the relationships we discussed in this lesson?
5. ☐ Describe something you learned or discovered in this lesson that you found particularly interesting.
6. ☐ What questions do you have about this lesson?

Looking Ahead

1. ☐ Complete the Prep Skills for Lesson 2-3.
2. ☐ Read the opening paragraph in Lesson 2-3 carefully then answer Question 0 in preparation for that lesson.

Answers to "Did You Get It?"

1. **a.** Amount of money is independent. How much you spend definitely affects the number of votes you get, but not the other way around.
   b. Amount of sugar is independent. That will affect how the drink tastes.
2. \$12.75
3. $A = 42{,}000 + 42{,}000(0.045)t$; \$64,680

Answers to "Prep Skills"

1. **a.** 243 **b.** 64 **c.** $\frac{1}{16}$ **d.** −25
2. **a.** 22 **b.** 17 **c.** 33
3. **a.** 70 **b.** −96 **c.** 24
4. **a.** 441 **b.** 3,591.71 **c.** 20.25 **d.** 1,764

2-2 Applications

Name ______________________________

1. If we were to find a formula that relates the distance it takes an average car to stop when the brakes are applied and the speed at which the car is traveling, which would be the input and which would be the output? Why?

2. The formula $D = R + 0.05R^2$ describes the total stopping distance in feet for an average car under ideal driving conditions, where R is the speed of the car in miles per hour. This distance includes the time needed for the driver to react and the time for the car to come to a full stop. Use the formula to fill in this table. I'd suggest using a spreadsheet or the table feature on a calculator, but you can do whatever you want. You're an adult.

Speed (mph)	Stopping Distance (ft)
5	
10	
15	
20	
25	
30	
35	
40	
45	
50	

Math Note

This formula was developed based on experimental data from drivingtesttips.biz. Later in this course we'll see how to develop formulas from experimental data.

3. If you're driving at 25 miles per hour through a school zone and a child runs out in front of you, how much distance will your car cover before stopping?

4. How slow would you have to go in order to decrease that stopping distance by 30 feet?

5. At the scene of an accident, a traffic investigator determines that the stopping distance for one of the cars involved was 103 feet. The speed limit in that area was 30 mph, and conditions were ideal. Was the driver speeding? Explain.

6. According to the U.S. Department of Transportation, in 2014 alone, 3,179 people were killed and 431,000 injured in motor vehicle crashes involving distracted drivers. If you're sending a text while driving on the freeway at 67 mph, how much stopping distance would you need to avoid an accident? (Legal disclaimer: NEVER EVER EVER EVER EVER EVER do this.)

7. What are some factors that would affect stopping distance? Remember, stopping distance factors in both the reaction time of the driver and the time needed to come to a complete stop after braking.

Lesson 2-3 Prep Skills

SKILL 1: RECOGNIZE QUANTITIES THAT CAN BE ADDED

Quantities can only be added if they're **like quantities.** For example, you can easily add \$10 and \$8, but you can't add \$10 and 8 bran muffins. The easiest way to decide if quantities are like is to cover up the number part: If what's left over is identical, the quantities are like. If there are any differences, then the quantities are NOT LIKE and CANNOT be added.

- 202 lb + 15 lb = 217 lb — Without the numbers, both quantities are pounds.
- 202 lb + 15 oz — Not like quantities (lb and oz) so can't be added.
- $12y + 22y = 34y$ — Without the numbers, both quantities are y.
- $12y + 22y^2$ — Not like quantities (y and y^2) so can't be added.

In some cases, quantities that aren't like can be changed into like quantities by changing units.

- 4 lb + 3 oz = 64 oz + 3 oz = 67 oz — 4 lb is 4 × 16 = 64 oz; now they're like quantities.
- \$3 + 50¢ = \$3 + \$0.50 = \$3.50 = 300¢ + 50¢ = 350¢ — Usually you can convert to whichever of the units you prefer.

SKILL 2: MULTIPLY POWERS OF A VARIABLE QUANTITY

We know that integer exponents represent repeated multiplication. For example, $5^4 = 5 \times 5 \times 5 \times 5$. The same applies to variable quantities, which can be represented by letters, like x, or units of measurement, like feet, meters, or others.

- $x \cdot x = x^2$
- $4y^2 \cdot 2y = 8y^3$ — Numbers and variables get multiplied separately.
- 4 in. × 2 in. = 8 in.^{2} — Numbers and units get multiplied separately.
- 6 m^2 × 10 m = 60 m^3

SKILL 3: WRITE A PROCEDURE AS A FORMULA

One of the most common and important uses of variables in math is writing formulas to represent mathematical procedures. For example, in order to find your average speed over a trip, you would divide the distance you went by the amount of time you traveled. (This is exactly why a common measure of speed is "miles per hour": It's a distance divided by a time.) Instead of writing that procedure out in words, we could more concisely write

$$s = \frac{d}{t}$$

where s represents the average speed, d represents the distance, and t represents the time.

PREP SKILLS QUESTIONS

1. Add the quantities if possible, or describe why they can't be added.

   a. 132 yd + 40 yd b. $-12t^3 + 4t$ c. $18y^2 + 10y^2$

2. If possible, rewrite the quantities so they can be added.

   a. 4 ft + 8 in. b. 12 yd + 16 sec c. 10 decades + 5 years

3. Perform each multiplication.

   a. $4y \cdot 6y$ b. (4 km)(6 km) c. $x^2(-3x)$ d. 8 yd² × 12 yd

4. To find the amount of simple interest earned, multiply the original amount, the interest rate, and the number of years. Write this procedure as a formula, and describe what each symbol in your formula represents.

5. The temperature in degrees Fahrenheit can be calculated by multiplying the temperature in Celsius by $\frac{9}{5}$ and adding 32. Write this procedure as a formula, and describe what each symbol in your formula represents.

Lesson 2-3 From Another Dimension

LEARNING OBJECTIVES

- ☐ 1. Determine units for area and volume calculations.
- ☐ 2. Use formulas to calculate areas and volumes.
- ☐ 3. Discuss important skills for college students to have.
- ☐ 4. Simplify expressions.

Even if you're on the right track, you'll get run over if you just sit there.

—Will Rogers

Courtesy Dave Sobecki

In a couple of lessons in Unit 1, we worked with lengths and distances. When you measure distance along a line, you're measuring in just one dimension. But unless you reside in South Park, you live in a three-dimensional world, and many measurements are used for two- and three-dimensional objects. How much carpet do you need for a certain room? That's an area—two dimensions. What's the cargo capacity of an SUV you're considering buying? Volume—three dimensions. How much paint do you need to buy when repainting a couple of rooms? Area. How large of a hot water tank do you need if you have five people in your house? Volume. Measurements like this play a very real role in everyday life, and understanding how to do calculations for measurement in multiple dimensions gives you a real perspective on size in our world.

0. Think of some units of measurement that are applied to areas, and to volumes.

2-3 Group

This activity is a bit less guided than most: The idea here is to promote some qualities that we'll study in the Class portion of this lesson, like problem solving, agility, adaptability, initiative, imagination, and (of course) collaboration, which is always one of the key goals in group work.

We'll begin by working with area and volume. **Area** is the measure of size for two-dimensional objects, like floors, walls, posters, fields, etc. **Volume** is the measure of size for three-dimensional objects, like beer mugs, swimming pools, buildings, and so on.

1. Notice that each of the small boxes in the figure here is a square that's one inch on each side. You could find the area of the large rectangle by adding up the number of boxes, but there's a quicker way. What is it? Use it to find the area.

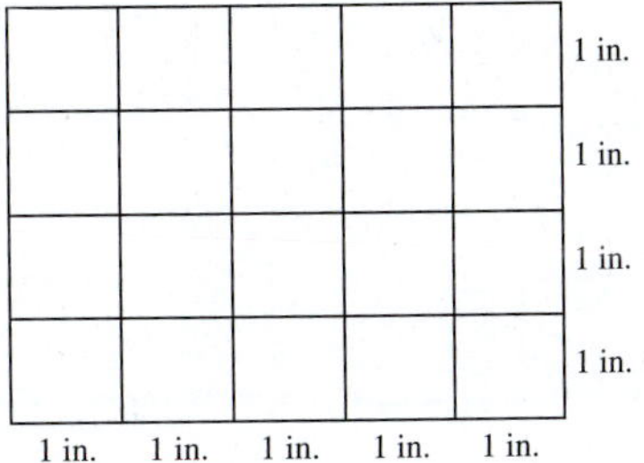

2. When length is measured in inches, the correct units for area are __________ __________.

3. Use your answer from Question 1 to write a formula for the area of a rectangle based on the length and the width.

4. This time, we have a three-dimensional figure made up of a bunch of little cubes. You could find the volume by counting all the little cubes, but you're more clever than that! Describe a quicker method, then use it to find the volume.

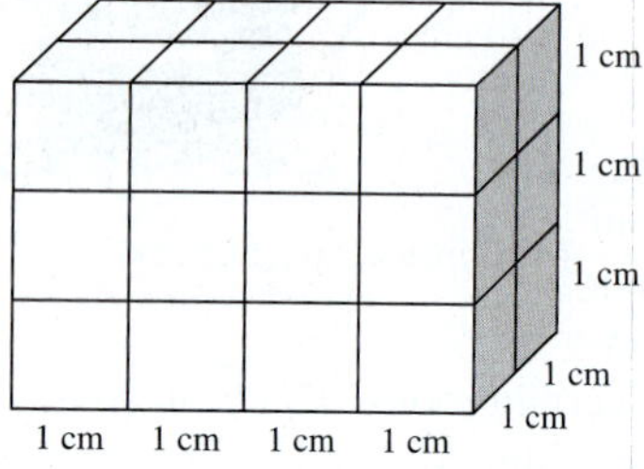

5. When length is measured in centimeters, the correct units for volume are ____________ ____________.

6. Use your answer from Question 4 to write a formula for the volume of a rectangular solid based on the length, width, and height.

Did You Get It

Try this problem to see if you understand the concepts we just studied. The answers can be found at the end of the Portfolio section.

1. Find the area of one figure and the volume of the other figure, whichever is more appropriate.

a.

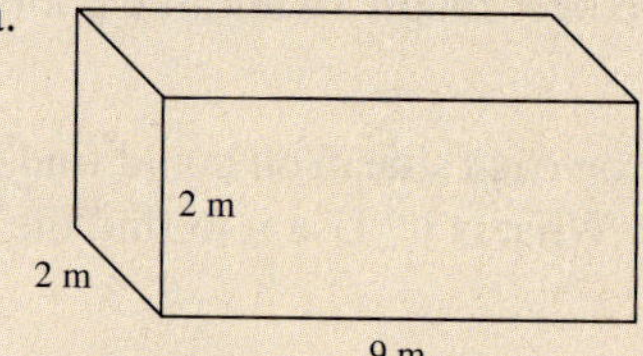

b.

2.5 yd

7 yd

7. My friend David Najar is an acclaimed artist. The value of a painting is to some extent determined by the area of the canvas. A Najar piece I recently acquired has dimensions as shown, and you can find the area using the formula $A = lw$, where l is the length and w is the width. What is the area?

Area = ________ × ________. (Include units on each number.)

Area =

The area units for the canvas are __________ __________ because there are ______ factors multiplied together, each of which has ________ as units.

Painting courtesy of David Najar, Photo courtesy of Dave Sobecki

8. The volume of a cylindrical oil storage tank can be found using the formula $V = \pi r^2 h$, where r is the radius of the top and bottom, and h is the height. Use the second diagram (not one of my best, but to be fair I'm a mathematician, not a graphic artist) to find the volume of the tank.

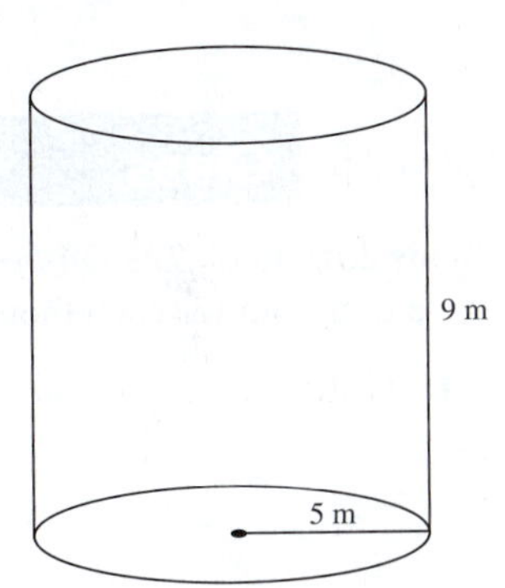

Volume = ________ × ________ × ________ × ________
(Write four separate factors, and again include units on all measurements.)

Volume =
(Round to the nearest tenth.)

The volume units for the oil tank are __________ __________ because there are ___ factors that have __________ as units.

In Questions 9–13, perform each calculation. Then describe whether the result is an area or a volume, and make an educated guess as to what type of figure you think the result measures. (If necessary, round results to one decimal place.)

> **Math Note**
>
> It would be a good idea to locate the π button on your calculator, and use it anytime a calculation involves pi. You'll get better results than if you use a decimal approximation like 3.14, and it's less work! Bonus.

9. (6 m)(4 m)

10. (3 ft)(5 ft)(10 ft)

11. $\dfrac{4\pi(4 \text{ in.})^3}{3}$

12. $\dfrac{4\pi}{3}(1.5 \text{ cm})^3 + \pi(1.5 \text{ cm})^2(3 \text{ cm})$

13. $\frac{1}{2}\pi(2\text{ ft})^2 + (8\text{ ft})(4\text{ ft})$

Did You Get It

2. For each calculation, describe if it's an area or a volume, then calculate the result, including units.
   a. $\frac{1}{2}(4\text{ mi} + 5\text{ mi})(3\text{ mi})$
   b. $13(14\text{ ft}^2)9\text{ ft}$
   c. $\frac{4}{3}\pi(7\text{ mm})(5\text{ mm})(11\text{ mm})$

2-3 Class

In his 2010 book *The Global Achievement Gap,* educational consultant Tony Wagner outlines seven survival skills that students have to master to reach their academic potential. Define what each of these skills means to you.

1. Critical thinking and problem solving

2. Collaboration across networks and leading by influence

3. Agility and adaptability

4. Initiative and entrepreneurialism

5. Accessing and analyzing information

6. Effective written and oral communication

7. Curiosity and imagination

Here's a key quote from Wagner's book: *"Today knowledge is ubiquitous, constantly changing, growing exponentially. . . . Today knowledge is free. It's like air, it's like water. It's become a commodity. . . . There's no competitive advantage today in knowing more than the person next to you. The world doesn't care what you know. What the world cares about is what you can do with what you know."*

8. What does this quote mean to you? Do you agree with it? Why or why not?

At this point, you should be pretty comfortable multiplying units of length to find the correct units for area and volume calculations. Multiplying two terms with variables is very similar to multiplying dimensions with units.

9. Perform each multiplication. For parts a and b, include units.

a. 2 ft × 3 ft = ______

b. 6 ft^2 × 5 ft = ______

c. $(2x)(3x) =$ ______

d. $(6x^2)(5x) =$ ______

When reviewing addition of similar quantities in Lesson 1-2, we saw that we can add the expressions $8x$ and $5x$ to get $13x$. Since $8x + 5x$ and $13x$ would give you the same output no matter what input you choose, we say that those expressions are **equivalent**.

When working with expressions, equations, and formulas it's often useful to simplify expressions. This involves replacing an expression with a simpler but equivalent one. As we saw in the previous paragraph, adding or subtracting like terms is one way to simplify an expression.

10. Use the formula $P = a + b + c$ to find the perimeter of this triangle. The length of each side is measured in inches. First write the formula, then substitute the expressions for each side of the triangle into the formula, then simplify the result. Make sure each step is a complete formula (has both a left side and a right side), and include units.

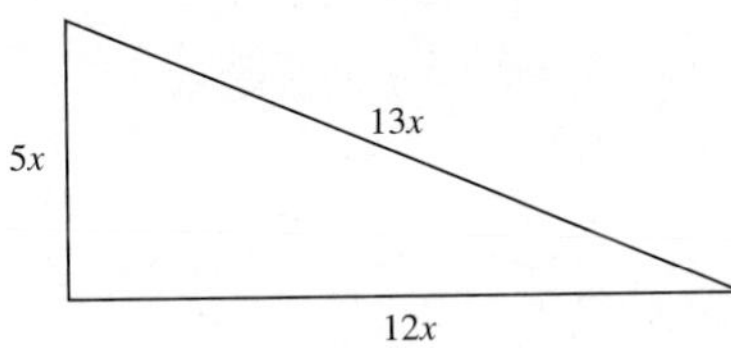

11. Repeat Question 10, but this time use the formula $A = \frac{1}{2}bh$ to find the area of the triangle. In this case, b represents the length of the triangle's base and h is the height.

12. Use your answers to Questions 10 and 11 to find the perimeter and area of the triangle if x represents 4 inches.

Did You Get It

3. Simplify each expression, and state whether each could be a volume, an area, or neither.
   a. $\pi(4x)^2 x$
   b. $lw + lw + wh + wh + lh + lh$
   c. $8y^2(10y^2)$

2-3 Portfolio

Name ______________________________

Check each box when you've completed the task. Remember that your instructor will want you to turn in the portfolio pages you create.

Online Practice

1. ☐ Include any written work from the online assignment along with any notes or questions about this lesson's content.

Applications

1. ☐ Complete the Applications problems.

Reflections

Type a short answer to each question.

1. ☐ Pick two of the seven survival skills from the Class portion of the lesson that you think are most important to you. Discuss why you chose them, and how strong you think you are in these skills.
2. ☐ Why are units so important in measurement? Giving some examples of measurements with no units should help.
3. ☐ What does it mean to say that two expressions are equivalent?
4. ☐ Name one thing you learned or discovered in this lesson that you found particularly interesting.
5. ☐ What questions do you have about this lesson?

Looking Ahead

1. ☐ Complete the Prep Skills for Lesson 2-4.
2. ☐ Read the opening paragraph in Lesson 2-4 carefully and answer Question 0 in preparation for that lesson.

Answers to "Did You Get It?"

1. **a.** 36 m^3 **b.** 17.5 yd^2
2. **a.** Area: 13.5 mi^2 **b.** Volume: $1{,}638 \text{ ft}^3$ **c.** Volume: $1{,}612.7 \text{ mm}^3$
3. **a.** $16\pi x^3$, volume **b.** $2lw + 2wh + 2lh$; area **c.** $80y^4$; neither

Answers to "Prep Skills"

1. **a.** 172 yd **b.** cannot be added; t^3 and t are not like **c.** $28y^2$
2. **a.** $4\frac{2}{3}$ ft or 56 in. **b.** cannot be added **c.** 10.5 decades or 105 years
3. **a.** $24y^2$ **b.** 24 km^2 **c.** $-3x^3$ **d.** 96 yd^3
4. $I = Prt$, where I represents the amount of interest earned, P is the original amount, r is the interest rate, and t is time in years.
5. $F = \frac{9}{5}C + 32$, where F is the temperature in degrees Fahrenheit and C is the temperature in degrees Celsius.

2-3 Applications

Name ______________________________

Tennis balls are packaged in a cylindrical can, and since they're round, that leaves a lot of empty space in the can. How much space? The radius of a tennis ball is 3.3 cm; the can has the same radius for a nice tight fit, and the height of the can is six times that radius, which is 19.8 cm.

©McGraw-Hill Education/ Ken Karp, photographer

1. The calculation below is intended to find the volume of the empty space in the can, but there's something wrong with it. Explain how you can tell, without doing any calculations or looking up any formulas, that the answer will be wrong. (Hint: The terms "units" and "like terms" might come into play.)

$$\text{Volume} = \underbrace{\pi(3.3\text{ cm})^2(19.8\text{ cm})}_{\text{Volume of can}} - 3\underbrace{\left(\frac{4}{3}\pi(3.3\text{ cm})^2\right)}_{\text{Volume of each ball}}$$

2. Find the fix needed to correct the mistake in the calculation. If you need help, do a Web search for "volume of a sphere." Then find the volume of empty space in the can. Round to the nearest hundredth.

3. Use your corrected formula to find the volume of the can, and the volume of all three balls in the can combined. Round each to the nearest hundredth.

4. What percentage of the volume of the can is filled by the tennis balls? When you set up your calculation, make sure to include units on all volumes. Round to the nearest hundredth.

5. After performing the division in Question 4, what units remain? Do you think this will always happen when finding percentages?

2-3 Applications

Name ____________________

6. Racquetballs have a radius of about 2.8 cm. If three balls are packaged in a can like we described for tennis balls earlier, find each volume. Round to the nearest hundredth.

   a. The volume of one ball.

©Comstock Images/Alamy RF

   b. The volume of three balls.

   c. The volume of the can. (Remember, the height is six times the radius.)

   d. The volume of empty space in the can.

   e. The percentage of the can filled by the racquetballs. How does it compare to the tennis ball percentage? Does this surprise you?

7. Golf balls also come in packages of three, but they are usually packaged in a rectangular box with a square bottom. The radius of each ball is about 2.2 cm. The length and width of the box's square base are twice the radius, and the height is six times the radius.

   a. Would you guess that there would be more or less empty space by percentage than there is in a tennis ball can? Why?

   b. Use your procedure from Question 4 to find the percentage of the golf ball box that is filled. Round volumes to the nearest hundredth.

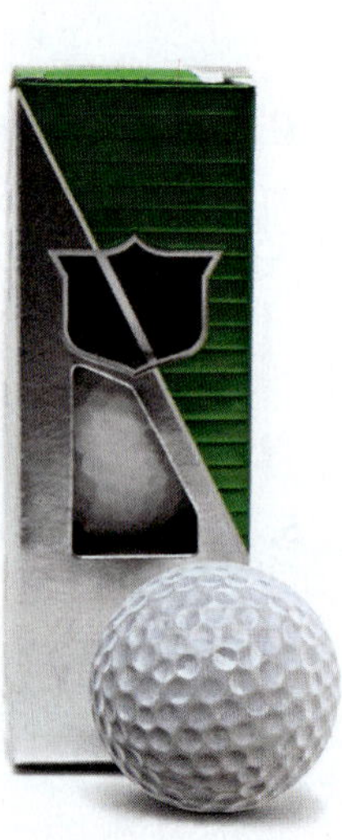

©Arina Zaiachin/123RF .com

Lesson 2-4 Prep Skills

SKILL 1: REDUCE FRACTIONS

The process of writing fractions in lowest terms is called **reducing fractions**. It's useful to do because it's easier to work with and interpret fractions when they're in lowest terms. For example, it's a lot easier to see that $\frac{5}{2}$ is two and a half than to see that $\frac{180}{72}$ is equal to two and a half. To reduce a fraction, divide BOTH the numerator and denominator by any common factors.

- $\frac{4}{6} = \frac{4 \div 2}{6 \div 2} = \frac{2}{3}$ The greatest factor common of 4 and 6 is 2.
- $\frac{30}{12} = \frac{30 \div 6}{12 \div 6} = \frac{5}{2}$ The greatest factor common of 30 and 12 is 6.

SKILL 2: BUILD UP FRACTIONS

Sometimes it's useful to do the opposite of reducing fractions: rewriting fractions so that they have a bigger numerator and denominator. This is called **building up** a fraction. Since it's the opposite of reducing fractions, we do exactly the opposite: Instead of dividing the numerator and denominator by the same number, we multiply the numerator and denominator by the SAME number.

- $\frac{3}{5}$ can be written as a fraction with denominator 20: $\frac{3 \cdot 4}{5 \cdot 4} = \frac{12}{20}$
- $-\frac{2}{3}$ can be written as a fraction with denominator 60: $-\frac{2 \cdot 20}{3 \cdot 20} = -\frac{40}{60}$

SKILL 3: MULTIPLY FRACTIONS

To multiply two fractions, you just multiply numerators and multiply denominators to find the fraction for the product.

- $\frac{2}{3} \cdot \frac{7}{11} = \frac{14}{33}$
- $\frac{5}{2} \cdot \frac{12}{7} = \frac{60}{14} = \frac{30}{7}$ Often, the result can be reduced.

If one of the factors isn't a fraction, you can make it a fraction by just giving it a denominator of 1.

- $100 \cdot \frac{3}{5} = \frac{100}{1} \cdot \frac{3}{5} = \frac{300}{5} = 60$

The same process can be used to multiply three or more fractions.

- $\frac{6}{1} \cdot \frac{1}{3} \cdot \frac{2}{1} \cdot \frac{4}{1} = \frac{48}{3} = 16$

SKILL 4: REDUCE FRACTIONS WITH VARIABLES

Variables in fractions can sometimes be a common factor. For example, in the expression $\frac{7a}{2a}$, we can divide the numerator and denominator by a to get $\frac{7}{2}$.

- $\frac{y(3-y)}{y} = 3 - y$ Numerator and denominator divided by y
- $\frac{12t}{t^2} = \frac{12}{t}$ Numerator and denominator divided by t

SKILL 5: RECOGNIZE UNIT EQUIVALENCES

You almost certainly know that 12 inches and 1 foot represent the same length. This is an example of a **unit equivalence**, which is a statement that two measurements in different units represent the same size. The following tables are a list of unit equivalences. The goal is certainly not to memorize these equivalences (although it wouldn't be a bad idea to learn some of them by heart). Just familiarize yourself with them so that you know where to refer back to when needing an equivalence throughout the remainder of the course.

English Measures and Equivalents

Length	Weight
12 inches (in.) = 1 foot (ft) 3 feet = 1 yard (yd) 5,280 feet = 1 mile (mi)	16 ounces (oz) = 1 pound (lb) 2,000 pounds = 1 ton (T)
Liquid Volume	**Time**
3 teaspoons (tsp) = 1 tablespoon (tbs) 8 fluid ounces (oz) = 1 cup (c) 2 cups = 1 pint (pt) 2 pints = 1 quart (qt) 4 quarts = 1 gallon (gal) 1 cubic centimeter (cm^3) = 1 mL	60 seconds (sec) = 1 minute (min) 60 minutes = 1 hour (hr) 24 hours = 1 day 7 days = 1 week 52 weeks = 1 year 365 days = 1 year

Metric System

Kilo	Hecto	Deka	Base Unit	Deci	Centi	Milli
km	hm	dam	**Length** Meter (m)	dm	cm	mm
kg	hg	dag	**Weight** Gram (g)	dg	cg	mg
kL	hL	daL	**Volume** Liter (L)	dL	cL	mL
Move decimal right →						
← Move decimal left						

Conversions between Systems

Length	Weight	Volume	Temperature
2.54 cm = 1 in. 1 m ≈ 3.28 ft 1.61 km ≈ 1 mi	28.3 g ≈ 1 oz 2.2 lb ≈ 1 kg	1.06 qt ≈ 1 L 3.79 L ≈ 1 gal	$F = \frac{9}{5}C + 32$ $C = \frac{5}{9}(F - 32)$ $C = K - 273.15$ $K = C + 273.15$

PREP SKILLS QUESTIONS

1. Write each fraction in lowest terms.

   a. $\frac{12}{18}$ b. $-\frac{5}{35}$ c. $\frac{20}{45}$

2. Write each fraction with the given denominator.

   a. $\frac{1}{3}$; 21 b. $\frac{7}{3}$; 18 c. $-\frac{3}{8}$; 80

3. Perform each operation, and reduce your answer if necessary.

   a. $\frac{2}{5} \cdot \frac{9}{2}$ b. $12 \cdot \frac{9}{8}$ c. $\frac{12}{1} \cdot \frac{1}{3} \cdot \frac{600}{1}$ d. $5 \cdot \frac{11}{5} \cdot \frac{12}{11} \cdot \frac{7}{12}$

4. Simplify each fractional expression by dividing out common factors.

   a. $\frac{8b}{3b}$ b. $\frac{5x(x+4)}{x}$ c. $\frac{4y^2}{8y}$

5. Use the table on the previous page to find an equivalent measure for each amount.

   a. 1 gal = ______ qts b. 1 in. = _____ cm c. 1 gal = _____ L

   d. 1 T = _____ lbs e. 1 kg = _____ lbs

Lesson 2-4 It Works Like Magic

LEARNING OBJECTIVES

- ☐ 1. Convert units using dimensional analysis.
- ☐ 2. Convert units within the metric system.
- ☐ 3. Convert rates of change.
- ☐ 4. Convert temperatures.

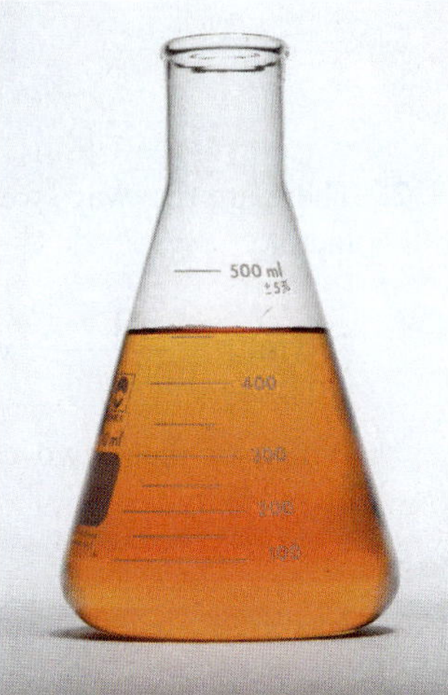

©David A. Tietz/Editorial Image, LLC RF

If you don't like change, you are going to like irrelevance even less.
—Eric Shinseki

Have you ever thought about the number of different ways we have to measure objects? Inches, feet, yards, miles, centimeters, meters, kilometers . . . and these cover only length; there are many more ways to measure length, as well as different units of measure for temperature, time, area, volume, mass, and other quantities. Why so many different measurements? The short answer is efficiency. Things come in so many different sizes that being stuck with just one unit of measurement would be very inconvenient. For example, it's about 5,892,480 inches from Chicago to Milwaukee. Clearly, measuring the distance in miles (93) is far more efficient.

On the other hand, this mysterious beaker is only about 0.00016 miles tall. Do you have even the first clue of how tall that is compared to, say, a house cat? Me neither. So measuring its height in inches (10) makes a whole lot more sense. In this lesson, we'll learn an incredibly valuable skill called **dimensional analysis** in the process of converting units. It sounds fancy, but don't worry: You'll be able to handle it just fine, and it'll help you out with a lot of day-to-day calculations that might have seemed puzzling before.

0. Think of a situation you've encountered where converting the units of some quantity did, or would have, come in handy.

2-4 Class

According to the unit equivalence table in the Prep Skills (which we'll use a TON in this lesson), 1 kg is equivalent to about 2.2 lb. Quick, without thinking about it too deeply, if you wanted to convert your weight from pounds to kilograms, would you multiply or divide by 2.2? Not so easy, is it? (Don't feel bad, it's confusing for me too, and I have all sorts of degrees and stuff.) Our first goal in this lesson is to develop a systematic procedure that takes thinking and guessing out of the business of converting units. It's based on an idea that's so brilliantly simple you should probably send us a sizeable cash donation: If measurements in two different units represent the same actual weight, like 1 kg and 2.2 lb, then a fraction formed from dividing those units is just a fancy way to say 1. In other words, 1 kg and 2.2 lb are different ways to express the exact same amount, so

$$\frac{1\text{ kg}}{2.2\text{ lb}} \text{ and } \frac{2.2\text{ lb}}{1\text{ kg}} \text{ are both equal to 1.}$$

We call a fraction of this type a **conversion factor**, and *multiplying any measurement by a conversion factor won't change the size of the measurement* because we're just multiplying by one; it will just change the units used to measure.

So here's our genius idea for converting units for a measurement: We'll multiply the measurement by a conversion factor in a way that makes the units we don't want divide out, leaving behind the units we do want. To see how this works, we'll return our attention to the beaker at the beginning of the lesson. It turns out that this particular beaker contains a revolutionary product, Dr. Phail's Magic Love Potion: guaranteed to lure the affection of that special someone, or your money back (minus a processing fee of just $29.95). The 450 mL of potion in the beaker can be converted to liters if we multiply by a convenient conversion factor.

1. Write two different conversion factors based on the fact that 1 liter is equivalent to 1,000 mL.

2. There are two ways we could multiply our measurement of 450 mL by a conversion factor:

$$450 \text{ mL} \cdot \frac{1 \text{ L}}{1{,}000 \text{ mL}} \quad \text{or} \quad 450 \text{ mL} \cdot \frac{1{,}000 \text{ mL}}{1 \text{ L}}$$

Which of these two calculations correctly eliminates milliliters and leaves behind liters? Why?

Math Note

If multiplying a number by a fraction, as in Question 2, bothers you at all, feel free to write the number as a fraction by giving it a denominator of 1.

3. What are the units for the other calculation? Do they have any meaning?

4. So how many liters of magic potion are in the beaker?

5. According to our handy-dandy conversion chart, 1 gallon is approximately equivalent to 3.79 L. Write two different conversion factors based on this fun fact.

6. Use one of the two conversion factors to convert your answer from Question 4 into gallons.

Did You Get It ?

Try this problem to see if you understand the concepts we just studied. The answers can be found at the end of the Portfolio section.

1. There are four quarts in a gallon. How many quarts are in the beaker at the beginning of the lesson? (You'll need your answer to Question 6.)

7. Now here's one of the great things about dimensional analysis: Notice that it took us two steps to get from 350 mL to gallons. First we converted to liters using one conversion factor, then from liters to gallons using another. If you're really clever, you can get there directly in one calculation by using BOTH of the conversion factors you used in Questions 4 and 6. Do it!

8. My house is just about 15 miles from the Gulf of Mexico on the Florida coast. How far is this in inches? Use two conversion factors in one calculation, like you did in Question 7.

Did You Get It 

2. There are 2 cups in a pint, and 8 pints in a gallon. If a large recipe calls for 7 cups of sweet cream, how many gallons is that?

Converting Units within the Metric System

When we converted 450 mL to L we used dimensional analysis, but it turns out that we could have just moved the decimal point. If you look at the conversion chart for metric units in the Prep Skills for this lesson, going from mL to L requires a move of 3 columns to the left. This corresponds to moving the decimal point 3 places to the left:

450. mL = 0.450 L

If, on the other hand, a unit you're converting to lives to the right on the chart, count the number of columns and move the decimal point that many places to the right.

9. Convert 0.44 L of love potion into dL and kL by shifting the decimal. Which is the more reasonable measurement in this case?

Did You Get It 

3. Use the metric conversion table to convert 275 grams into cg and dag.

10. Dr. Phail's potion also comes in a 0.75-gallon container. Convert this to cubic centimeters (cc) using equivalances in the Prep Skills table and dimensional analysis.

Now here's a key idea: Another way to write cubic centimeters is cm^3, or cm · cm · cm. We can use this idea to convert units of volume.

11. According to the table of equivalences, 1 in. = 2.54 cm. What is the result of the following calculation? Pay close attention to the units.

$$1{,}000\ \text{cm}^3 \cdot \frac{1\ \text{in.}}{2.54\ \text{cm}} \cdot \frac{1\ \text{in.}}{2.54\ \text{cm}} \cdot \frac{1\ \text{in.}}{2.54\ \text{cm}}$$

12. Use the idea from Question 11 to convert your answer from Question 10 into cubic inches.

13. Based on what you just did, fill in the blank to create a shortcut to convert between cubic centimeters and cubic inches.

$1\ \text{in.}^3 =$ ____________ cm^3

Converting Area and Volume Units

You can use known conversion factors involving length to convert units in two dimensions (area) and in three dimensions (volume). You'll just need to use them enough times to be sure the area or volume units you started with have been completely divided away.

For example, my bedroom has an area of 196 ft^2. This area can be converted to square inches by using the fact that 1 ft = 12 in.

$$\frac{196\ \text{ft}^2}{1} \cdot \frac{12\ \text{in.}}{1\ \text{ft}} \cdot \frac{12\ \text{in.}}{1\ \text{ft}} = 28{,}224\ \text{in}^2$$

Did You Get It

4. Use the procedure for converting units of area or volume to convert 4 square miles into square feet.

2-4 Group

Rates of Change

A **rate** is a ratio that compares two quantities. Examples of rates include $\frac{75 \text{ mi}}{1 \text{ hr}}$, $\frac{10 \text{ mg}}{1 \text{ mL}}$, and $\frac{\$15.00}{2 \text{ tickets}}$. The first two are also called **unit rates** because they have a 1 in the denominator.

1. Explain what is meant when we say a car is traveling $\frac{75 \text{ mi}}{1 \text{ hr}}$ (more commonly written as 75 mph).

2. The entire 450 mL of Dr. Phail's Love Potion (from the beginning of the lesson) is supposed to be slowly dripped into a mixture over a 5-hour period. (Kind of a pain, but hey—love doesn't come easy.) Express this quantity as a rate.

> **Math Note**
>
> You know how sometimes we write fractions using a slanted line, like 5/2? DON'T do that when using dimensional analysis. The whole process relies on clearly recognizing which units are in the numerator, and which are in the denominator.

3. Simplify the rate you wrote in Question 2 so that it's a unit rate.

4. It might seem like it would be really complicated to convert this rate into liters per minute, but dimensional analysis will save the day. If we're going to end up with liters per minute, we'll need to get rid of mL in the numerator and replace it with L, and we'll need to get rid of hr in the denominator and replace it with min. This can be accomplished in one calculation if we multiply by two different conversion factors that eliminate the unwanted units, just like we did in the class portion. Do that now. Focus on putting the units you want to eliminate on the OPPOSITE side of the fraction when writing your conversion factors. Write your answer in decimal and scientific notation.

Did You Get It

5. By several recent estimates, global sea levels are rising by at least 0.32 cm per year. Convert this rate to inches per century. (A century is 100 years.)

Converting Temperatures

Unlike other unit conversions, temperature conversions are done using a formula. (We'll get a quick look at why in the Applications.) There are two common temperature units for everyday use: degrees Fahrenheit in the English system, and degrees Celsius in the metric system. There's a third unit often used in science, particularly chemistry, known as degrees Kelvin.

5. While in production, the love potion needs to be kept below 400 K at all times. The thermometer on the most recent batch currently reads 200°F. One of the formulas provided will convert that temperature to degrees Celsius. Which formula would you use? Why?

$$F = \frac{9}{5}C + 32 \qquad C = \frac{5}{9}(F - 32)$$

Math Note

In the Celsius and Fahrenheit scales, we use the degree symbol (as in 32°F or 21°C), but in the Kelvin scale, we just use a K (as in 200 K).

6. Use the appropriate formula to convert 200°F to Celsius.

7. Use one of the two formulas provided to convert that temperature to Kelvin. Explain why you chose that formula. Will the batch of love potion be ruined?

$$C = K - 273.15 \qquad K = C + 273.15$$

8. Use a calculator or spreadsheet along with an appropriate formula to complete the table, which compares Fahrenheit and Celsius temperatures.

Degrees Celsius	Degrees Fahrenheit
0	
10	
20	
30	
40	
50	
60	
70	
80	
90	
100	

Did You Get It

6. The hottest temperature ever reliably recorded was 129.2°F in Death Valley, California, in June 2013. The lowest was −89.2°C at Vostok Station, a Soviet research installation on Antarctica, in July 1983. Convert the high temperature to Celsius and the low temperature to Fahrenheit.

We'll close this lesson with a bit more practice on dimensional analysis problems, coming to us from the world of nursing, where I promise you it comes in handy. We'll look at some related examples in the Applications.

9. According to the online dosage guide for one brand of insulin, the recommended delivery is 0.2 units of insulin per kilogram of weight each day. How many daily units would be indicated for a 178-pound patient?

10. A dosage of 1,700 mg of cefadroxil is to be delivered to a patient in an IV drip over 75 minutes. It's supplied in a solution that contains 1 gram of the drug in 20 cm^3 of solution. The pump used to deliver the drug uses units of cc's per hour. At what rate should it be set?

2-4 Portfolio

Name ______________________________

Check each box when you've completed the task. Remember that your instructor will want you to turn in the portfolio pages you create.

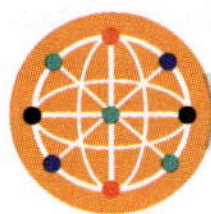

Technology

The method of dimensional analysis is all about multiplying by an appropriate factor, and spreadsheets are really good at doing multiplication, right? So it should be possible to build a spreadsheet that does unit conversions. Starting with the template in the online resources for this lesson, build a spreadsheet that converts lengths in meters to each of feet, inches, yards, and miles. To figure out how to write the formulas, it'll be helpful to first do a couple of sample conversions by hand, then apply that knowledge to the spreadsheet. See the template for further instructions.

Online Practice

1. ☐ Include any written work from the online assignment along with any notes or questions about this lesson's content.

Applications

1. ☐ Complete the Applications problems.

Reflections

Type a short answer to each question.

1. ☐ Describe the process we developed for converting units; focus on the fact that we're multiplying by one.

2. ☐ What's a conversion factor? Where do they come from? What are they used for?

3. ☐ What's a rate? What about a unit rate?

4. ☐ How can we use length conversion factors to convert areas or volumes?

5. ☐ Name one thing you learned or discovered in this lesson that you found particularly interesting.

6. ☐ What questions do you have about this lesson?

Looking Ahead

1. ☐ Complete the Prep Skills for Lesson 2-5.

2. ☐ Read the opening paragraph in Lesson 2-5 carefully and answer Question 0 in preparation for that lesson.

Answers to "Did You Get It?"

1. 0.48 qt

2. 0.4375 gal

3. 275 g = 27,500 cg = 27.5 dag

4. 111,513,600 ft^2

5. About 12.6 $\frac{\text{in.}}{\text{century}}$

6. High: 54°C; Low: −128.56°F

Answers to "Prep Skills"

1. a. $\frac{2}{3}$ **b.** $-\frac{1}{7}$ **c.** $\frac{4}{9}$

2. a. $\frac{7}{21}$ **b.** $\frac{42}{18}$ **c.** $-\frac{30}{80}$

3. a. $\frac{9}{5}$ **b.** $\frac{27}{2}$ **c.** 2,400 **d.** 7

4. a. $\frac{8}{3}$ **b.** $5(x + 4)$ **c.** $\frac{y}{2}$

5. a. 4 qts **b.** 2.54 cm **c.** 3.79 L **d.** 2,000 lbs **e.** 2.2 lbs

2-4 Applications

Name ______________________________

For Questions 1–5, use dimensional analysis with conversion factors to convert each measurement to the given units. You'll need unit equivalences from this section and from page 192 to build your conversion factors. Round to two decimal places if necessary.

1. How many miles is a 10-kilometer race?

2. How many feet is a 10-kilometer race?

3. The average offensive lineman in the National Football League weighs 141.3 kilograms. How many pounds is that?

4. A surgical patient needs to have her liquid intake carefully monitored. How many mL of fluid did she consume if she drank $\frac{1}{2}$ L of water, a pint of milk, and 8 oz of juice?

5. The earth moves at about 98,000 feet per second as it revolves around the sun. How fast is that in miles per hour? (Recall that 1 mile is 5,280 feet.)

©Getty Images/Digital Vision RF

Glaciers are large masses of ice that flow like rivers across the ground. Really, really slow rivers—did I mention that they're ice? Most move less than a foot per day. At one point, the San Rafael glacier in Chile was moving 203 millimeters per day.

6. How fast was it moving in inches per hour? (1 inch is 25.4 mm.)

Source: Photograph by Bruce F. Molnia, U.S. Geological Survey

7. Find the speed in miles per hour, then explain why that's a silly unit of speed in this case. Write your answer in scientific notation.

8. How long would it take the glacier to move the length of a football field (100 yards)? Write your answer in decimal notation rounded to the nearest tenth of an hour, then in scientific notation.

9. A patient needs 175 mg of a low-dose pain reliever. The pharmacy sends pills marked 0.35 g. How many should be administered?

Lesson 2-5 Prep Skills

SKILL 1: LOCATE NUMBERS ON A NUMBER LINE

When numbers live exactly at one of the labels on a number line, it's pretty easy to decide where to put them, as in this example:

- Locate −15 on the number line

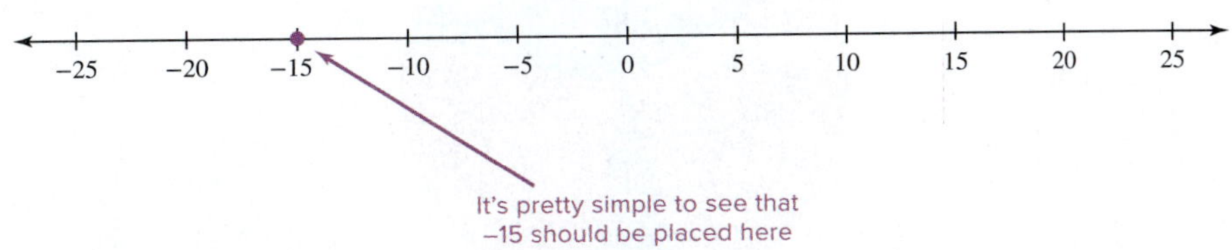

It's a more interesting question, and typically involves some estimation, when locating numbers that live in between labels.

- Locate 3, 8.2, 13.4, and 19.9 on the number line

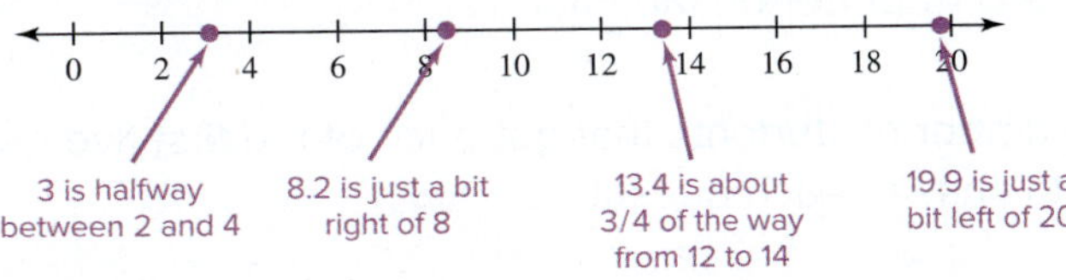

SKILL 2: READ BAR GRAPHS, PIE CHARTS, AND SCATTER PLOTS

These skills have been previously covered in this book. Cool! Bar graphs and pie charts are in Lesson 1-1, and scatter plots were introduced in Lesson 1-5.

PREP SKILLS QUESTIONS

1. Locate −40, 15, −15, 27, and −2 on the number line.

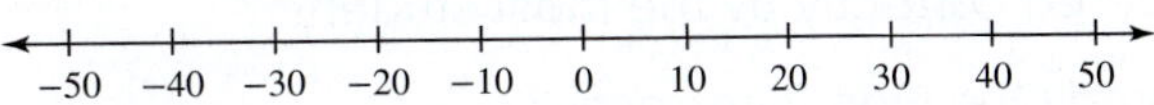

2. The pie chart shows the distribution of grades on a final exam I just finished grading for a Calc 2 class with 31 students. Use it to answer each question.

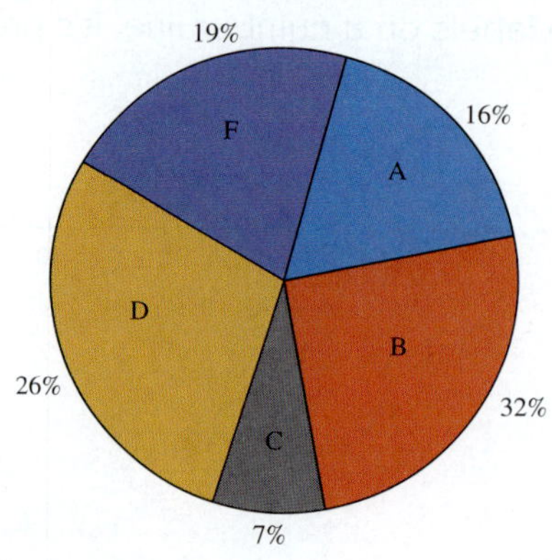

a. What percentage of students got a B on the final?

b. How many students got a C on the final?

c. How many students got a D or better on the final?

3. The bar graph shows the number of students that got each of the first five questions on the final completely correct. Use it to answer each question.

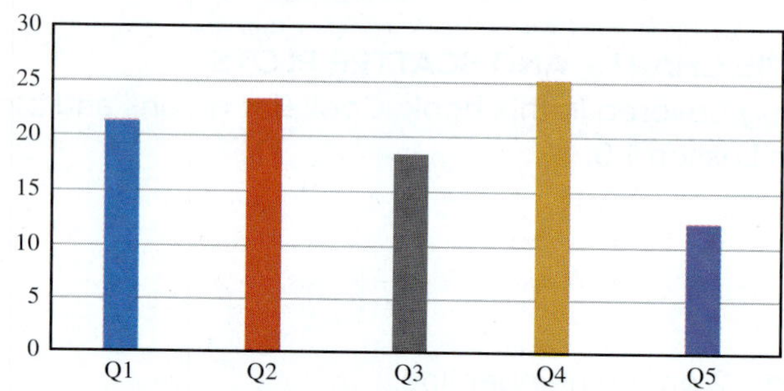

a. Which question was answered correctly by the most students?

b. How many people answered Question 2 correctly?

c. What percentage of the 31 students taking the test got Question 5 completely right?

4. The scatter plot shows the final grades for all 31 students. On the horizontal axis is the student's position alphabetically on the class roster. The vertical axis shows their score out of 160 on the final.

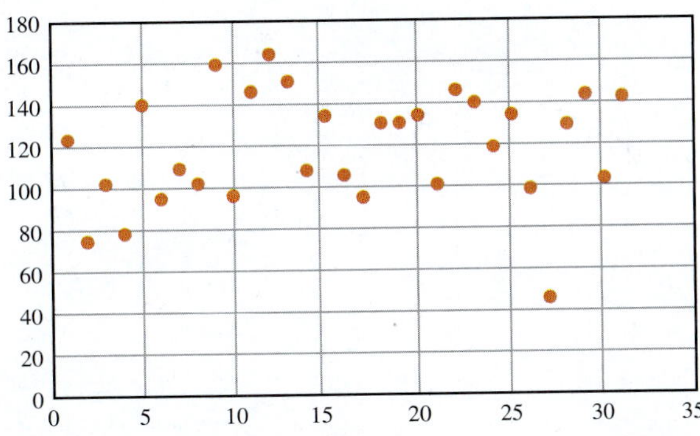

a. How many students scored less than 80 points?

b. How many students scored 140 or more points?

c. About what was the lowest score on the test?

Lesson 2-5 Take a Guess!

©Rich Legg/Getty Images RF

LEARNING OBJECTIVES

- ☐ 1. Identify the steps in a systematic problem-solving procedure.
- ☐ 2. Make educated guesses.
- ☐ 3. Compare numbers using inequality symbols.

Don't be afraid to fail. Be afraid not to try.
—Michael Jordan

Consider this scenario: You're a contestant on a game show, and you're given a trivia question worth $50,000. Pressure! The question is "Which state was first to secede from the union in 1860?" Do you know the answer? If so, terrific—drinks are on you. But if you don't know the answer, you'd have to guess. Would your list of guesses include the Pacific Ocean, Justin Bieber, Twinkies, and a goat? Of course not. Those would be completely ridiculous, and you're not a goof.

The point is that "guessing" doesn't usually mean just throwing out some random answer and praying for a miracle. Instead, we use information we know to be true and some reasoning ability to make an **educated guess**. In this section, we'll try to gain some confidence with educated guessing, which sometimes goes by a different name in math: **estimation**.

From an instructor's point of view, it seems like in math classes, a lot of students are afraid to make educated guesses. But sometimes that's a completely reasonable step in problem solving. Nobody likes to be wrong, but if you're not willing to at least answer a question, you have no chance at ever being right. Not being afraid of failure is one of the characteristics shared by almost everyone that's very successful in their chosen field.

Think Michael Jordan never failed? Do a Google search sometime for "Michael Jordan high school team." If one of the greatest athletes of all time can overcome the occasional fail, then surely we can, too. So let's get over the fear of being wrong and gain the confidence to make educated guesses.

0. Describe a situation, either in the classroom or out, where having the confidence to make an educated guess would be helpful.

2-5 Class

1. Write a brief description of two or three problems that you've had to solve outside of school.

2. What are some methods that you use to solve problems in your life? Do you think that they apply to solving problems in college classes?

We're now well into the second unit of this course. So what is it really about? Math would be the easy answer, but we're sure hoping that you've recognized by now that this isn't a typical math course. A better answer to that question would be "problem solving." Hopefully you've noticed that we haven't done any abstract math: Every topic we've covered has been in an attempt to solve some problem. So by now, we've gotten some pretty good experience in solving problems. And without even thinking about it, we've carved out some general guidelines for attacking problems.

A Hungarian mathematician named George Pólya published a book on problem solving in 1945. In it, he described a problem-solving procedure that he felt was common to most of history's greatest thinkers. His book has been translated into at least 17 languages and is still a big seller on Amazon, so obviously he must have been on to something! Pólya's procedure isn't exactly earth shattering: Its brilliance really lies in its simplicity. Pólya's basic steps are listed in the colored box, followed by our spin on each step.

Polya's Four-Step Procedure for Solving Problems

Step 1: Understand the problem. The best way to start any problem is to write down relevant information as you read it. Especially with longer word problems, if you read the whole problem at once and don't DO anything, it's easy to get panicked. Instead, read the problem slowly and carefully, writing down information as it's provided. That way, you'll always at least have a start on the problem. Another essential step: Carefully identify AND WRITE DOWN what it is they're asking you to find; this almost always helps you to devise a strategy.

Step 2: Devise a plan to solve the problem. Good planning is the key to any successful operation! This is where problem solving is at least as much art as science—there are many, many ways to solve problems. A short list of common strategics includes making a list of possible outcomes; drawing a diagram; trial and error; finding a similar problem that you already know how to solve; and using arithmetic, algebra, geometry, or good old-fashioned common sense.

Step 3: Carry out your plan to solve the problem. There's no point in making a plan if you don't carefully execute it. If your original plan doesn't work, don't settle for failure and give up—try a different strategy! There are many different ways to attack problems. Be persistent!

Step 4: Check your answer. In ANY problem-solving situation, you should think about whether or not your answer is reasonable. In many cases you'll be able to use math to check your answer and see if it's exactly correct. Even if you can check an answer, before you start solving a problem (or after you finish it) the ability to make an estimate to decide whether your answer is reasonable is a valuable skill.

Estimation is a valuable skill when it comes to reading information from charts and graphs. In fact, we used that skill (without even realizing it!) several different places in Unit 1. For example, when reading a bar graph, if the length of a bar falls between two lines, we have to estimate its height. We can practice this useful skill by looking at number lines.

This number line is a timeline of important advances in medical science. Use it to answer Questions 3–10.

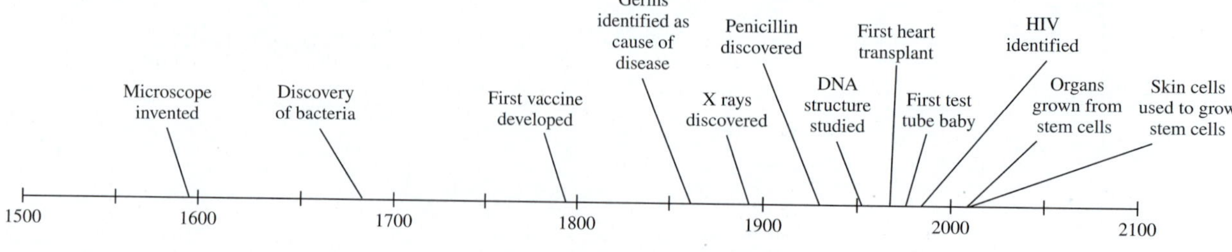

3. Estimate the year that the microscope was invented, and describe how you got your answer.

4. Estimate the year that germs were identified as the cause of disease, and describe how you got your answer.

5. Which happened first: the first test tube baby, or the first heart transplant? Estimate the year for each.

6. Which happened first: organs grown from stem cells, or the identification of HIV? Estimate the year for each.

There are several places in this course where comparing the sizes of numbers plays a key role. So it's worth a bit of our time to review some of the symbols and terminology that go along with these comparisons. In some cases our example comparisons will involve just numbers; in others, we'll be using letters to represent variable quantities.

Comparing Numbers

Verbal Description	In Symbols	Relationship on a Number Line
10 and $\frac{20}{2}$ are equal.	$10 = \frac{20}{2}$	10 and $\frac{20}{2}$ are the same point.
x is approximately equal to 10.	$x \approx 10$	The quantity x is close to 10, but they aren't necessarily the same point.
10 is not equal to 15.	$10 \neq 15$	10 and 15 are different points.
10 is less than 12.	$10 < 12$	10 is to the left of 12.
y is less than or equal to 10.	$y \leq 10$	The quantity y is either to the left of or on 10
10 is greater than 6.	$10 > 6$	10 is to the right of 6.
n is greater than or equal to 10.	$n \geq 10$	The quantity n is either to the right of or on 10

7. Use your answer from Question 5 to write an inequality involving $<$ that compares the dates for the advances mentioned in that question.

> **Math Note**
>
> A statement that involves one of the symbols $>$, $\geq$ $<$, or $\leq$ is called an **inequality**.

8. Use your answer from Question 6 to write an inequality involving $>$ that compares the dates for the advances mentioned in that question.

9. The two most recent events on the timeline are organs grown from stem cells, and skin cells grown into stem cells. How do the dates compare?

Your first instinct might be that growing stem cells from skin cells came later because the words are set farther right, but their position on the graph indicates that they might also have occurred in the same year. When comparing two quantities that may be equal, we can use the symbol $\leq$, which is read "less than or equal to," or $\geq$, which is read "greater than or equal to."

10. Use one of the symbols introduced in the previous paragraph to describe the relationship between the dates for organs being grown from stem cells, and stem cells being grown from skin cells. Do so using verbal descriptions of those quantities first, then assign a letter to represent each quantity and write a second inequality with your variables.

Did You Get It

Try this problem to see if you understand the concepts we just studied. The answers can be found at the end of the Portfolio section.

1. Use the timeline to make an estimate of the year in which the first vaccine was developed. Then write an inequality that compares that date to the date when penicillin was discovered.

Now we'll practice estimation using a bar graph. The next graph shows the average monthly cost of homeowners insurance in five states that are common destinations for people hoping to move to a warmer climate. Use it for Questions 11–15.

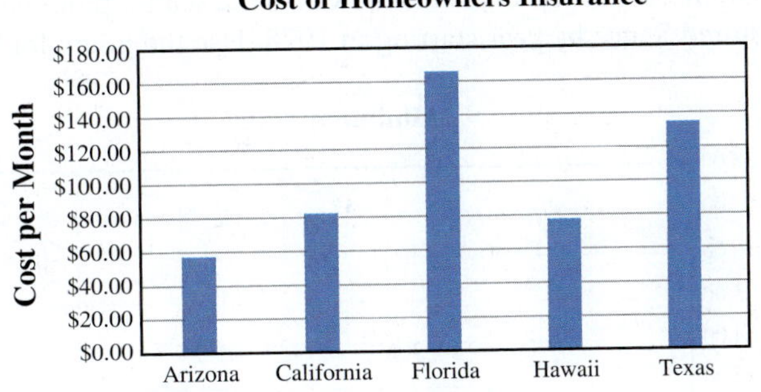

11. Estimate the monthly cost of homeowners insurance in California.

12. Do you think your estimate may be too high, or too low? Why?

13. Using your answers to Questions 11 and 12, write an inequality using either $\geq$ or $\leq$ that describes the cost of homeowners insurance in California. Use letter C to represent the cost.

14. Suppose that you were planning a move to Arizona and wanted to budget a monthly amount for homeowners insurance. Would you prefer an overestimate or an underestimate? Explain.

15. What estimate would you use for the monthly cost of homeowners insurance in Arizona?

Did You Get It

2. Use the bar graph on insurance to fill in the blanks: The cost of homeowners insurance in Florida $\geq$ ________ per month, which means my estimate was an ________________ (overestimate or underestimate).

Another type of graph that we often have to estimate values from is the scatter plot, like the one here. This graph illustrates the federal minimum wage in the United States by year starting in 1978. Use the graph for Questions 16–18.

16. Estimate the minimum wage in 2009.

17. Do you think your estimate may be too high, or too low? Explain.

18. Suppose that a high school graduate in 2009 was trying to figure out how to budget for college while working a minimum wage job. Would she have preferred an underestimate of the minimum wage, or an overestimate? Why?

The following box describes some words in English that are used to describe the inequality symbols we've been working with.

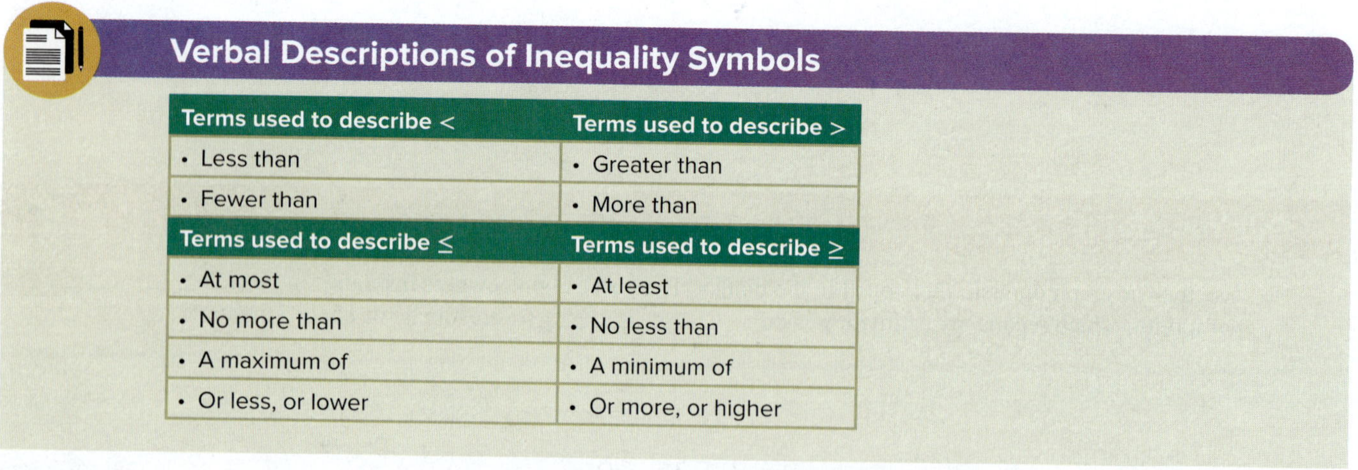

Verbal Descriptions of Inequality Symbols

Terms used to describe $<$	Terms used to describe $>$
• Less than	• Greater than
• Fewer than	• More than
Terms used to describe $\leq$	**Terms used to describe $\geq$**
• At most	• At least
• No more than	• No less than
• A maximum of	• A minimum of
• Or less, or lower	• Or more, or higher

In Questions 19–22, write the statement as an inequality, then give an example of an actual value that would match each description. For example, if I say, "Man, I need to get at least a 2.2 GPA this semester or I'll be on double-secret probation," then 2.5 would be a value matching the description. Assign a variable to each quantity.

19. We can fit at most 12 people on the Slow Ride pedal bus.

20. It's estimated that there are at least 50 ways to leave your lover.

21. I get a text alert from my bank if the balance in my checking account drops below 100 bucks.

22. Any amendment to the U.S. Constitution will only be enacted if more than 37 states vote to accept it.

Did You Get It

3. Write each statement as an inequality. Assign a variable to each quantity.
   a. If I don't get 80% or higher in this class, my parents are going to sell me to a traveling circus.
   b. I took a volcano tour in Hawaii Volcanoes National Park that allowed no more than 10 people near the crater at any one time.

2-5 Group

An **isotherm** is a curve on a map that connects locations where the temperature is the same. By looking at a large map with isotherms, you can get a rough idea of what the temperature is like in many different locations. Using the map, estimate the temperature in each city.

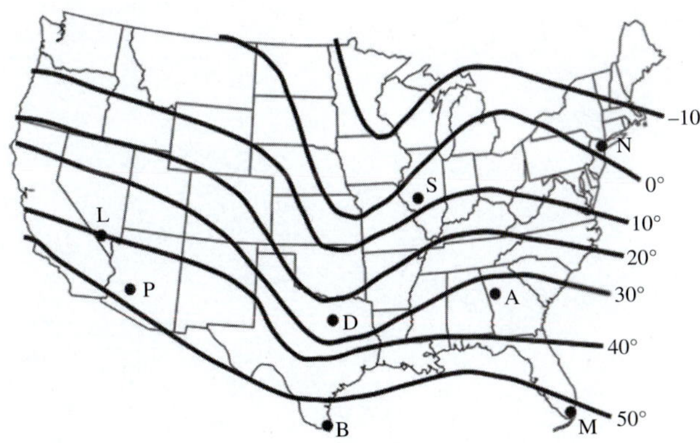

1. Las Vegas, NV (L) ________
2. Phoenix, AZ (P) ________
3. Dallas, TX (D) ________
4. Springfield, IL (S) ________
5. Atlanta, GA (A) ________
6. Miami, FL (M) ________
7. New York, NY (N) ________

Did You Get It

4. The southernmost city in Texas is Brownsville, located at point B on the map. Estimate the temperature there. Why is this one harder than the cities in Questions 1–7?

At this point, we've talked about estimation, and learned a little bit about Pólya's problem-solving method. This is a theme that we'll carry on as we work through the course: Being able to estimate an answer to a problem is *always* a valuable skill, but better still if you can then solve to find an exact answer. For now, we'll focus on doing some estimates, and we'll revisit many of these questions later on.

Spend some time thinking about Questions 8–12. After talking with your group, make your best guess and explain your reasoning. Remember, these are supposed to be EDUCATED guesses, not shots in the dark, so think about using some basic calculations to help you. No using phones or computers to look anything up yet!

8. Estimate the monthly payment if we borrow $10,000 at 7% interest for 3 years to buy a car. If we were planning a budget, would we want an overestimate or an underestimate? Why?

9. An average cup of plain old coffee (not one of those designer blends in a monster cup that you could bathe a small dog in) has about 80 mg of caffeine. How long would it take for that caffeine to leave your system? Based on your guess, how many mg do you think will be in your system in half that time? What are you assuming when you make that estimate?

10. About how much would you spend in a year if you bought two $2.80 cups of coffee each weekday (Monday through Friday)? Does this amount surprise you? Does it sound worth the expense?

11. If you put $1,000 into an Individual Retirement Account (IRA) right now, how much would it be worth when you retire? What are some of the things you need to factor in to make a reasonable estimate here?

12. How much money are you completely wasting every time you skip a college class? What information would you need to calculate this exactly?

2-5 Portfolio

Name ______________________________

Check each box when you've completed the task. Remember that your instructor will want you to turn in the portfolio pages you create.

Online Practice

1. ☐ Include any written work from the online assignment along with any notes or questions about this lesson's content.

Applications

1. ☐ Complete the Applications problems.

Reflections

Type a short answer to each question.

1. ☐ What kinds of things would you think about when making an educated guess in a math problem?
2. ☐ Describe at least one situation where estimating is good enough, and one where it isn't.
3. ☐ Describe the four steps in Pólya's problem-solving method. Then write about a time in this course that you've used each step.
4. ☐ Name one thing you learned or discovered in this lesson that you found particularly interesting.
5. ☐ What questions do you have about this lesson?

Looking Ahead

1. ☐ Complete the Prep Skills for Lesson 2-6.
2. ☐ Read the opening paragraph in Lesson 2-6 carefully and answer Question 0 in preparation for that lesson.

Answers to "Did You Get It?"

1. The first vaccine was developed around 1795; the year when the first vaccine was discovered $<$ the year when penicillin was discovered.
2. $160, underestimate
3. **a.** $G \geq 80\%$ where G is the grade needed
   b. $P \leq 10$ where P is the number of people allowed near the crater
4. It's definitely more than 50°, but we can't gauge how much more because there's no 60° isotherm on the map.

Answers to "Prep Skills"

1. −50 −40 −30 −20 −10 0 10 20 30 40 50

2. **a.** 32 **b.** 2 students **c.** 25 students

3. **a.** Question 4 **b.** about 23 **c.** about 39%

4. **a.** 3 **b.** 9 **c.** about 47 out of 160

2-5 Applications

Name ______________________________

1. According to a survey reported by the Nielsen Corp., in the fourth quarter of 2014, the average American adult spent just about 11 hours a day using electronic media of some sort. The pie chart displays the percentage of time spent using various types of media. Use the chart to make an educated guess on the amount of time spent using the media types in the table, and briefly explain how you made your estimates.

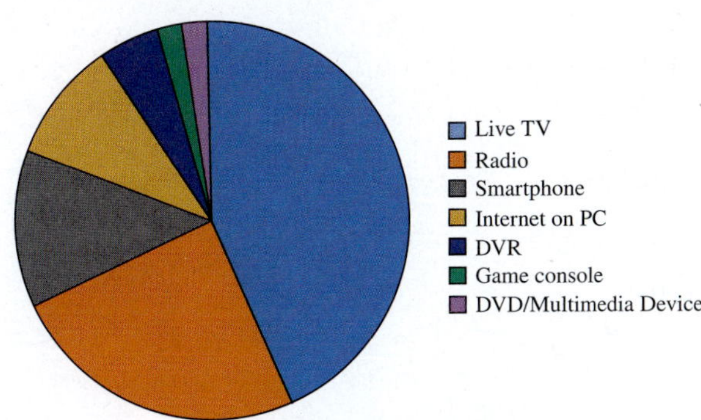

Media	Time
Live TV	
Radio	
Smartphone	
DVD/Multimedia	

2. Write the set of described numbers using the roster method.
   a. The number of people in a boat that can hold at most 7 people.

   b. The number of days you'll work out next week if you plan to work out at least twice.

Shopping provides a really good example of the value of estimation. Most people have a rough idea of how much they have to spend in a given shopping trip. While you could pull out the old cell phone calculator and add up all of your purchases to the cent, who has time for that? Instead, it's helpful to round prices to easy-to-use numbers and keep rough track of the total. Use this idea to estimate the total cost of each group of items in Questions 3–5. Round each price to the nearest dollar, then try to estimate the overall cost in your head.

3. Four boxes of mac and cheese (89 cents per box)

A gallon of orange juice ($3.79)

A loaf of bread ($1.79)

Two pounds of ground turkey ($2.97 per pound)

4. Two pairs of jeans ($24.95 and $32.95)

A three-pack of socks ($7.99)

Three workout shirts ($11.99 each)

5. Two Xbox games (regular price $29.50 each, on sale for 50% off)

iPhone case ($18.95)

Headphones ($22.49)

6. Estimation is useful for tipping when you're dining out, too. If the service is good, 20% of the overall bill is a nice tip to leave. Here's a clever way to find that: Move the decimal of the total amount one place to the left: that gives you 10%. Round this value to the nearest dollar and double the amount, and voila! Twenty percent. Use this to find a reasonable tip for good service on each bill.

a. $31.45

b. $129.80

This graph shows the violent crime rate in 2014 for five different American cities, measured in violent crimes per 100,000 citizens. Use it for Questions 7–10.

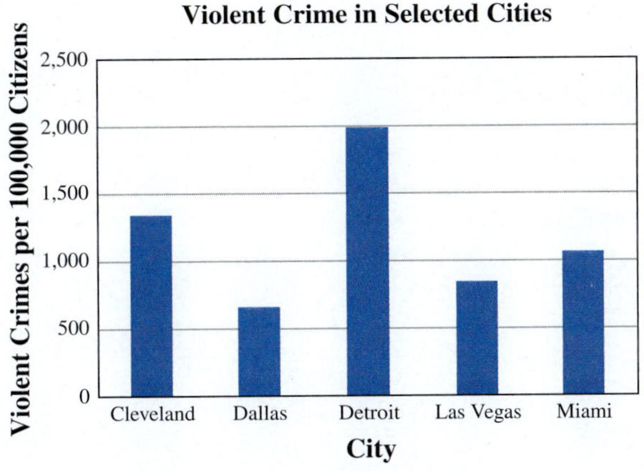

7. Estimate the number of violent crimes per 100,000 citizens in Detroit.

8. Do you think your estimate may be too high, or too low? Why?

9. Write an inequality that uses your estimate from Question 7 and your answer to Question 8. Assign a variable to the quantity you're illustrating.

10. Describe the number of violent crimes per 100,000 citizens in Miami using either the words "at least" or "at most."

Lesson 2-6 Prep Skills

SKILL 1: CONVERT FRACTIONS INTO PERCENT FORM

To find the percentage that corresponds to a given fraction, we can do these two steps: (1) Divide the numerator by the denominator, and (2) convert the resulting decimal into percent form by moving the decimal point two places to the right (multiply by 100).

- $\frac{38}{50} = 0.76$; moving the decimal two places to the right, we get 76%.
- $\frac{8-12}{12} \approx -0.333$; moving the decimal two places to the right, we get −33.3%.

SKILL 2: FIND THE AREA OF A RECTANGLE

We covered area of rectangles in Lesson 2-3: Recall that the area is the length times the width. It's very important that the two measurements are in the same units!

- The area of a rectangle with length 15 in. and width 3 ft can be calculated in one of two ways. We could convert 15 in. to 1.25 ft, and get area (1.25 ft)(3 ft) = 3.75 sq. ft. Or we could convert 3 ft into 36 inches and get area (15 in.)(36 in.) = 540 sq. in. We could then do a unit conversion to show that these two areas are the same.

SKILL 3: CONVERT UNITS OF AREA AND VOLUME

This skill was covered in Lesson 2-4. We use dimensional analysis, keeping in mind that we'll have to multiply by the same conversion factor either twice or three times when the units we're trying to eliminate are squared or cubed.

- $27\text{ in.}^2 \cdot \frac{1\text{ ft}}{12\text{ in.}} \cdot \frac{1\text{ ft}}{12\text{ in.}} = 0.1875\text{ ft}^2$
- $4\text{ cm}^3 \cdot \frac{1\text{ in.}}{2.54\text{ cm}} \cdot \frac{1\text{ in.}}{2.54\text{ cm}} \cdot \frac{1\text{ in.}}{2.54\text{ cm}} \approx 0.244\text{ in.}^3$

PREP SKILLS QUESTIONS

1. Convert each fraction into percent form.

   a. $\frac{12}{40}$ b. $-\frac{0.3}{1.6}$ c. $\frac{58.3-60}{60}$

2. Find the area of a building made up of three rectangular rooms that are 12 ft by 18 ft, and a hallway that's 22 ft long and 4 ft wide.

3. Using only the fact that 1 yd = 3 ft, convert 10 cubic yds into cubic feet.

4. Using only the fact that 12 in. = 1 ft, convert 5,000 sq. inches into sq. feet.

Lesson 2-6 It's All Relative

LEARNING OBJECTIVES

- ☐ 1. Compare change to relative change.
- ☐ 2. Apply percent error.

Source: Library of Congress, Prints and Photographs Division [LC-USZ62-60242]

When you sit with a nice girl for two hours you think it's only a minute, but when you sit on a hot stove for a minute you think it's two hours. That's relativity.
—Albert Einstein

In a nutshell, Einstein's theory of relativity says that even the passage of time is not an absolute, but rather is relative to your frame of reference. In this lesson, we'll study appropriate ways to compare numbers, focusing on the importance of relativity. Consider these two scenarios: 1) I owe you $20, and I give you $10 instead. Will that work for you? 2) You win $5 million in a lottery, but you only get paid $4,999,990. Would that ruin your day? You're out ten bucks either way, but clearly that's a much bigger deal when the amount owed is $20. If that makes sense to you, then you have an instinctive understanding of the concept of relative change.

0. How does Einstein's quote about relativity connect to the examples of being shorted by $10? And how cool is Einstein's hair?

2-6 Class

When asked to compare the sizes of two numbers, most people think of subtraction, but the example above shows that this isn't necessarily the best choice. The **change** between two quantities is found by simply subtracting the new value from the old. This is sometimes referred to as **actual change**.

The Change in a Quantity

$$\text{Change} = \text{New value} - \text{Original value}$$

If we use x_n to represent the new value and x_o to represent the original value, we get

$$\text{Change} = x_n - x_o$$

In both scenarios in the opening paragraph, the change in what was owed and what was paid is $10, which leads us to conclude the significance of that $10 is the same in each case. But doesn't it seem clear that the $10 shortfall is more significant when the total amount owed is $20? That's where relative change comes in. The **relative change** between two quantities measures the change as a fraction of the original value.

The Relative Change in a Quantity

$$\text{Relative change} = \frac{\text{Change}}{\text{Original value}} = \frac{\text{New value} - \text{Original value}}{\text{Original value}}$$

Using the same variables as in our definition of change,

$$\text{Relative change} = \frac{x_n - x_o}{x_o}$$

From our first scenario:

Relative change $= \frac{\$10 - \$20}{\$20} = -\frac{1}{2}$ or -50%. (The negative indicates that you got shorted. Bummer.)

This shows that the change between the amount owed and the amount received is half of the amount owed. That's a big deal!

1. Find the relative change if you win \$5,000,000 and get paid \$4,999,990. Write as a fraction, not a decimal.

> **Math Note**
>
> Relative changes are often written as percentages, but they don't have to be. Fraction or decimal form is fine, too.

Did You Get It

Try this problem to see if you understand the concepts we just studied. The answers can be found at the end of the Portfolio section.

1. A stock that was worth \$65 per share yesterday is now trading at \$38 per share after the CEO was arrested for fraud. Find the actual change and the relative change in the value of the stock.

2. Compare the relative change for the two scenarios in the opening paragraph from this lesson. Does this match your intuition about the scenarios?

3. Your instructor tells you that you're going to get an additional 2 points on a recent test or quiz. Describe when this would be great news, and when it wouldn't help very much.

For Questions 4 and 5, write a sentence that describes in your own words what exactly the given statement means. For example, when we found that the relative change from \$20 owed to \$10 paid was $-1/2$, this means the amount paid was too low by an amount half of what it was supposed to be.

4. The relative change in my budget was 7% compared to last year.

5. The relative change in enrollment this year compared to last year at my college was −3.5%.

For Questions 6 and 7, write a statement using the term "relative change" or "percent change" to describe the information provided. This is pretty much the opposite of what you did in Questions 4 and 5.

6. Apple's profit in 2016 was 14.4% lower than it was in 2015.

7. The value of my retirement fund was 5.5% more at the end of this year than it was at the end of last year.

Did You Get It

2. I'm at the Cincinnati airport (CVG) right now, and I just learned that the number of passengers flying in or out of CVG was about 7% higher in 2016 than it was in 2015. Write a sentence using the phrase "relative change" that describes this statistic. Make sure your answer has a − or + in it.

If you want some interesting reading, look up "baseline budgeting." Currently, the federal budget has an automatic annual increase of 7% to many programs.

8. Suppose that $435 million is spent for improvements to interstate highways for 2017. If the automatic increase of 7% applies to this budget, how much is expected for that program in 2018? (Round to the nearest million.)

9. Suppose that only $440 million is budgeted for that program, not the expected amount that you found in Question 8. Find the actual change and the relative change in the amount spent from 2017 to this budgeted amount of $440 million for 2018. Is it positive or negative? What does this change mean?

10. A news article reports that this program has been cut by 5.4%. Does this match what you found in Question 9? What values have been used to come up with that calculation? (Hint: It is still relative change.)

11. Write a sentence or two describing the 5.4% cut and what is misleading about it.

2-6 Group

Next up is a really useful topic that's related to relative change. We've already talked quite a bit about estimation in this course. In most cases, when some quantity is estimated there's going to be some error in that estimate. When comparing an estimate to an exact value, the **error** or **actual error** in the estimate can be expressed by subtracting the actual value from the estimated value. That is,

$$\text{Error} = \text{Estimated value} - \text{actual value}$$

The percent error is calculated just like the relative change. The idea is to measure how far off the estimate is as a percentage of the actual value.

The Percent Error in an Estimate

When an estimate is made of some quantity, the percent error in the estimate is given by

$$\text{Percent error} = \frac{\text{Estimated value} - \text{actual value}}{\text{Actual value}}$$

If we use x_e to represent the estimated value and x_a to represent the actual value,

$$\text{Percent error} = \frac{x_e - x_a}{x_a}$$

This calculation gives you percent error in decimal form; we'll usually want to rewrite that as a percentage.

Some sources use an absolute value in the numerator of this formula so that the percent error is always positive. But since we're all about interpretation in this course, we'll want to preserve the sign so that we can tell if the estimate is too high (positive percent error) or too low (negative percent error).

If you're ever in the market for some new carpet, you might consider searching the Web for the Carpet Buyer's Handbook. The site has many useful resources on buying and installing carpet. It turns out that figuring out how much you're going to spend is more complicated then you might think. Carpet is sold by the square foot, and you can measure the square footage of a room. But you can't buy exactly a certain number of square feet because carpet comes on rolls of a certain width (usually

12 feet). If you need a 2 ft × 3 ft piece for a closet, you can't buy just 6 square feet: You'd have to buy 2 feet off of the roll, which would be 24 square feet.

In addition, there are certain guidelines that installers will adhere to in order to get the best result. These govern how the pieces off the roll can be situated, leading to more potential waste. The bottom line is that when a professional installer measures a room to estimate the amount of carpet needed, the result will usually be quite a bit different from the actual square footage that will be covered.

1. When you're buying carpet, do you want an overestimate or an underestimate? Explain.

A carpet installer estimated that $24\frac{2}{3}$ square yards of carpet would be needed for the bedroom and closet shown in the diagram.

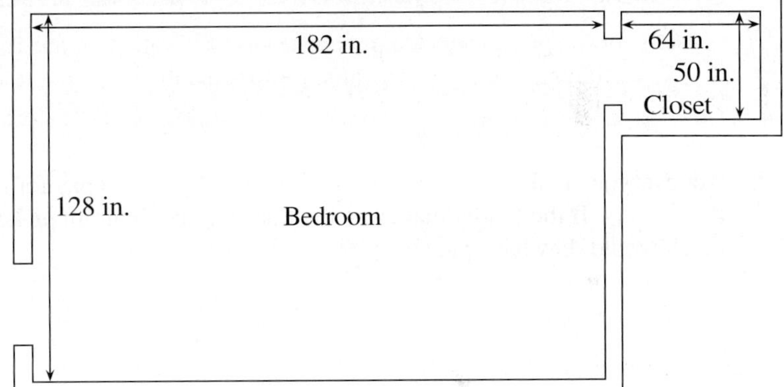

2. Find the area of the bedroom only in square inches. (Recall that the area of a rectangle is length times width.)

3. Find the area of the closet in square inches.

4. Each of the doorways shown is 30 inches wide, and the walls are 4 inches thick. Carpet is needed for the entire doorway between the bedroom and closet, and in half of the doorway into the bedroom. How many square inches does this add to the total carpeted area? What is the total carpeted area?

5. Convert the total carpeted area to square yards. Round to one decimal place. (You may find it useful to look back at the colored box on converting area and volume measurements on page 198.)

6. Write a sentence or two describing how the professional estimate compares to the actual carpeted area.

7. Find the percent error in the estimate. Round to one decimal place. Write a sentence explaining what this tells us about how the estimate compares to the actual carpeted area.

Did You Get It

3. One of the guest rooms in my house has two walls that are 15′6″ by 8 ft, and two others that are 12′3″ by 8 ft. When buying paint for the walls, I estimated the wall area to be about 450 sq. ft. Find my percent error.

8. Not everyone realizes this, but when you buy carpet, you pay for the estimated amount, not the actual amount covering your floors. If the family that owns the house in the diagram picked carpet that costs $28 per square yard installed, how much would they have paid?

9. Use the actual error in the estimate to find the dollar value of wasted carpet.

10. According to the Internet (which knows everything), a good rule of thumb is to plan for 5–10% waste when buying carpet. How did the estimator do?

11. One suggestion for estimating carpet is to measure each room in inches and round up to the nearest half foot (and ignore the doorways). Using the measurements for the closet and the bedroom, make an estimate for the square feet needed. How does this estimate compare to the actual area?

12. List some situations where relative change is a much more useful number than actual change.

2-6 Portfolio

Name ______________________________

Check each box when you've completed the task. Remember that your instructor will want you to turn in the portfolio pages you create.

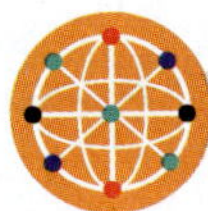

Technology

1. ☐ The first few rows of a spreadsheet designed to compute change and relative change are shown here. Your job is to download the template in online resources, then fill in formulas in row 3 for computing the change and relative change from one row to the next for each account. Then copy those formulas down to fill in the rest of the sheet. Finally, write any observations you have based on the result.

	A	B	C	D	E	F	G
1	**Account #1**	**Change**	**Relative Change**		**Account #2**	**Change**	**Relative Change**
2	$ 10,000.00				$ 10,000.00		
3	$ 10,700.00				$ 10,500.00		
4	$ 11,400.00				$ 11,025.00		

Online Practice

1. ☐ Include any written work from the online assignment along with any notes or questions about this lesson's content.

Applications

1. ☐ Complete the Applications problems.

Reflections

Type a short answer to each question.

1. ☐ How would you explain to someone you know that relative change is often much more meaningful than actual change?
2. ☐ Both Shawna and John found out that they're getting a $1,000 annual raise this year. John went out to celebrate, while Shawna yawned and said "Yeah, whatever." Use the topic of this section to describe what you think might be likely to account for this discrepancy in their reactions to the news.
3. ☐ Name one thing you learned or discovered in this lesson that you found particularly interesting.
4. ☐ What questions do you have about this lesson?

Looking Ahead

1. ☐ Complete the Prep Skills for Lesson 2-7.
2. ☐ Read the opening paragraph in Lesson 2-7 carefully and answer Question 0 in preparation for that lesson.

Answers to "Did You Get It?"

1. Actual: −$27; Relative: about −41.5%
2. The relative change in the number of passengers from 2015 to 2016 was +7%.
3. About 1.4% error

Answers to "Prep Skills"

1. **a.** 30% **b.** −18.75% **c.** about −2.8%
2. 736 ft^2
3. 270 ft^3
4. about 34.7 ft^2

2-6 Applications

Name ____

After moving into a new house, a couple wants to have a concrete patio poured to support a hot tub, because . . . well, because it's a hot tub. The plans call for a 14 ft by 16 ft slab of concrete 4 inches thick. Round any calculations to two decimal places if you need to.

1. How many cubic feet of concrete will be needed?

2. Convert the volume of concrete to cubic yards.

3. The coolest thing about ordering ready-mix concrete is that a concrete mixer will show up at your house. Most companies require orders by the cubic yard. Our couple's patio was estimated at 3 cubic yards. How much extra concrete was ordered?

4. Find the percent error in the estimate compared to the actual amount of concrete needed. Write a sentence explaining exactly what the relative error means.

5. In 2016, an average cost for concrete was $140 per cubic yard. How much would be spent on concrete for the patio at that price?

6. Find the dollar value of the wasted concrete.

7. Suppose that you've budgeted \$250 per month for a new car, but the salesperson goes into sales mode and talks you into one with a few extra snazzy features. Next thing you know, you have a car payment of \$265 per month. Find the actual change and the relative change needed for our car payment budget to accommodate our impulsive decision to go with the fancy car.

8. If the total of all payments for the original budgeted amount is \$12,000, how much extra would you end up paying for the snazzy features? Do you think that would be worth it? Discuss.

9. A certain pain medication has a recommended dosage of 0.3 mg per pound of body weight, and is reported to only be effective if the dosage administered is within 5% of the recommendation. Would a dosage of 50 mg be effective for a 175-pound patient?

Lesson 2-7 Prep Skills

SKILL 1: FIND A MEAN

Mean, or what is commonly called "average," is one of the main topics of Lesson 2-1. In short, the mean of a list of numbers is computed by adding all of the numbers and dividing by how many numbers there were on the list. But you already knew that, didn't you?

SKILL 2: INTERPRET INFORMATION FROM A DIAGRAM

Identifying information from a diagram plays an important role in this lesson. The skill you'll need is pretty similar to reading info from a pie chart, like this example:

- The chart represents the final grades for 51 students in two sections of a math class. Students that earned a C or better were eligible to move on to the next class. If we add the percentages for the A, B, and C categories, we find that 74% of students qualified to move on. This represents (0.74)(51), or 38 students.

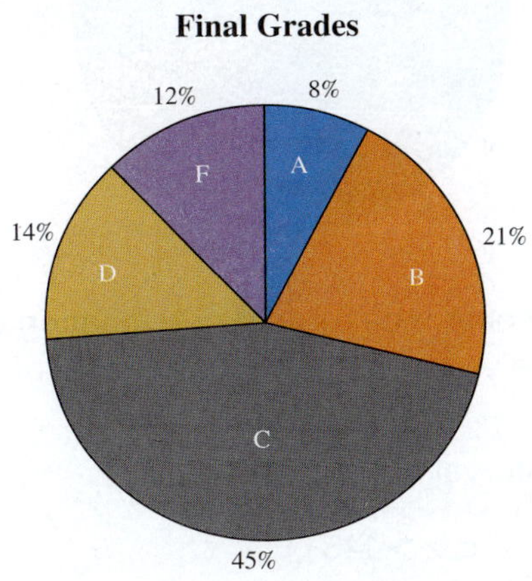

SKILL 3: FIND PROBABILITY FROM A PERCENT CHANCE

Recall that probability is really the same thing as the percent chance that some event will occur. The only difference is that probability is typically written in fraction or decimal form.

- If there's a 41% chance that some event will occur, we find the probability of it occurring by converting 41% to decimal form. Move the decimal point 2 places to the left to get a probability of 0.41.

SKILL 4: WRITE INTERVALS IN INEQUALITY NOTATION

The inequality symbols $<$ and $>$ can be used to describe an interval of numbers. For example, if we want to write an inequality that represents all of the real numbers that are less than 12, we could write $x < 12$. The letter that we choose to represent the variable quantity is immaterial: we could have used any letter.

- The set of all real numbers that are more than −8 is written $x > -8$.
- The set of all real numbers that are no more than 5 is written $x \le 5$.
- The set of all real numbers that are no less than 100 is written $x \ge 100$.

PREP SKILLS QUESTIONS

1. Find the mean of the numbers listed.

   23 34 45 56 67 78 89 11 22 33

2. The pie chart represents the breakdown of customers by day for a breakfast restaurant in a downtown area. There were 2,431 customers for the week.
   a. What percentage of customers visited on the weekend?
   b. How many customers visited on a weekday?

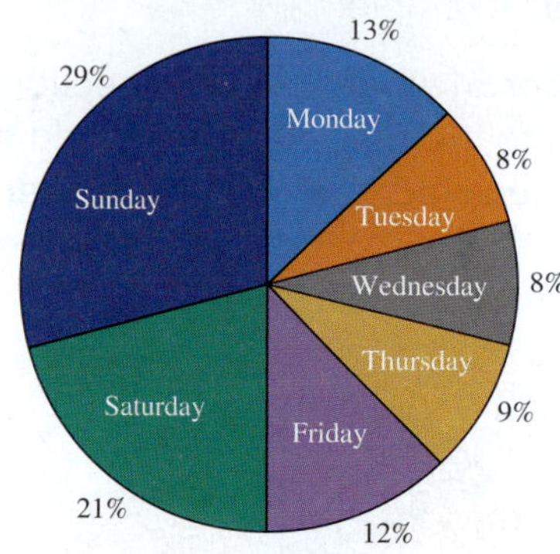

3. In an average year, there is at least some measurable precipitation on about 42% of days in Seattle, WA. If you choose a random day during the year, what's the probability of precipitation?

4. Write a verbal description of each interval described.

   a. $x \geq -11$ b. $x < 9$ c. $x > 143$

5. Write an inequality that describes each interval.
   a. The set of all real numbers that are less than 40.
   b. The set of all real numbers that are more than 3.2.
   c. The set of all real numbers that are at least 5.

Lesson 2-7 Is That Normal?

©Dserov/Shutterstock RF

LEARNING OBJECTIVES

- ☐ 1. Identify the steps in computing standard deviation, and describe why they lead to a measure of variation.
- ☐ 2. Compute and interpret standard deviation.
- ☐ 3. Use a normal distribution to find probabilities.

The only normal people are the ones you don't know very well.
—Alfred Adler

When I was in college, I had a button on my favorite jacket that said "Why be normal?" Some people tend to want to be different, while others would prefer to blend in. But what exactly does "normal" mean? Is it what's most common? Most accepted? If you have "just a normal day," is that a good thing or a bad thing? I suppose that a normal day would be a day where things happen like they usually do, and that's not a bad description of how we'll be using the word "normal" in this lesson. We've already learned a lot about measures of average, but it turns out that a data set often has more of a story to tell than simply what the average is (whatever that means). We'll begin this lesson by thinking about whether a lot of the values in a data set are "normal" (meaning similar to the others), or if there are a lot of values that do their best to stand out from the crowd.

0. After reading the opening paragraph, how do you think we're going to apply the word "normal" to studying sets of data?

2-7 Group

Two golfers were competing in a televised competition for a scholarship to play on a college golf team. The scores for each golfer over all six rounds of the competition are shown in the table. If you're not familiar with the sport, in golf a lower score is better.

Player	Rd. 1	Rd. 2	Rd. 3	Rd. 4	Rd. 5	Rd. 6
Brittany	75	80	72	81	75	77
Ji-Min	69	77	71	75	80	83

1. Compute the mean score for each player. Round to one decimal place.

2. Find the range of scores for each golfer. (The **range** of a data set is calculated by subtracting the highest value minus the lowest value.)

3. Which golfer do you think was the better player in this competition? Which do you think was more consistent? Explain your reasoning.

4. Can you make up four exam scores that have a mean of 80, and a highest score of 100? Are the scores close together, or spread out?

5. Can you make up four exam scores that have a mean of 80 and a highest score of 84? Are the scores close together or spread out?

6. What do the previous two questions tell you about how there's more to the story told by a data set than just the measures of average?

Did You Get It

Try these problems to see if you understand the concepts we just studied. The answers can be found at the end of the Portfolio section.

1. If nine of the ten houses on one street are all worth about $210,000 and the one at the end of the street is valued at $1,200,000, how does the mean price on the street compare to a similar neighborhood that only has the nine houses worth $210,000?
2. Is the range a meaningful measure of how spread out the home values are in the first neighborhood?

2-7 Class

As we saw in the Group portion of this lesson, while the mean is a very useful tool in analyzing data, there's more to the story than just the mean because it can't factor in how spread out a group of values is. The range is one way to analyze spread, but it totally ignores all but the highest and lowest values. So if there's just one value that's really low or high compared to the others, the range will make it seem like the data are more spread out than they actually are.

For this reason, we'll study a more detailed measure of spread known as the **standard deviation**. In short, standard deviation, which is often represented by a lowercase Greek sigma (σ), is a measure of how far on average the values are from the mean. A large standard deviation means there are a lot of values far away from the mean. A small standard deviation means that most of the values are close to the mean. Here are the steps for computing standard deviation, using Brittany's golf scores from the Group portion as an example: 75, 80, 72, 81, 75, 77.

Step 1: Find the mean. Standard deviation is all about measuring distance from the mean, so we better start by finding the mean. We already found the mean in group Question 1: It's 76.7.

Step 2: Subtract the mean from each data value in the data set. This is where we're really measuring the spread.

1. Fill in the second column of the table below.

Step 3: Square the differences from Step 2. This is a simple way to eliminate the signs: We're trying to measure how far away from the mean the values are, not whether they're bigger or smaller.

Score	Score − Mean	(Score − Mean)^2
75		
80		
72		
81		
75		
77		

2. Fill in the third column of the table by squaring each value in the second column.

Step 4: Find the mean of the squares. This is almost finding the average of the distances of each data value from the mean: The only problem is that we're really averaging the squares. We'll fix that soon.

3. Find the mean of the values in the third column of the table.

Step 5: Find the square root of the mean from Step 4. In essence, this is "undoing" the squares used to eliminate the signs.

4. Find the standard deviation by computing the square root of your answer from Question 3.

Using Technology: Computing Standard Deviation

TI-84 Plus Calculator

1. Enter the data in list **L1**; press STAT then ENTER.
2. Enter the list of values under **L1**.
3. We want choice 1 under the **STAT-CALC** menu, which you get to by pressing STAT ▶ ENTER.
4. Choose **1: 1-VarStats.** Among the information displayed will be the mean, which is denoted $\bar{x}$, and the standard deviation (σ_x).
5. If appropriate, select Calculate and press ENTER.

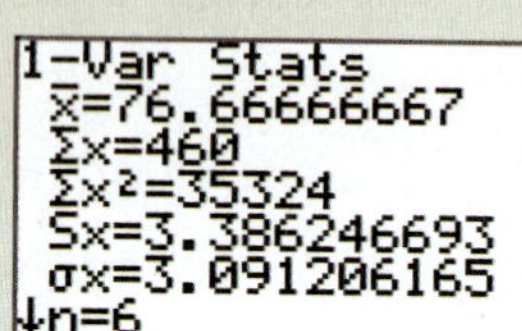

Excel Spreadsheet

B8 | f_x =STDEV.P(B2:B7)

	A	B	C
1		Score	
2		75	
3		80	
4		72	
5		81	
6		75	
7		77	
8	Standard Deviation	3.1	
9			

1. Enter the values in a row or column.
2. Enter = STDEV.P(B2: B7); note that this is the range of cells containing the data values.
3. Format the cell where you entered the formula to display the number of decimal places you want.

See Lessons 2-7-1 and 2-7-2 Using Tech videos in class resources for further instruction.

The Normal Distribution

How many people do you know that are taller than 6′5″, or shorter than 5′1″? For most people, the answer to that question is "not very many." The heights of humans tend to exhibit a phenomenon shared by many characteristics of living things: There are a lot of values close to the mean, and less and less as you get farther from the mean. The picture to the right is an example of 100 maple seed pods that I collected from my back yard, arranged by length; note the pattern that appears.

5. Describe the pattern made by the seed pods, and explain what it says about their lengths.

Courtesy of Dave Sobecki

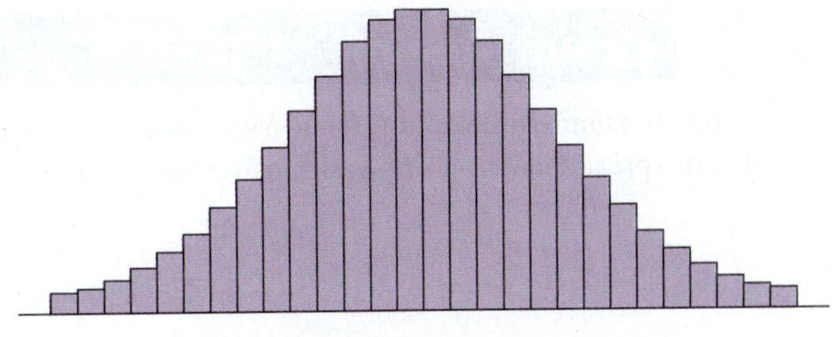

This phenomenon is so common, in fact, that data sets that follow a similar pattern are said to be **normally distributed**. Things like sizes of individuals, IQ scores, weights of packaged products, and lifespans of batteries or lightbulbs are often normally distributed.

I only collected 100 pods, out of thousands (maybe millions) on the tree. If I had collected a lot more of them, the picture would most likely have started to look a lot like the nice, symmetric diagram to the right. When a group of data is normally distributed, and we know the mean and standard deviation, there's a rule that allows us to estimate how many data values fall within certain ranges. This is known as the **empirical rule,** and it's illustrated by the following diagram.

The empirical rule says that when a data set is normally distributed, about 68% of all values will fall within 1 standard deviation of the mean; about 95% will fall within 2 standard deviations of the mean; and about 99.7% will fall within 3 standard deviations of the mean.

An Illustration of the Empirical Rule

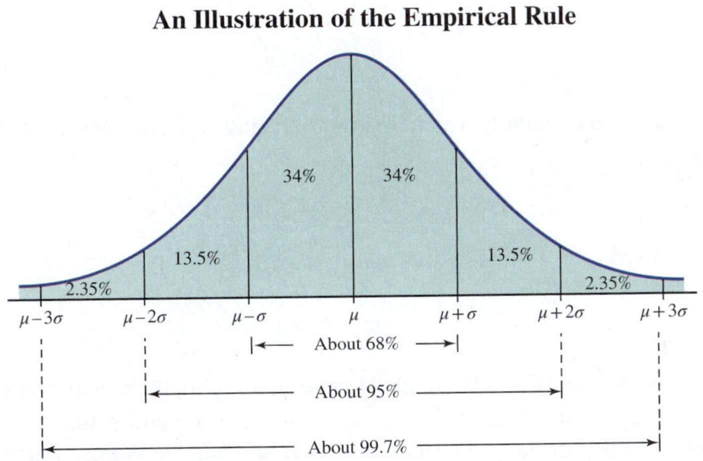

Math Note

The mean for data that are normally distributed is often represented by the lowercase Greek letter mu (μ).

For example, the heights of American men are normally distributed with mean 5 feet 9.3 inches and standard deviation 2.8 inches. That is, μ is 5 feet 9.3 inches, and σ is 2.8 inches.

6. Use the information just given about μ and σ to fill in the blanks on the next empirical rule diagram with heights, using the formulas below the blanks for guidance.
7. Based on your diagram, about what percentage of American men fall into the height range from 5 feet 6.5 inches to 6 feet 0.1 inch?

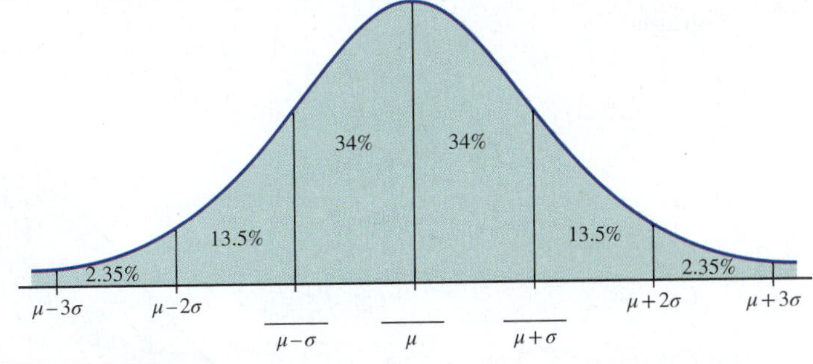

Did You Get It ?

3. About what percentage of American men are between 5 feet 9.3 inches and 6 feet 0.1 inch tall?

2-7 Group (Again)

1. Find the standard deviation for Ji-Min's golf scores (provided on page 243). Compute first by hand, then using a calculator or spreadsheet to check your answer. Round to one decimal place.

2. Discuss how comparing the two standard deviations can help you to decide which of the two golfers is a more consistent player.

A standard package of Oreos is supposed to contain 510 grams of chocolatey goodness. But there's variation in just about anything, including production and packaging, so some packages will contain more and some will contain less. In fact, this is exactly the sort of quantity that tends to be normally distributed. The folks that run Nabisco aren't stupid, and they know that customers won't be very happy if they weigh a package of cookies and find that it contains less than the labeled amount. The typical approach to keep that from happening is to design the packaging process so that the mean is something more than 510 grams, with a standard deviation that guarantees that the vast majority of packages contain 510 grams or more.

3. Let's say that the mean is 518 grams and the standard deviation is 4 grams.
   a. Fill in all of the blanks on the empirical rule diagram with weights in grams. Use the formulas below the blanks for reference.
   b. What percent of all Oreo packages would contain between 514 and 522 grams?

©McGraw-Hill Education/John Flournoy, photographer

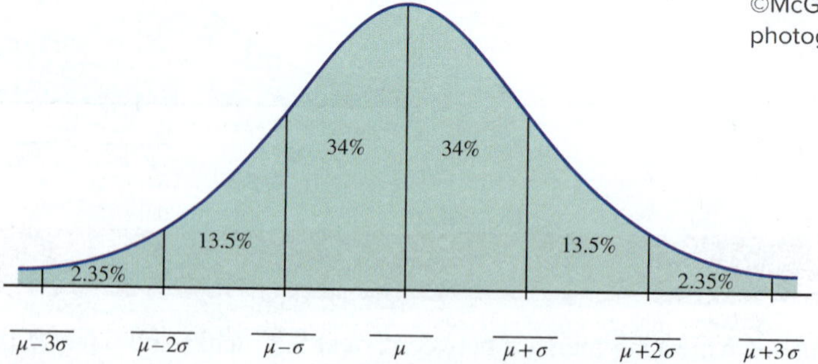

4. Using what we know about the connection between probability and percent chance, what's the probability that a randomly chosen package contains between 514 and 522 grams?

The probability of a package of Oreos containing between 514 and 522 grams can be written as $P(514 < x < 522)$, where x represents the weight. This uses a *compound inequality*, which is a combination of two inequalities. In this case, $514 < x < 522$ means that x is both greater than 514 (that's the $514 < x$ part) AND less than 522 (which is the $x < 522$ part). *In other words, x is between 514 and 522.*

Writing Compound Inequalities

There are many occasions where we'd like to represent an interval between two specific numbers. To do this concisely, we use a **compound inequality**. An inequality of the form

$$a < x < b$$

describes the set of all numbers that are between a and b. When writing an interval in this form, we only use the *less than* symbol. The first number is the lower boundary of the interval we're describing, and the last number is the upper boundary.

5. Write a compound inequality with variable x that describes the set of all weights between 510 and 526 grams.

6. Write an expression using your answer to Question 5 that describes the probability of a randomly selected Oreo package containing between 510 and 526 grams of joy (cookies, actually).

7. Use the empirical rule to find the probability in Question 6.

Did You Get It

4. Write an expression using a compound inequality that represents the probability of a randomly selected Oreo package weighing between 514 and 530 grams, then use the empirical rule to find that probability.

8. Write a description of the probability represented by the expression $P(x < 510)$, where x represents the weight of a randomly selected package of Oreos. Shoot for a description that one of your classmates could easily understand.

9. Find the probability described in Question 8. (This will require some interpretation of the diagram illustrating the empirical rule.)

10. Based on your answer to Question 9, if 1,000 Oreo packages are sampled, how many will have less than 510 grams of cookies?

Did You Get It

5. Use the empirical rule to find the probability that a randomly selected Oreo package weighs more than 514 grams.

11. Is it unusual for a package to weigh more than 530 grams? Explain.

2-7 Portfolio

Name ______________________________

Check each box when you've completed the task. Remember that your instructor will want you to turn in the portfolio pages you create.

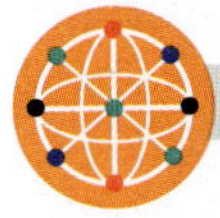

Technology

1. ☐ Using the Internet as a resource, find a data set that you find interesting, then use a spreadsheet to compute the mean, median, and standard deviation. Then write a brief report about what these measures of average and spread tell you about the data. Make sure that you list the source of your data.

Online Practice

1. ☐ Include any written work from the online assignment along with any notes or questions about this lesson's content.

Applications

1. ☐ Complete the Applications problems.

Reflections

Type a short answer to each question.

1. ☐ Describe exactly what the standard deviation of a data set tells us. More detail, as usual, is better.
2. ☐ What can we learn from analyzing the standard deviation for a data set that we couldn't learn from just looking at measures of average?
3. ☐ Look back at your answer to Question 0 at the beginning of the lesson. How did you do? Would you change your answer now?
4. ☐ Name one thing you learned or discovered in this lesson that you found particularly interesting.
5. ☐ What questions do you have about this lesson?

Looking Ahead

1. ☐ Complete the Prep Skills for Lesson 2-8.
2. ☐ Read the opening paragraph in Lesson 2-8 carefully and answer Question 0 in preparation for that lesson.

Answers to "Did You Get It?"

1. The mean for the second neighborhood is obviously \$210,000. For the first it's \$309,000.
2. The range is huge—almost \$1,000,000—which makes it seem like the values are very spread out. But they're not since all but one are identical.
3. About 34%.
4. $P(514 < x < 530) = 0.8385$
5. 0.84

Answers to "Prep Skills"

1. 45.8
2. **a.** 50% **b.** 1,215 or 1,216
3. 0.42
4. **a.** The set of all real numbers that are no less than –11
   b. The set of all real numbers that are less than 9
   c. The set of all real numbers that are more than 143
5. **a.** $x < 40$ **b.** $x > 3.2$ **c.** $x \geq 5$

2-7 Applications

Name ______________________________

1. In Lesson 2-1, you found the mean of the exam scores shown in the table. Find the standard deviation for these scores, using whatever method you prefer. Then describe what the mean and standard deviation tell you about the data.

	A	B	C
1	**Student**	**Exam 1 (%)**	**Exam 2 (%)**
2	Michael	80	89
3	Andy	77	93
4	Pam	68	84
5	Jim	81	88
6	Dwight	96	91
7	Stanley	54	75
8	Phyllis	75	54
9	Kevin	81	86
10	Creed	71	0
11	Darryl	89	83
12	Gabe	56	64
13	Toby	81	64
14	Holly	92	74

2. According to numerous online resources, the mean height for American women is 5 feet 5 inches, with a standard deviation of 3.5 inches. Use the empirical rule to find the probability that a randomly chosen American woman is between 5 feet 1.5 inches and 5 feet 8.5 inches.

3. Write an expression of the form *P*(inequality) that represents the probability you found in Question 2. Use h to represent the height of a randomly chosen woman, and write heights in inches.

4. In a group of 500 women, how many would you expect to be taller than 6 feet? (You'll need to interpret the diagram that illustrates the empirical rule.)

2-7 Applications

Name ____________________

5. On one campus, about 95% of students work between 6 and 12 hours per week, and the number of hours worked is normally distributed. What is the mean number of hours worked likely to be? What is the standard deviation? (This one requires a little bit of ingenuity.)

For each of the quantities in Questions 6–9, decide whether you think each is likely to be normally distributed, and explain why or why not.

6. The SAT scores for all high school juniors in Texas.

7. The amount of tax revenue taken in by the United States government over the last 60 years.

8. The length of daylight hours over the course of a year.

9. The amount of time it takes all students in a class to finish a given homework assignment without a time limit.

Lesson 2-8 Prep Skills

SKILL 1: COMPUTE A BASIC PROBABILITY

The simplest way to compute a probability is to divide the number of outcomes in a certain event by the total number of outcomes possible. For example, when flipping a coin, you can either get heads or tails: There are two outcomes. Exactly one of those satisfies getting tails, so the probability of getting tails is ½.

- A board game spinner has 8 spaces, numbered from 1 to 8. For each spin, there are 8 possible outcomes. Three of those 8 are more than 5, so the probability of getting a number more than 5 is

$$\frac{\text{Number of ways to get more than 5}}{\text{Total possible outcomes}} = \frac{3}{8}$$

SKILL 2: CONVERT FROM PERCENT FORM TO DECIMAL FORM

When we're given a percentage, to convert it into decimal form, we execute two steps: (1) Move the decimal point two places to the left (divide by 100); (2) drop the percent symbol.

- 90% = 0.90
- 12.5% = 0.125
- 4% = 0.04

PREP SKILLS QUESTIONS

1. There are 52 cards in a standard deck, equally divided among hearts, clubs, diamonds, and spades. Find the probability of drawing a club with one randomly selected card, and the probability of not drawing a spade.
2. When you roll one six-sided die, what's the probability of getting a number less than 3?
3. Convert each percentage to decimal form:

   a. 87% b. 33.5% c. 9% d. 1.3%

Lesson 2-8 Meeting Expectations

LEARNING OBJECTIVES

- ☐ 1. Estimate expected value experimentally.
- ☐ 2. Compute expected value.
- ☐ 3. Compute weighted grades and GPA.

Don't lower your expectations to meet your performance. Raise your level of performance to meet your expectations.

—Ralph Marston

©McGraw-Hill Education/Mark Steinmetz, photographer

If you bought 10 $1 scratch-off lottery tickets that have prizes ranging from "you get nothing and like it" to $10,000, how much money would you expect to end up winning or losing? If that question makes sense to you, great—it means you have an understanding of what the topic of expected value is about. When some game or experiment involving probability has numeric outcomes, like the amount of money won or lost when buying 10 lottery tickets, expected value is used to describe what would be expected to happen, on average, if you repeated that experiment a boatload of times. It's a tremendously useful calculation in many different settings: It's not hard to think of many situations where you'd like to have some idea of whether or not the end result is likely to be favorable. Games of chance are obvious applications, but there are tons of others: Buying insurance, investing, job decisions, and what items a store should put on sale are four that come to mind. By the way, if you guessed that you'd lose somewhere between $6 and $7, for most instant lottery games, you'd be right. Now let's jump in and learn about expected value. I expect that you'll do fine.

0. Describe what you think it means to say that the expected value of buying 10 $1 lottery tickets is $7.

2-8 Class

Expected Value Lab

Supplies needed: At least one 6-sided die for each group. (A phone app will work just fine.)

The point: The expected value of an event is the value that can be expected, on average, to occur. In this case, you'd probably guess that the expected value is one of the numbers from 1 to 6. But does it have to work out that way? We shall see!

The procedure: One really good way to gain some perspective on what expected value is all about is to play a game and see how the results compare to what we might expect. It's a pretty simple game: Roll one die, and you win the dollar amount of the number facing up—$1 for a 1, $2 for a 2, $3 for a 3, etc. So each group should roll their die 100 times, and record the result of each roll. You can either split the rolls among the group, or have a roller and a recorder.

Record results in the following table. If you have access to a computer, you might want to record the results in a spreadsheet instead, which will help make the calculations we need to do easier.

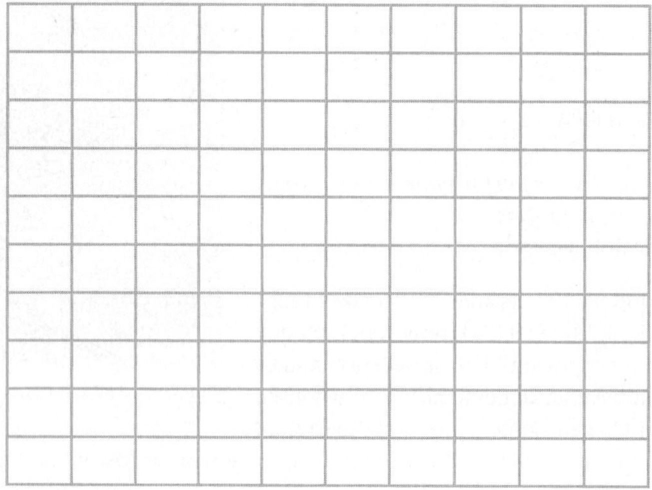

1. Add the numbers across each of the 10 rows, and write the sum to the right. Then add up all of those sums. How much money would you have won on those 100 rolls?

2. On average, how much did you win per roll?

3. Add the answers from Question 2 for every group in your class. It's a good idea to have each group write their answer on the board. Then divide by the number of groups to find the average amount won for the entire class.

4. Congratulations: You've just found the expected value for one roll in this game, which is the average amount you'd expect to win per roll if you played this game a LOT. Is it possible to win this amount on one roll? Explain.

5. What is the probability of getting a 1 when rolling one die?

6. What is the probability of getting any individual number from 1 to 6 when rolling one die?

7. Complete the second column of this table by writing in the probability of each outcome. Then complete the third row by multiplying each outcome by its probability.

Outcome (*X*)	Probability *P*(*X*)	Product *X* · *P*(*X*)
$1		
$2		
$3		
$4		
$5		
$6		

Math Note

Do all of your probabilities add up to one? If not, you must have made a mistake.

8. Find the sum of the values in the third column. How does it compare to the expected value for your class?

9. Obviously, playing this game for free would ROCK: You'd always win something for nothing. How likely is that to happen in real life, though? (Hint: Not at all.) Instead, you'd typically pay to play. What do you think would be a fair price to charge to play this game? Explain your choice.

10. Let's say you were setting up this game somewhere on your campus. Would you expect to make money if you were charging students $2 to play? What about $5?

Finding Expected Value

To compute the expected value of a probability experiment:

Step 1: List out all possible numeric outcomes.

Step 2: Write the probability of each outcome.

Step 3: Multiply each numeric outcome by its probability.

Step 4: Add the results. The sum is the expected value.

11. If you were charging $2 to play, the outcomes for each roll would be different. For example, instead of winning $1 for a roll of 1, you'd lose $1 because you paid $2 to play, and the outcome would be –$1. This in turn changes the expected value. We've revised the table you filled in earlier to reflect this change in the outcome for each roll. Use the new table to find the expected value for the player. Does your answer agree with your prediction in Question 10?

Roll	Outcome (X)	Probability P(X)	Product X · P(X)
1	–$1	$\frac{1}{6}$	$-\$\frac{1}{6}$
2	$0	$\frac{1}{6}$	$0
3	$1	$\frac{1}{6}$	$\$\frac{1}{6}$
4	$2	$\frac{1}{6}$	$\$\frac{2}{6}$
5	$3	$\frac{1}{6}$	$\$\frac{3}{6}$
6	$4	$\frac{1}{6}$	$\$\frac{4}{6}$

Did You Get It

Try these problems to see if you understand the concepts we just studied. The answers can be found at the end of the Portfolio section.

1. Suppose that you buy travel insurance after booking a trip, at a cost of $49. If you take the trip as scheduled, you've lost that $49, which would be an outcome of –$49. If something happens that forces you to miss the trip, you get back the $1,400 you originally paid, which would be an outcome of +$1,351. Let's say there's a 0.98 probability that you'll be able to go on the trip and a 0.02 probability that you won't. Find the expected value.
2. Will the company offering the insurance make or lose money in the long run? Why?

2-8 Group

Just about every college syllabus contains information about a grading policy and a grading scale. Are you aware of the policies for this course? Do you understand the policies? If not, make sure that you talk to your instructor.

Here's a potential grading scale that could be used in this course. A scale like this is called a **weighted scale**, because some assignments count for a bigger portion of the overall grade than others. We'll refer to the percentage of the overall grade corresponding to a given category as the **weight** for that category.

<table>
<tr>
<td>
Grading:
20% Homework
<ul>
<li>Technology</li>
<li>Skills</li>
<li>Applications</li>
<li>Reflections</li>
</ul>
20% Group work and class participation
<ul>
<li>Group projects and presentations</li>
<li>Exam reviews</li>
<li>Attendance and participation</li>
<li>Binder check</li>
<li>Final portfolio assignment</li>
</ul>
40% Unit exams
20% Final exam
</td>
<td>
Grading Scale:
A 90%
B 80%
C 70%
D 60%
F Below 60%
</td>
</tr>
</table>

Let's try to figure out the final grade for a fictional student named Christine. She earned 220/320 points on homework assignments (boo!!), 365/410 points on group work and class participation, 77/100 on exam 1, 82/100 on exam 2, 71/100 on exam 3, 85/100 on exam 4, and 162/200 on the final. This table can be used to organize all of her scores.

Category	Student's Average (as a percent)	Weight (in decimal form)	Points for category (out of 100)
Homework			
Group work and class participation			
Unit exams			
Final exam			
		Grade =	

1. Using the scores provided, find the percentage of points earned in each category and put them in the Student's Average column.

2. Use information on the grading scale to fill in the weights for each category. Remember, the weight is the percentage of overall points corresponding to a category. Make sure you're writing the weights in decimal form, not percent form.

3. The points out of 100 earned for each category is the student's average multiplied by the weight: Use this to fill in the last column.

4. The overall grade, out of 100, is the sum of all the points earned. Write this total in the bottom right-hand corner.

5. What letter grade did Christine earn?

6. Which category would you say had the most negative effect on Christine's grade? Discuss why you feel that way. Provide evidence, not gut feelings.

7. As a check on the grading system, what should all of the weights add up to? Is that the case here? What would it mean if they didn't add up to that total?

Did You Get It

3. Find the final course grade for this history student.

Category	Student's Average (as a percent)	Weight (in decimal form)	Score (as a percent)
Quizzes	70	0.2	
Exams	75	0.3	
Final paper	87	0.5	
		Grade =	

Now that we know how to compute weighted grades, let's do something more useful. It's certainly a good idea to figure out your grade and make sure your instructor got it right. But it's even a better idea to figure out where you stand at some point in a course, and use that information to decide on how well you need to do the rest of the way to get a grade you're shooting for.

So let's say that you have the following scores in the week before your final exam.

Category	Student's Average (as a percent)	Weight (in decimal form)	Score (as a percent)
Homework	95.0	0.1	
Quizzes	88.0	0.15	
Unit exams	89.0	0.5	
Final exam		0.25	
		Grade =	

8. Complete the score column as much as you can.

9. If you don't even show up for the final, what would be your final grade?

10. How many points on a 200-point final would you need in order to get an A in the course? Feel free to use trial and error, trying out some different percentages at the bottom of the first column. You can assume that the professor would round your average to the nearest whole number using standard rounding rules.

Hopefully you noticed that the calculations we did for computing a weighted grade average are the same as the calculations we did to find an expected value. In each case, a value is multiplied by a percentage in decimal or fraction form (the weight, or the probability of an outcome), and the results are added. Cool. We'll close the Group portion with another expected value calculation to make sure you've got it down.

In a classic carnival game, you have to plunk down $2 to spin the lucky wheel, shown below. If the pointer lands on red, you're the proud owner of $3 (which means the net result is a gain of $1). If it lands on yellow, you get to spin again for free (net gain $0). If it lands on purple, you my friend are out two American dollars (net loss of $2), and if it lands on blue, you get paid $5. Rock on!

11. What's the net gain if you land on blue? Write your answer at the bottom of the Gain column next to Blue.

Color	Gain (X)	Probability $P(X)$	Product $X \cdot P(X)$
Red	$1.00		
Yellow	$0.00		
Purple	−$2.00		
Blue			
		Expected value =	

12. Compute and record the probability for each gain (X). You'll need the picture of the spinner.

13. Calculate and record the product for each $X*P(X)$.

14. What's the expected value? Record in the table, then describe exactly what the result means.

15. A game of chance is called a **fair game** if the expected value is 0. Why do you think that is?

16. A game favors the player if the player's expected value is positive, and favors the house (a nickname for those running the game) if the expected value is negative. Is the spinner game fair? Does it favor one side? Explain.

17. If you played this game 100 times, how much would you expect to win or lose?

Did You Get It

4. Let's make a minor change to the spinner game: If landing on blue pays out \$6 instead of \$5, what's the new expected value? Is the game fair, or does it favor one side?

2-8 Portfolio

Name ______________________________

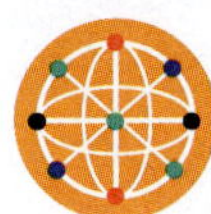

Technology

1. ☐ Can a spreadsheet roll a die? Physically, no. Numerically, yes. The command "=RANDBETWEEN(1,6)" generates a random number from the set 1, 2, 3, 4, 5, 6, which is exactly the same thing as rolling a die. That means you can repeat the experiment that started this lesson many, many times by just copying and pasting—if you're clever. The technology template for this lesson will get you started. The goal is to repeat the experiment enough times that the average expected value gets really close to the theoretical expected value that you calculated in Question 8 of the Class portion.

Online Practice

1. ☐ Include any written work from the online assignment along with any notes or questions about this lesson's content.

Applications

1. ☐ Complete the Applications problems.

Reflections

1. ☐ For all but one bet in standard American roulette, the expected value on a \$1 bet is −\$0.053. Explain what that means.

2. ☐ Betting on 0, 00, 1, 2, and 3 in roulette has an expected value of −\$0.079. Is that a smarter bet, or a dumber one? Explain.

3. ☐ What makes a grading system weighted?

4. ☐ Name one thing you learned or discovered in this lesson that you found particularly interesting.

5. ☐ What questions do you have about this lesson?

Looking Ahead

1. ☐ Complete the Prep Skills for Lesson 3-1.

2. ☐ Read the opening paragraph in Lesson 3-1 carefully and answer Question 0 in preparation for that lesson.

Answers to "Did You Get It?"

1. The expected value is −\$21.
2. The company will make money because in the long run, customers will lose \$21 each time they buy.
3. 80%
4. The expected value is now \$0, and the game is fair.

Answers to "Prep Skills"

1. $\frac{1}{4}; \frac{3}{4}$
2. $\frac{1}{3}$
3. **a.** 0.87 **b.** 0.335 **c.** 0.09 **d.** 0.013

2-8 Applications

Name ____

The goal of the first few questions will be to compute a grade point average, which is a good example of weighted grading: Courses with more credit hours contribute more to the GPA than those with less.

Below are the grades that our friend Christine earned last semester. You'll be filling in the rest of the table as you work through Questions 1–4.

Course	Grade	Grade Value	Credit Hours	Grade Points
English	A		3.0	
Math	C		4.0	
History	B		3.0	
Science	D		5.0	
		Totals:		

1. In most GPA systems, an A is worth 4 points, a B is worth 3 points, a C 2 points, and a D 1 point. Fill in the appropriate values in the Grade Value column.
2. In a GPA system, the number of credit hours act like the weights in a weighted grading system: The number of grade points for each course is the product of the grade value and the credit hours. Fill in the Grade Points column.
3. The GPA is calculated by dividing the number of grade points by the total number of credit hours. Fill in the Totals row, then use the results to find Christine's GPA. GPAs are almost always rounded to two decimal places.
4. What would her GPA have been if all courses were weighted the same? How did you decide?

Next, we'll study a traditional carnival game that we just now made up. The player rolls a single die. If the roll is 1, you win \$1. You lose \$2 if you roll a 2, 3, or 4. You win \$3 if you roll a 5 or 6.

5. Without doing any calculations, does this game sound fair? Would you be willing to play it? Explain.
6. Find the expected value of this game. The table, which we're generously providing at no added cost, should help.

Result	Outcome (X)	Probability $P(X)$	Product $X \cdot P(X)$
1			
2, 3, 4			
5, 6			

7. Would you prefer to be the player in this game, or the person running the game? Why?

8. What should the player expect to happen if he overstays his welcome and plays the game 500 times?

9. The grading structure for my Calc 2 class last semester is described here, and one student who shall remain nameless (because otherwise he or she could sue me back to the Stone Age and probably get me fired as well) had the scores shown. What was this student's overall average?

4 Tests: 40% total	Test scores: 78/100, 85/100, 68/100, 92/100
6 Quizzes: 20% total	Quiz score (overall): 140/150
Final: 20%	Final score: 175/200
Homework: 10%	HW score: 88/94
Projects: 10%	Project scores: 38/50, 50/50

Unit 2 Language and Symbolism Review

Carefully read through the list of terminology we've used in Unit 2. Consider circling the terms you aren't familiar with and looking them up. Then test your understanding by using the list to fill in the appropriate blank in each sentence.

$A = P(1 + r)^t$
$A = P + Prt$
area
change
compound inequality
constant
conversion factor
dependent variable
dimensional analysis
empirical rule
equation
equivalent
error
estimation
evaluate
expected value
expression
fair game
formula
future value
greater than
greater than or equal to
independent variable
inequality
input
interest rate
less than
less than or equal to
mean
median
mode
normally distributed
output
percent error
principal
range
rate
relative change
standard deviation
time
unit rate
variable
volume
weighted scale

1. The ______________ of a set of numbers is found by adding all of the numbers, then dividing by how many numbers there are in the list.
2. The ______________ of a list of numbers is the value that lives right in the middle of the set if it's arranged in order.
3. The ______________ of a list of numbers is the value that appears most often.
4. A ______________ is a quantity that is able to change, or vary.
5. A ______________ is a quantity that can't vary.
6. Course grade is an example of a ______________ because other factors cause your course grade to vary.
7. The amount of time you spend studying is an example of an ______________ because the amount of time spent studying can cause changes in your course grade.
8. An ______________ is a combination of variables and constants using mathematical operations and grouping symbols.
9. An ______________ is a statement that two quantities are equal, built using expressions and an equal sign (=).
10. A ______________ is an equation with multiple variables that is used to calculate some quantity that we're interested in.
11. The independent variable is sometimes called the ______________ and the dependent variable is sometimes call the ______________.
12. To ______________ an expression (or a formula), replace all occurrences of the variable or variables with specific numbers. This allows us to find a numeric value.
13. The future value of an account using simple interest is given by ______________.
14. The future value of an account using compound interest is given by ______________.
15. In Questions 13 and 14, the new value of an account A is known as the ______________, P represents the original amount or ______________, r represents the ______________ in decimal form, and t represents the number of ______________ periods.
16. Two expressions that would give you the same output no matter what input you choose would be considered ______________ expressions.

17. ____________ is the measure of size for two-dimensional objects, like floors, walls, posters, fields, etc.

18. ____________ is the measure of size for three-dimensional objects, like beer mugs, swimming pools, buildings, and so on.

19. ____________ is the incredibly valuable skill we use in the process for converting units.

20. Multiplying any measurement by a ____________ won't change the size of the measurement because we're just multiplying by one; it will change the units used to measure.

21. A ____________ is a ratio that compares two quantities.

22. A ____________ is a rate that has a 1 in the denominator.

23. ____________ is another name for making an educated guess.

24. A statement that involves one of the symbols $>$, $\geq$, $<$, or $\leq$ is called an ____________.

25. A phrase used to describe $<$ is ____________.

26. A phrase used to describe $>$ is ____________.

27. A phrase used to describe $\leq$ is ____________.

28. A phrase used to describe $\geq$ is ____________.

29. The ____________ between two quantities is found by simply subtracting the new value from the old, $(x_n - x_o)$.

30. The ____________ between two quantities measures the change as a fraction of the original value, $\frac{x_n - x_o}{x_o}$.

31. When comparing an estimate to an exact value, the ____________ in the estimate can be expressed by subtracting the actual value from the estimated value.

32. The ____________ is calculated just like relative change. The idea is to measure how far off the estimate is as a percentage of the actual value.

33. The ____________ of a data set is calculated by subtracting the highest value minus the lowest value.

34. ____________, which is often represented by a lowercase Greek sigma (σ) is a measure of how far on average the values are from the mean.

35. Things like sizes of individuals, IQs, weights of packaged products, and lifespans of batteries or lightbulbs are often ____________.

36. When a group of data is normally distributed, and we know the mean and standard deviation, the rule that allows us to estimate how many data values fall within certain ranges is known as the ____________.

37. A combination of two inequalities is known as a ____________.

38. The ____________ of an event is the value that can be expected, on average, to occur.

39. A grading scale where some assignments count for a bigger portion of the overall grade than others is called a ____________.

40. A game of chance is called a ____________ if the expected value is 0.

Unit 2 Technology Review

This is a short review of the technology skills we've used in this unit. In each case, rate your confidence level by checking one of the boxes, and then answer the question. If you feel like you're struggling with these skills, consult the online resources for extra practice.

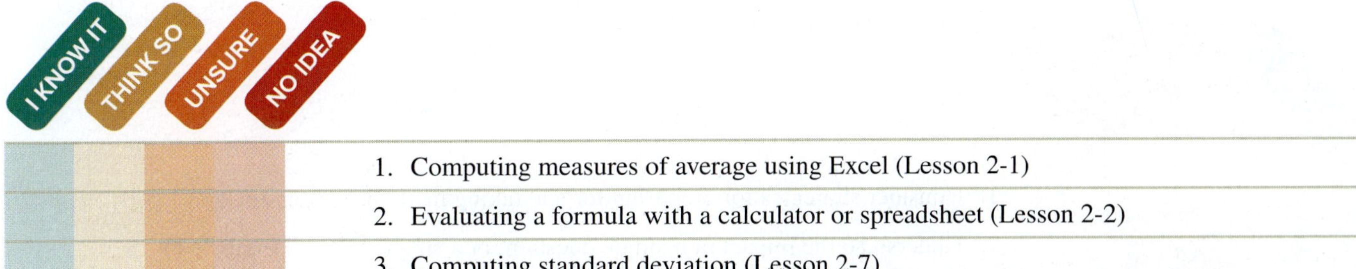

1. Computing measures of average using Excel (Lesson 2-1)
2. Evaluating a formula with a calculator or spreadsheet (Lesson 2-2)
3. Computing standard deviation (Lesson 2-7)

1. Compute the mean, median, and mode for the lengths of my 10 favorite Arnold Schwarzenegger action movies in the list.

	A	B
1	**Movie**	**Length (min)**
2	Collateral Damage	108
3	Commando	90
4	Conan The Barbarian	129
5	Predator	107
6	Teminator 2	137
7	Terminator	107
8	Terminator 3	109
9	Terminator Genisys	126
10	The Running Man	101
11	True Lies	141
12		
13		
14	Mean	
15	Median	
16	Mode	

2. A prepaid calling card starts with a value of \$40. It costs \$0.06 per minute to use the card. The formula $V = 40 - 0.06x$ gives the value remaining on the card after x minutes of talking time. Use a calculator or a spreadsheet to complete this table.

x (min)	V (value remaining)
0	
60	
120	
180	
240	
300	

3. Compute the standard deviation for the lengths of my 10 favorite Arnold Schwarzenegger action movies in the list.

	A	B
1	**Movie**	**Length (min)**
2	Collateral Damage	108
3	Commando	90
4	Conan The Barbarian	129
5	Predator	107
6	Teminator 2	137
7	Terminator	107
8	Terminator 3	109
9	Terminator Genisys	126
10	The Running Man	101
11	True Lies	141
12		
13		
14	Standard Deviation	

Unit 2 Learning Objective Review

This is a short review of the learning objectives we've covered in this unit. In each case, rate your confidence level by checking one of the boxes, and then answer the question. If you feel like you're struggling with these skills, consult the lesson referenced next to the objective and see the online resources for extra practice.

1. Consider strategies for preparing for and taking math tests. (Lesson 2-1)
2. Understand the impact of a single question, or a single exam. (Lesson 2-1)
3. Calculate, interpret, and compare measures of average. (Lesson 2-1)
4. Distinguish between inputs (independent variables) and outputs (dependent variables). (Lesson 2-2)
5. Evaluate expressions and formulas. (Lesson 2-2)
6. Write and interpret expressions. (Lesson 2-2)
7. Determine units for area and volume calculations. (Lesson 2-3)
8. Use formulas to calculate areas and volumes. (Lesson 2-3)
9. Discuss important skills for college students to have. (Lesson 2-3)
10. Simplify expressions. (Lesson 2-3)
11. Convert units using dimensional analysis. (Lesson 2-4)
12. Convert units within the metric system. (Lesson 2-4)
13. Convert rates of change. (Lesson 2-4)
14. Convert temperatures. (Lesson 2-4)
15. Identify the steps in a systematic problem-solving procedure. (Lesson 2-5)
16. Make educated guesses. (Lesson 2-5)
17. Compare numbers using inequality symbols. (Lesson 2-5)
18. Compare change to relative change. (Lesson 2-6)
19. Apply percent error. (Lesson 2-6)
20. Identify the steps in computing standard deviation, and describe why they lead to a measure of variation. (Lesson 2-7)
21. Compute and interpret standard deviation. (Lesson 2-7)
22. Use a normal distribution to find probabilities. (Lesson 2-7)
23. Estimate expected value experimentally. (Lesson 2-8)
24. Compute expected value. (Lesson 2-8)
25. Compute weighted grades and GPA. (Lesson 2-8)

After you've evaluated your confidence level with each objective and gone back to review the objectives you weren't sure about, use the problem set as an additional review.

1. How did things go for the first exam? What do you plan to do the same for the second exam? What do you plan to change?

Question #	Points Possible
1	3
2	2
3	6
4	5
5	4
6	6
7	8
8	6
Total	40

2. The table shows the number of points possible on an 8 question quiz. What is the highest percentage a student could earn if he skipped the question worth the lowest number of points? What if he skipped the question worth the highest number of points instead?

Consider the table showing points earned on exam 1 for a group of students as you answer Questions 3–6.

	Exam 1
Beavis	28
Butthead	28
Hank	78
Bart	84
Lisa	100
Homer	71
Marge	92

3. Determine the mean of these scores.

4. Determine the median of these scores.

5. Determine the mode of these scores.

6. Which measure of average would you use if you wanted to brag about how well these students did on the exam?

7. According to http://www.boeing.com/, the Boeing 747 consumes about 5 gallons of fuel per mile. If a plane starts with 63,500 gallons of fuel and consumes fuel at that same rate, which expression would best represent the number of gallons of fuel **remaining** in the tanks of a plane after x miles had been flown? Circle the correct choice, and then explain why you believe it is correct.

©imageshop–zefa visual media uk ltd/ Alamy RF

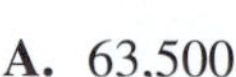

A. 63,500 **B.** $63{,}500 - x$

C. $63{,}500 - 5x$ **D.** $12{,}700x$

Explain:

8. Use your calculator or a spreadsheet to complete the table where the miles flown is x and the gallons remaining is given by the correct expression from question 7.

Miles flown	Gallons remaining
0	
250	
500	
750	
1000	
1250	
1500	
1750	
2000	

9. How many gallons would be remaining after 1,000 miles were flown?

10. How many miles were flown if there were 56,000 gallons of fuel remaining?

11. Write a sentence or two using the terms **input, output, independent,** and **dependent** to describe the relationship between the miles flown and the gallons remaining.

12. Which is most likely the correct way to compute the volume of a cone? Circle the correct choice and clearly explain your reasoning using the units (not by looking up a formula).

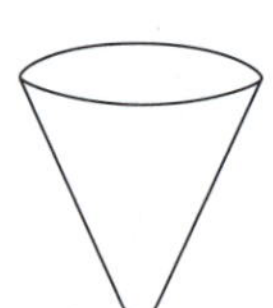

A. **Volume** $= \frac{1}{3}\pi(1.2 \text{ in.})^2(3 \text{ in.})$

B. **Volume** $= \frac{1}{3}\pi(1.2 \text{ in.})(3 \text{ in.})$

13. The formula $A = \pi r^2$ can be used to find the area of a circle with radius r. Find the area of the circle in the figure. Round to the nearest tenth.

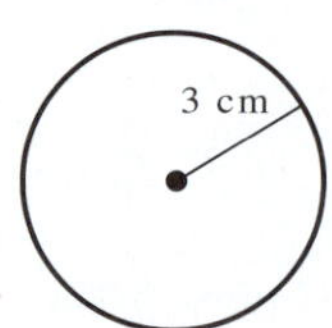

14. The formula $V = \pi r^2 h$ can be used to find the volume of a cylinder with radius r and height h. Find the volume of the cylinder in the figure. Round to the nearest tenth.

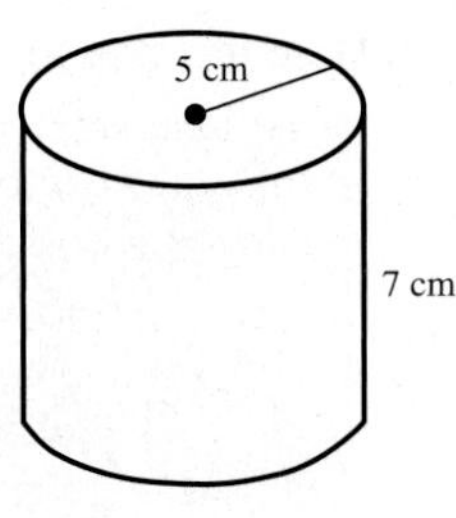

In Questions 15 and 16, simplify each expression.

15. $2(3x)(4x)$

16. $5x + 6y + 2x + 3y$

17. Determine the number of cups that are in a 2-liter bottle of Dr Pepper.

18. Determine the speed in inches per second of a space-shuttle crawler that travels at 2 miles per hour.

19. A 5-k race is 5 kilometers long. How long is that in millimeters?

20. Convert 68° Fahrenheit to Celsius.

21. Write a brief summary of Polya's 4-step problem-solving procedure.

22. Use the map to estimate the distance from St. Louis to New York, and explain your reasoning.

Distance:

Reasoning:

23. If in a typical month your cell-phone bill is around \$54, you might want to budget at least \$60 for cell-phone expenses. Write an inequality to state that the amount budgeted *c* for cell-phone expenses was at least \$60.

24. A seven-passenger van can transport at most seven people safely. Write an inequality stating that the number of passengers *n* is at most seven.

25. On March 1, 2017, the Dow Jones Industrial Average closed at \$21,115.55. The next day it closed at \$21,002.97. What was the actual change in value of the Dow from March 1 to March 2, 2017? What was the relative change?

26. If a jet was fueled with 68,000 lbs of fuel and was supposed to be given 72,000 lbs of fuel, what's the percent error? Then write a sentence using your answer to describe the impact this mistake would have on the amount of time the plane could remain flying.

Runners often keep a training log to keep track of the distance they run over a period of time. Nikki's running log is displayed here showing total miles run by month for 2015 and 2016.

	Miles in 2015	Miles in 2016
Jan	40	29
Feb	43	48
March	38	22
April	49	21
May	47	44
June	55	80
July	41	75
Aug	35	10
Sept	42	49
Oct	48	45
Nov	29	43
Dec	35	38

27. In which year did Nikki run more miles?

28. Compute the standard deviation for the miles run for each year. Feel free to use your favorite form of technology to assist with the calculations.

29. Write a sentence or two describing the meaning of the standard deviations you just computed.

30. It has been reported that credit card debt for college seniors is normally distributed with a mean of \$3,262 and a standard deviation of \$1,100. Fill in the missing values under the normal curve.

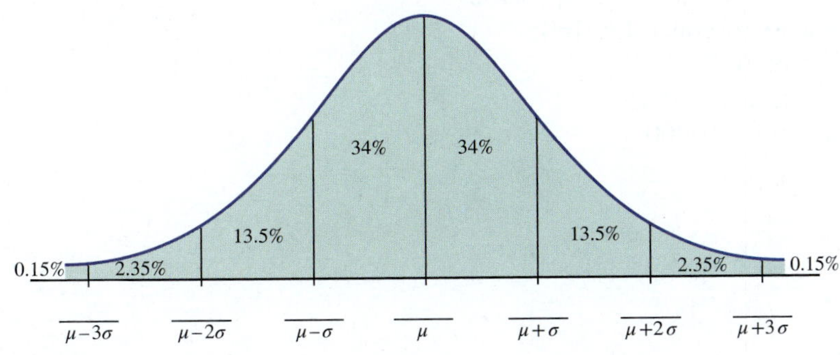

31. Write an expression that describes the probability of a randomly selected college senior having credit card debt over \$4,362, and use the empirical rule to find this probability.

32. Use the empirical rule to find $P(2{,}162 < x < 5{,}462)$ and write a sentence explaining what it means.

Suppose you decide to set up a simple dice-rolling game at a carnival and you plan to pay $1 for a 1 or 2, $2 for a 3 or 4, and $3 for a 5 or 6. If each player pays you $2 for each roll, YOU would make $1 if the player rolled a 1 or 2, $0 if the player rolled a 3 or 4, and lose $1 if the player rolled a 5 or 6. Those outcomes have been completed in the table.

Outcome (*X*)	Probability *P*(*X*)	Product *X* · *P*(*X*)
$1		
$0		
−$1		

33. Complete the associated probabilities for each outcome and then complete the third column and use it to find the expected value of this game.

34. Is this a fair game? Would it make sense for you to offer this game? Explain.

35. The grading structure for a Stats course is described here, and one student had the scores shown. What was the student's overall average? If necessary, round the student's scores to the nearest full percent.

3 Tests: 30% total	Test scores: 65/100, 88/100, 93/100
8 Quizzes: 20% total	Quiz score (overall): 139/160
Homework: 5%	HW score: 65/70
Project: 15%	Project scores: 48/50
Final: 30%	Final score: 191/200

Unit 3
Thinking Linearly

©Brand X Pictures/PunchStock RF

Outline

Lesson 3-1 Prep Skills

SKILL 1: USE DIMENSIONAL ANALYSIS

Dimensional analysis was the main skill we learned in Lesson 2-4: It's the process we used to convert measurements from one unit to another. If you don't remember the process very well, it would be an excellent idea to review that lesson before starting Lesson 3-1. Try the Prep Skills Questions first to see how you do, then refer back to Lesson 2-4 if necessary.

SKILL 2: BUILDING UP FRACTIONS

Sometimes it's useful to do the opposite of reducing fractions: rewriting fractions so that they have a bigger numerator and denominator. This is called **building up** a fraction. Since it's the opposite of reducing fractions, we do exactly the opposite: Instead of dividing the numerator and denominator by the same number, we multiply the numerator and denominator by the SAME number.

- $\frac{3}{5}$ can be written as a fraction with denominator 20: $\frac{3}{5} \cdot \frac{4}{4} = \frac{12}{20}$
- $-\frac{2}{3}$ can be written as a fraction with denominator 60: $-\frac{2}{3} \cdot \frac{20}{20} = -\frac{40}{60}$

SKILL 3: DRAWING A SCATTER PLOT

This important skill was introduced in Lesson 1-5. The key things to remember are (1) write the given information as ordered pairs; (2) put the information corresponding to the first coordinates on the horizontal axis and the information corresponding to the second coordinates on the vertical axis; and (3) put an appropriate scale on those axes so that all of the points on your plot fit in the space you have, and don't leave a ton of empty space with no points. In almost every case, you'll want the scale on the vertical axis to begin at zero so that the heights of points don't give a misleading view of differences in the data.

SKILL 4: INTERPRETING INFORMATION FROM GRAPHS

Interpreting graphs is the most important skill we covered in Lesson 1-5. As we pointed out at the time, if you can't use a graph to get information, then it's just really bad art. In essence, this is the exact opposite of drawing a scatter plot. When you draw a plot, you're taking two corresponding pieces of information and drawing a point that represents them. When you interpret a graph, you're looking at a point and recognizing what those two pieces of information are. In both cases, the key is the same: carefully identifying exactly WHAT INFORMATION is represented by the scale on each axis.

PREP SKILLS QUESTIONS

1. Use dimensional analysis to perform each conversion. The unit equivalence reference tables can be found on page 192.

   a. 1,100 yards to feet b. 93 km to meters

   c. 2 miles to yards d. 32 hours to seconds

2. Rewrite each fraction with the given denominator.

   a. $\frac{5}{3}$; 21 b. $-\frac{2}{5}$; 60 c. $\frac{14}{4}$; 10

3. Draw a scatter plot of the data provided, which are the global sales (in millions) for each of Metallica's first 9 albums, as of January 1, 2016.

Album number	1	2	3	4	5	6	7	8	9
Sales (millions)	5.5	7.1	7.8	9	23	8.5	7	4.3	2.3

4. The graph shows the number of hours of daylight over the course of a year in Portland, Oregon. Use it to answer the questions.

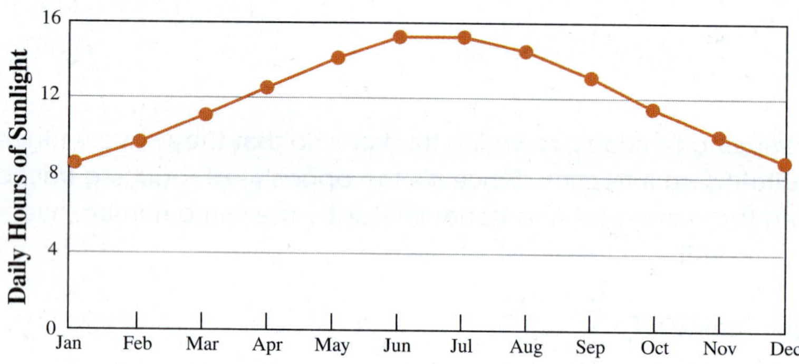

a. About how many hours of daylight could I expect on my birthday, July 8?

b. About when does the daylight time get longer than 12 hours in the spring?

c. What are the highest and lowest amounts of daylight during the year?

Lesson 3-1 88 Miles Per Hour!

LEARNING OBJECTIVES

- ☐ 1. Interpret a rate of change.
- ☐ 2. Predict a future value from a rate of change.
- ☐ 3. Calculate a rate of change.
- ☐ 4. Find the intercepts of a line.
- ☐ 5. Interpret the meaning of the intercepts of a line.

©Universal Pictures Photographer: Ralph Nelson, Jr./Photofest

The future is something which everyone reaches at the rate of 60 minutes an hour, whatever he does, whoever he is.
—C.S. Lewis

Anyone familiar with the classic movie *Back to the Future* instantly recognizes the significance of 88 miles per hour: It's the speed the DeLorean time machine needed to reach in order to travel through time. Speeds (like 88 miles per hour) are familiar examples of the topic of this lesson, **rates of change**. Our world is a dynamic, ever-changing place, so studying the rate at which things change is an excellent way to study the world around us. The amount your grade changes as you spend more time studying? That's a rate. Your hourly wage if you work a part-time job? Rate. A waiter's decrease in tips as less customers eat at a restaurant? Rate. They're everywhere! Fortunately, studying rates will tie together some of the skills we've already practiced in this course, meaning we're in a good position to make the most of this useful topic.

0. Think of an example of a rate of change different from the ones listed in the opening paragraph.

3-1 Class

A rate is a ratio that compares two quantities. By rate of change, we mean a rate that compares the change in one quantity to the change in another. A speed like 88 miles per hour qualifies, because we can write it this way:

$$\frac{88 \text{ mi}}{1 \text{ hr}}$$

This quite literally compares a change in distance (88 miles) to a change in time (1 hour). That's exactly why we measure speeds in miles per hour.

The word **equivalent** is an important word in math, and one that we'll use often in this course. Make sure you clearly understand what that word means. For our purposes, we'll say that two quantities are equivalent if they have the same value or are interchangeable even if they look different.

1. If you make \$100 a day for doing a certain job, how much money would you make in a 5-day work week?

2. Are $\frac{\$100}{\text{day}}$ and $\frac{\$500}{5 \text{ days}}$ equivalent rates? What does that mean?

Question 2 demonstrates a useful fact about rates of change: They can be scaled up or scaled down to fit a given situation. Scaling up a fraction is done by writing an equivalent fraction (that is, one that has the same value) using larger numbers. Scaling down is writing an equivalent fraction with smaller numbers. These processes are done by either multiplying BOTH the numerator and denominator by the same number, or dividing BOTH the numerator and denominator by the same number.

3. Write a fraction equivalent to $\frac{60\text{ mi}}{4\text{ hr}}$ that has 2 in the denominator.

4. Write a fraction equivalent to $\frac{60\text{ mi}}{4\text{ hr}}$ that has 1 in the denominator. This is known as a **unit rate**.

5. Write a fraction equivalent to $\frac{\$320}{4\text{ days}}$ that has 10 in the denominator. (Hint: First scale down, then up.)

6. If one bicyclist is pedaling at a rate of 40 miles in 3 hours, and another at a rate of 50 miles in 4 hours, which is faster? (Scaling either up or down can be used!)

Did You Get It

Try this problem to see if you understand the concepts we just studied. The answer can be found at the end of the Portfolio section.

1. If I leave my home and drive for 2 hours, covering 128 miles, what was my rate?

Most experienced runners get to a point where they can comfortably jog long distances at a consistent pace. This "pace," of course, is another way to say "rate of change," because speed is the rate at which distance changes compared to time. One particular runner jogs one lap around a 400-meter track in 2 minutes. In Questions 7–16, you can assume that the runner can maintain this pace for a long time.

7. For every _______ minutes, the runner jogs _______ meters.

8. Complete the table, continuing the pattern for times and finding the associated distances for this runner.

Time (Min)	Distance (m)
2	400
4	

9. Notice that the rate at which the runner's distance changes is constant. When the rate of change is constant, a quantity illustrates what type of growth (that we encountered in Unit 1)?

10. Write the runner's rate as a fraction using meters in the numerator.

Math Note

Using rates of change to calculate sizes of quantities is one of the most important applications of dimensional analysis.

11. Write this rate as a unit rate in meters per minute.

12. Convert your rate from Question 11 to meters per second using dimensional analysis. (Round to two decimal places.)

13. What distance will the runner cover in 40 minutes?

14. What distance will he cover in 40 seconds?

15. Based on the previous two questions, can you write a formula for finding the distance traveled by an object when you know the speed and amount of time traveling at that speed?

Did You Get It?

2. Based on the rate you found in Did You Get It 1, what distance could I cover in 3.5 hours driving at that rate?

16. What would this runner's total time be for a 10k race? (Ten kilometers, that is. Recall that 1 km = 1,000 m. Your answer to Question 11 will help.)

3-1 Group

At this point, we know that a rate of change measures how some quantity is changing. Now ponder this question: Do rates of change change? That might look like a typo, but it's not. The rate at which some quantity changes either stays constant, or changes. For example, if you make $9 per hour at a job, the rate at which your pay changes as you work more hours is always the same: Your pay grows at the rate of $9 per hour. But if you're driving in traffic, the rate of change of your position (what we commonly call speed) probably changes quite a bit. Our next order of business is to study situations where the rate of change of a quantity stays constant: In that case, modeling that situation with an equation or a graph is very manageable.

1. According to taxifarefinder.com, a cab ride from the airport in Las Vegas will cost you an initial charge of $5.10, plus $2.60 per mile. Use this information to fill in the rest of the table describing the total cost for various distances. Then create a scatter diagram, using the distance as x coordinate and the cost as y coordinate.

©Stockbyte/Punchstock RF

Distance	Cost ($)
0	$5.10
1	$7.70
2	$10.30
3	
4	
5	

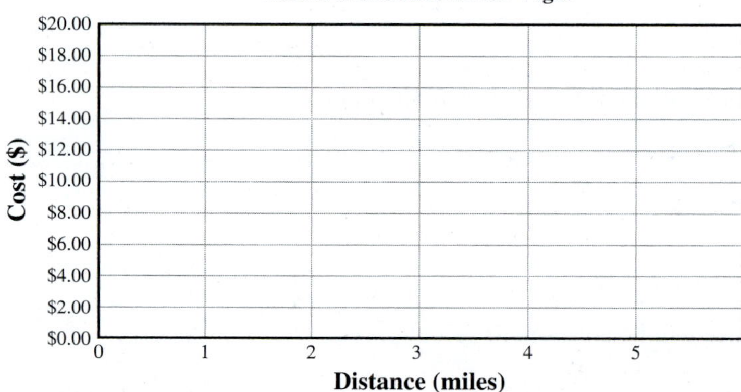

2. What do you notice about the pattern that the points on your scatter diagram are making? Use that observation about a pattern to connect the points, drawing a graph that represents the cost of a cab ride based on distance.

3. The point where any graph crosses the y axis is called the ***y* intercept** for the graph and a point where the graph crosses the x axis is called an ***x* intercept**. Write the coordinates of the y intercept for the graph that you drew. More importantly, what information does the y intercept represent? Does the graph have an x intercept? Why or why not?

4. How much more will you pay for a 3-mile ride than for a 2-mile ride?

©Glow Images RF

5. How much more will you pay for a 5-mile ride than a 2-mile ride? Explain why that makes sense based on your answer to Question 4.

6. Pick any two points on the graph that you drew, and subtract the second coordinates. Then divide the result by the difference of the first coordinates. Each person in your group should pick a different pair of points. What does the result represent about the cab ride?

In dividing the difference of the two costs by the difference of the two distances in Question 6, you found the rate of change of the cost as distance changes. When applied to the graph of a line, we call this number, which describes how steep the line is, the **slope** of the line.

The Slope of a Line

The slope of a line is a number that describes how steep a line is. This is accomplished by comparing the change in height over some span to the change in horizontal position.
As a formula, if the two points are (x_1, y_1) and (x_2, y_2), the slope (usually represented by the letter m) is

$$m = \frac{y_2 - y_1}{x_2 - x_1} = \frac{\text{Difference of second coordinates}}{\text{Difference of first coordinates}}$$

The key to interpreting slope is to recognize that each of the numerator and denominator is a change in some quantity. So slope compares the change in two quantities, which makes it a rate of change. Specifically, slope is the rate at which the second coordinate changes compared to the first coordinate.

Did You Get It

3. Find the slope of a line that connects the points (3, 10) and (8, 30).

7. If you were to connect the points (1, 7.70) and (5, 18.10) on the graph in Question 1 with a triangle, it would look something like this. Label the coordinates of the two given points on this triangle.

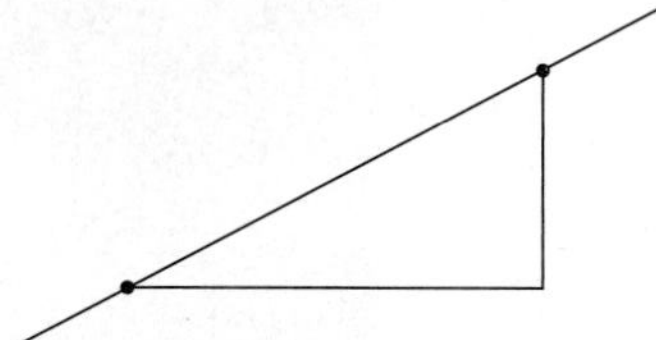

8. How much more distance was added to the ride going from (1, 7.70) to (5, 18.10)? Label this value on the triangle.

9. How much cost was added to the ride going from (1, 7.70) to (5, 18.10)? Label this value on the triangle.

10. Find the slope of the line using your answers to Questions 8 and 9. What do you notice?

11. How much would 4 miles add to the cost of a taxi ride? Would it matter if the change was from 1 mile to 5 miles or if it was 20 miles to 24 miles?

12. The thing that makes a line a line is the fact that the slope never changes. We now know that the slope of a line describes the rate at which the y coordinate changes compared to the x coordinate. Based on this, how can you decide if the relationship between two quantities might be modeled well by a straight line?

13. Estimate the distance of a \$16 cab ride. You can use either the table or the graph, but make sure you explain how you got your answer.

Everyone knows that after you buy a car, in most cases its value decreases as it gets older. Suppose you bought a used car for $15,000 a few years ago, and its value has been decreasing at the rate of $2,000 per year since then. **Make sure you answer each of the remaining questions with a full sentence,** with punctuation and everything!

©Vladimiroquai/iStock/Getty Images RF

Time (yrs)	Value ($)
0	$15,000
1	$13,000
2	$11,000
3	$9,000
4	$7,000
5	$5,000

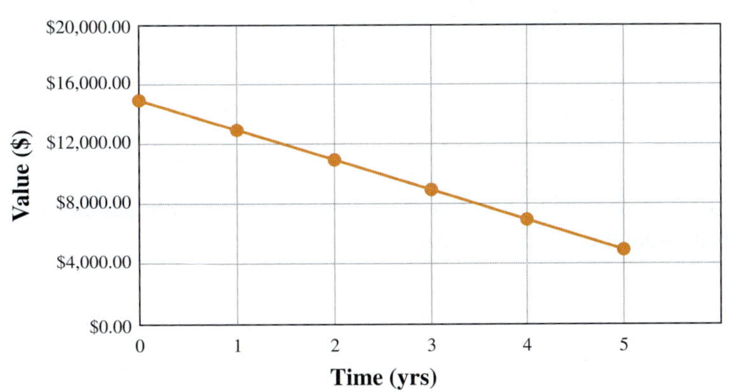

14. Find the y intercept of the graph and explain what it means. (Remember, the y intercept is a point, not a number.)

15. Find the slope of the line and explain what it means. In particular, what's the significance of the sign?

16. Estimate the value of the car after 42 months. (Careful about time units!)

17. Estimate the number of years it took for the car to reach a value of $6,000.

Math Note

The process of an object losing value as time passes is called **depreciation**. This concept is used by businesses in calculating net assets for tax purposes.

Did You Get It 

4. Estimate the value of the car 18 months after it was purchased.
5. About how long after it was purchased was the car worth $10,000?

18. Give a description of how to find the slope of a line, and why it represents a rate of change. Be specific!

19. Give a description of how to find the *y* intercept of a graph, and the *x* intercept as well. Include a definition in your own words of what the *y* and *x* intercepts of a graph are.

3-1 Portfolio

Name ______________________________

Check each box when you've completed the task. Remember that your instructor will want you to turn in the portfolio pages you create.

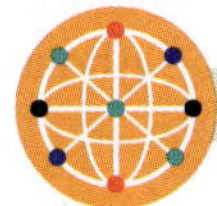

Technology

1. ☐ When you know the rate at which a quantity changes, you can build a spreadsheet to calculate the size of that quantity. Build a spreadsheet that calculates pay given hours worked and hourly rate, like the one shown here. You should be able to enter any rate you like in dollars per hour, and have a formula calculate the pay corresponding to any time you enter. Also enter a formula to calculate total pay for all employees. A template to help you get started can be found in the online resources for this lesson.

D11 fx

	A	B	C	D
1	**Employee**	**Time (hrs)**	**Rate ($/hr)**	**Pay ($)**
2	Napoleon	10	$8.75	$87.50
3	Deb	20		
4	Rico	32		
5	Me	40		
6			**Total Pay**	
7				

Online Practice

1. ☐ Include any written work from the online assignment along with any notes or questions about this lesson's content.

Applications

1. ☐ Complete the Applications problems.

Reflections

Type a short answer to each question.

1. ☐ In this section, we learned that the speed of an object is a rate of change; specifically, the rate at which distance traveled changes as time changes. Think of at least three other rates of change, and describe specifically what they measure. If you want to impress your instructor, try to come up with one that doesn't have a unit of time in it.
2. ☐ Explain in your own words what the slope of a line is. You should discuss both what it means graphically and what the practical significance is.
3. ☐ What are the intercepts of a graph?
4. ☐ Name one thing you learned or discovered in this lesson that you found particularly interesting.
5. ☐ What questions do you have about this lesson?

Looking Ahead

1. ☐ Complete the Prep Skills for Lesson 3-2.
2. ☐ Read the opening paragraph in Lesson 3-2 carefully and answer Question 0 in preparation for that lesson.

Answers to "Did You Get It?"

1. 64 mi/hr 2. 224 miles 3. 4 4. About $12,000 5. About 30 months

Answers to "Prep Skills"

1. a. 3,300 ft b. 93,000 m c. 3,520 yd d. 115,200 sec

2. a. $\frac{35}{21}$ b. $-\frac{24}{60}$ c. $\frac{35}{10}$

3.

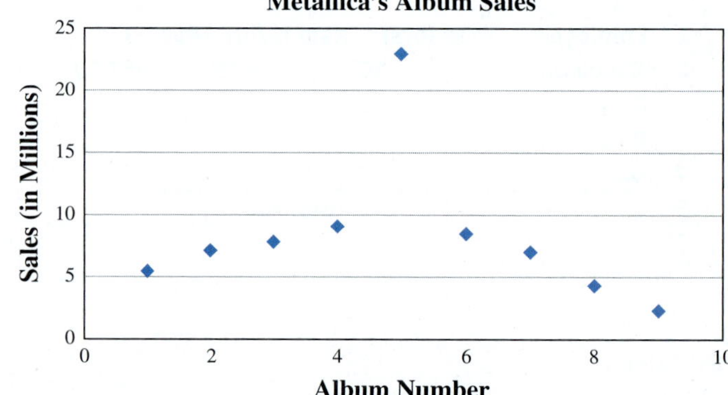

4. a. about 15 hours b. around March 28 c. about 15 hours; about 8.5 hours

3-1 Applications

Name ______________________________

When advertising special financing deals, car makers often use rates to describe what your monthly payment will look like. (This is because the amount you need to borrow varies widely based on the car you choose and the down payment you make.) In January 2013, Hyundai offered 2.9% financing on the Genesis coupe which would result in monthly payments of $22.09 per $1,000 borrowed for 48 months.

Courtesy of Dave Sobecki

1. For every ________ borrowed, the monthly payment will be ________.

2. Complete the table of loan payments using your answer to Question 1.

Amount Borrowed	Loan Payment
$1,000	$22.09
$2,000	
$3,000	
$4,000	
$5,000	
$6,000	
$7,000	
$8,000	
$9,000	
$10,000	

3. Write the rate Hyundai advertised as a fraction.

Use your answer to Question 3 and dimensional analysis for Questions 4 and 5.

4. What would your monthly payment be if you borrowed $18,000?

5. If your monthly payment is $331.35, how much did you borrow?

6. If you agree on a price of $23,900 (including taxes and fees) for a Genesis and the dealership offers you $7,500 in trade for your old car, how much would your monthly payment be?

3-1 Applications

Name ______________________________

A manufacturing company buys a new stamping machine for $28,000. The maker of the machine informs the company's CEO that on average, it depreciates in value according to the schedule shown in the table.

Months	Value
0	$28,000
6	$24,500
12	$21,000
18	$17,500

7. If the depreciation continues at the same rate, how long will it take until the machine has no value?

8. Based on the pattern you see in the table, how do you know that the graph will be a straight line?

9. Pick an appropriate scale for each axis, and graph the value of the machine until it reaches zero. Use the number of months as first coordinate.

10. Find the slope of the graph and explain what it means.

11. Find the intercepts of the graph, and describe what each means.

Lesson 3-2 Prep Skills

SKILL 1: ADD OR SUBTRACT LIKE TERMS

In Lesson 1-2, we saw that quantities can only be added if they're like quantities, which means they're identical except possibly for the numeric part. For example, $8n$ and $12n$ are like quantities because they're both just n without the 8 and the 12, while $8n$ and $12n^2$ are not like quantities because n and n^2 aren't identical.

The individual pieces of an expression containing addition and subtraction are called **terms**. In this case, we use the phrase *like terms* rather than like quantities. Deciding which terms can be added or subtracted is an important part of simplifying algebraic expressions.

- $10k + 8 - 3k + 4$ $10k$ and $-3k$ are like terms; 8 and 4 are like terms
  $= 7k + 12$
- $4n - 7n^2 + 11n + 7 - 4n^2$ $4n$ and $11n$ are like terms; $-7n^2$ and $-4n^2$ are like terms
  $= 15n - 11n^2 + 7$

SKILL 2: FIND SLOPE AND *Y* INTERCEPT

Slope was one of the key concepts we learned about in Lesson 3-1. It's discussed in the Group portion of that lesson, with a reminder on how to compute slope in the colored box on page 287. Finding and interpreting the *y* intercept of a graph is also in that Group portion. In short, the *y* intercept is the point where a graph crosses the *y* axis. It always tells you the value of the output for an equation when the input is zero.

SKILL 3: EVALUATE EXPRESSIONS

This was one of the most important skills we studied in Lesson 2-2. When we use an algebraic expression to model some quantity, it's very common to calculate the value of that quantity by replacing an input variable with some number. This is what we called evaluating an expression. The only thing you need to think about is replacing every single occurrence of the variable or variables with the given number(s).

- For the future value formula $A = P(1 + rt)$, if we want to know the future value of a \$5,000 initial investment at 6% interest for 5 years, we're given values of $P = 5{,}000$, $r = 0.06$, and $t = 5$. So we replace each letter in the formula with the associated value:

  $A = \$5{,}000(1 + 0.06(5)) = \$6{,}500$

SKILL 4: DRAW A SCATTER PLOT

This important skill was introduced in Lesson 1-5 and reviewed in the Prep Skills for Lesson 3-1.

PREP SKILLS QUESTIONS

1. Perform each addition or subtraction.

   a. $3x + 8x - 12$ b. $14y + 6 - 2y - 10$ c. $11a + 12b - 13 + 7a - 20b$

2. The graph on the next page shows the value of an antique painting based on the number of years since it was found in someone's attic. Find the slope of the line and write a description of what it tells us about the painting.

3. What is the *y* intercept of the graph? What does it tell us?

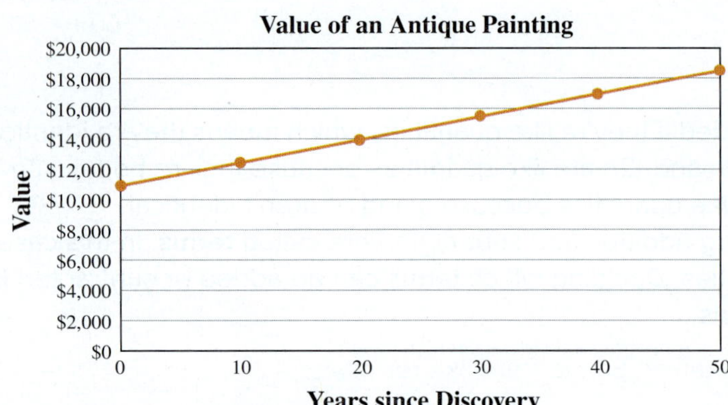

4. Evaluate each expression for the given value of the variable(s).

a. $K = C + 273.15$; $C = 35$

b. $A = lw$; $l = 10$, $w = 35$

c. $C = 4{,}200 + 1.35x$; $x = 675$

Lesson 3-2 A Snow Job

LEARNING OBJECTIVES

- ☐ 1. Write expressions based on given information.
- ☐ 2. Interpret algebraic expressions in context.
- ☐ 3. Evaluate and simplify expressions.

"Between calculated risk and reckless decision-making lies the dividing line between profit and loss."

—Charles Duhigg

©Richard Ellis/Alamy

Most folks that start a business do so with the goal of making money. And running a business isn't easy—there are tons of things you need to keep track of. Where is your money being spent? Is your investment of time and money worthwhile? How many customers do you have, and how much money are you bringing in? If you plan to just start a business without a plan for carefully monitoring every aspect of the company, you have about the same chance of success as an entrepreneur planning to sell snow cones at the South Pole. Before this course, you may have thought that algebra was just an abstract thing that pointy-headed intellectuals do to amuse themselves, but we've already seen a ton of ways that algebra is used to make sense of things in our lives. In this lesson we're going to show how some of the skills we've been working on are useful in monitoring a business. And that's no snow job.

0. In your own words, what is algebra?

3-2 Class

If you were running a business, like one that sells snow cones, you'd probably have a lot of data related to your costs and the amount of money you can make. And if you had also completed the first two units of this course, you'd know a lot about organizing and working with data, so let's put some of the stuff we've learned to work by studying the operation of a snow cone stand. First, we'll need to know some terminology from business and economics.

The amount of money that a business needs to spend for their operation is known as their **costs**. The amount of money that customers pay them for their goods or services is called the **revenue**. The difference between these two numbers is the amount of money that the company makes or loses, which is called the **profit**.

1. When a company is making money—which is the goal—the profit will be positive. What is the relationship between revenue and costs when this happens?

The spreadsheet describes the costs associated with a snow cone business, as well as the revenue and profit.

	A	B	C	D	E	F	G	H	I	J	K	L
1	**Snow cones sold**	**Cups**	**Cost of cups**	**Syrup (oz)**	**Cost of syrup**	**Ice (oz)**	**Cost of ice**	**Combined supply costs**	**Fixed costs**	**Total costs**	**Revenue**	**Net profit**
2	0	0	$0.00	0	$0.00	0	$0.00	$0.00	$245.00	$245.00	$0.00	-$245.00
3	100	100	$1.00	200	$4.00	500	$7.00	$12.00	$245.00	$257.00	$75.00	-$182.00
4	200	200	$2.00	400	$8.00	1000	$14.00	$24.00	$245.00	$269.00	$150.00	-$119.00
5	300	300	$3.00	600	$12.00	1500	$21.00	$36.00	$245.00	$281.00	$225.00	-$56.00
6	400	400	$4.00	800	$16.00	2000	$28.00	$48.00	$245.00	$293.00	$300.00	$7.00
7	500	500	$5.00	1000	$20.00	2500	$35.00	$60.00	$245.00	$305.00	$375.00	$70.00

2. If 300 snow cones are sold, how many ounces of syrup are needed? What about if 500 snow cones are sold?

3. A formula was typed into cell D5 to obtain the result you used to answer Question 6. What formula would give the correct result? (The formula should use cell A5.)

4. Write a verbal description of the relationship between the number of snow cones sold and the number of ounces of syrup that will be needed.

5. Complete the following sentence. If x represents the number of snow cones sold, then ________ represents the number of ounces of syrup used.

When writing an expression that represents some quantity, we'll often turn that expression into an equation (or formula, if you like) by using a letter to represent the value of the expression.

6. Using the letter S to represent the amount of syrup used, write a formula that describes the ounces of syrup needed in terms of the number of snow cones sold.

Courtesy of Greg Stiff

How do we know so much about the snow cone business? Because one of us used to own one!

Did You Get It

Try these problems to see if you understand the concepts we just studied. The answers can be found at the end of the Portfolio section.

1. A hot dog vendor makes \$0.60 on each hot dog sold. Write an expression to describe how much she makes, in dollars, from selling x hot dogs.
2. Using the letter D for the amount she makes, write a formula that describes the amount she makes in terms of number of hot dogs sold.

7. Does the number of snow cones sold depend on the amount of syrup used, or does the amount of syrup used depend on the number of snow cones sold?

8. In your formula from Question 6, which is the independent variable (input), and which is the dependent variable (output)?

9. Is the relationship between the amount of syrup used and the number of show cones sold linear, exponential, or neither? Explain.

10. If x represents the number of snow cones sold, what quantity described in the spreadsheet would be represented by $5x$?

11. If 70 snow cones are sold, the value of $5x$ is ________. Write a sentence or two explaining exactly what that result tells us about the snow cone business.

12. If you were to graph the data in columns A and J from the spreadsheet, with number of snow cones on the horizontal axis and total costs on the vertical axis, would the graph be a line? How can you tell?

13. Now plot points corresponding to columns A and J. Do they form a straight line? What is the y intercept?

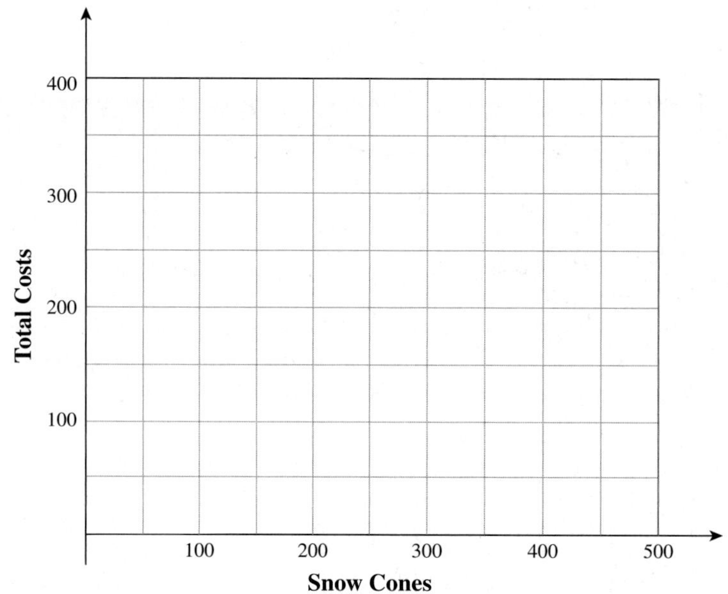

14. Write a unit rate that describes the combined costs of all supplies needed per snow cone sold.

15. Pick two points from the graph (remember, the points are coming from the table) and find the slope of the line connecting them. What do you notice?

16. If each cup costs $0.01, write an expression for the total cost in dollars of the cups based on x, the number of snow cones sold.

17. If $0.04 is spent on syrup for each snow cone sold, write an expression for the total cost in dollars of the syrup based on x, the number of snow cones sold.

18. If $0.07 is spent on ice for each snow cone sold, write an expression for the total cost in dollars of the ice based on x, the number of snow cones sold.

19. The total combined supply costs for making one snow cone looks like this:

cost of cups + cost of syrup + cost of ice

Write an algebraic expression for this, and then simplify your expression. Your previous three answers will surely help.

20. The total cost of selling snow cones is the sum of the variable cost for supplies (which you just found) and the **fixed costs**: These are costs that aren't affected by the number of snow cones sold, like rent, insurance, licensing, etc. Look back at the table to find the fixed costs for this business, then use that to write an equation of the form $C =$ _______ that represents the total cost based on x, the number of snow cones sold.

Did You Get It

3. A company that makes paper plates has fixed costs of $750 and it costs them $0.43 to make each package of plates. Write an equation of the form $C =$ _______ that represents the total cost based on x, the number of packages of plates they make.

3-2 Group

Next, we'll study the profit made by the business, beginning with a graph. The graph was drawn using information from the data table at the beginning of the lesson.

Answer each question with a full sentence or face consequences too dire to mention in polite company.

©RoongsaK/Shitterstpcl RF

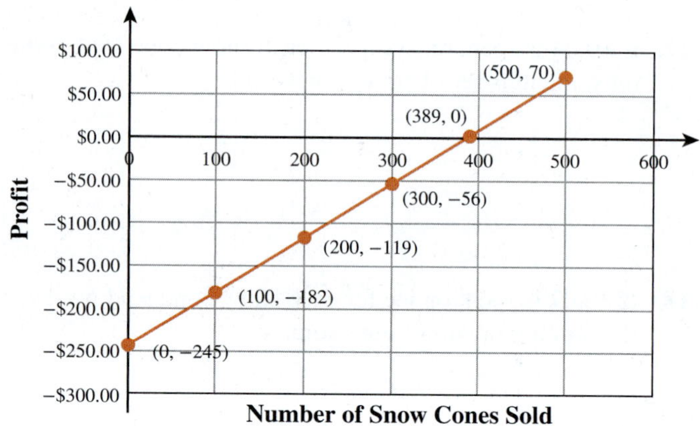

1. Find the y intercept of the graph and explain what it means.

2. Find the x intercept of this graph and explain what it means.

3. Find the slope of the line and explain what it means.

4. Estimate the net profit if 150 snow cones are sold. What's the significance of the sign?

5. Estimate the number of snow cones that need to be sold to make a profit of $50.

6. Fill in the blanks: The profit starts at ________ when no snow cones are sold, and goes up by ________ for each sale.

7. Use your answers to the previous question to write an equation of the form $P =$ ______ that gives the profit made from selling x snow cones.

Now that we're representing data with equations, we can use a calculator to draw graphs. This can be useful in checking an equation that you've written: If its graph matches the data that you started with, then you know you have the right equation.

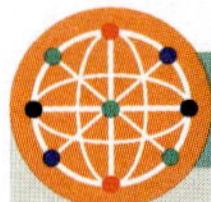

Using Technology: Graphing an Equation with a Graphing Calculator

Starting with an equation with two variables x and y, with y alone on one side:

1. Press [Y=] to get to the equation editor screen.
2. Enter the side of the equation containing x, using [X,T,θ,n] for the variable.
3. Press [WINDOW] to get to the window screen.
4. Enter the lowest value you want to display along the x axis after Xmin, and the highest value you want to display along the x axis after Xmax. After Xscl, enter the distance that you want between each tick mark on the x axis. If you make this too small, the tick marks will be indistinct on your graph.
5. Enter the lowest value you want to display along the y axis after Ymin, and the highest value you want to display along the y axis after Ymax. After Yscl, enter the distance that you want between tick marks on the y axis.
6. Press [GRAPH] to display the graph. If you're not seeing a graph, you probably need to adjust the values you entered in the window screen.

See the Lesson 3-2 Using Tech video in class resources for further information.

8. Use a graphing calculator to graph the equation you wrote in Question 7. Does it match the graph provided earlier? Draw a quick sketch of what your calculator screen looks like.

Use the spreadsheet at the beginning of the lesson to answer the following questions.

9. Find the total cost from selling 400 snow cones.

10. Find the revenue from selling 400 snow cones.

11. A formula that uses cells J7 and K7 was typed in cell L7. What is that formula? (Focus on the dollar amounts in cells J7 and K7.)

12. Write a verbal explanation of the relationship between the profit made from selling a certain number of snow cones and the revenue and total costs from selling that number of snow cones.

13. If we use P to represent profit, C to represent total costs, and R to represent revenue, write a formula that describes the profit in terms of C and R.

14. In Question 17 of the Class portion, you represented total costs with the expression $0.12x + 245$. Using the same line of reasoning, find an expression that describes the revenue from selling x snow cones. (Recall that we used a unit rate to find the expression for total costs.)

15. Is $5 - (7 - 3)$ equal to $5 - 7 - 3$, or $5 - 7 + 3$? Perform each calculation to decide, then try to explain why it worked out that way.

16. Subtract the expression $0.12x + 245$ from the one you wrote in Question 13. Use what you discovered in Question 14 to perform the subtraction. Then write a description of what exactly this new expression represents.

17. Use your expression to find the profit made if the company sells 8,000 snow cones at a festival.

Did You Get It

4. If the company from Did You Get It 3 that makes paper plates sells their plates for $2.49 per package, write and simplify an expression that describes the company's profit in terms of the number of packages sold.
5. Would the company make or lose money if it sells 250 packages? How much money?

3-2 Portfolio

Name ______________________________

Check each box when you've completed the task. Remember that your instructor will want you to turn in the portfolio pages you create.

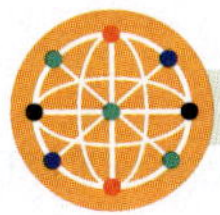

Technology

1. ☐ The spreadsheet on page 298 is in the online resources for this lesson. Make a copy of the spreadsheet, then use the fill-down feature to extend the existing pattern to show up to 2,100 snow cones sold. This should be in row 23 if all goes well.

2. ☐ Now create a scroll bar that changes the values in cell A23. (We learned how to make scroll bars in the online resources for Lesson 2-1.) The settings should be Current = 2,000, minimum = 0, maximum = 2,000. Adjust the scroll bar to the point where the profit is as close to zero dollars as possible. Record the value on the scroll bar, and describe what that tells you about the business.

Online Practice

1. ☐ Include any written work from the online assignment along with any notes or questions about this lesson's content.

Applications

1. ☐ Complete the Applications problems.

Reflections

Type a short answer to each question.

1. ☐ Describe how we used algebra in this lesson to study the operation of a snow cone business.
2. ☐ Describe what each of costs, revenue, and profit mean in terms of operating a business.
3. ☐ If a company's costs start out at $3,000 when no items are produced, and it costs $1.25 to produce each item, describe how you can write an expression that provides the cost of making x items. Don't just write the expression: Describe how you got it.
4. ☐ Name one thing you learned or discovered in this lesson that you found particularly interesting.
5. ☐ What questions do you have about this lesson?

Looking Ahead

1. ☐ Complete the Prep Skills for Lesson 3-3.
2. ☐ Read the opening paragraph in Lesson 3-3 carefully and answer Question 0 in preparation for that lesson.

Answers to “Did You Get It?”

1. $0.6x$ **2.** $D = 0.6x$ **3.** $C = 750 + 0.43x$

4. $2.06x - 750$ **5.** They’d lose \$235.

Answers to “Prep Skills”

1. a. $11x - 12$ **b.** $12y - 4$ **c.** $18a - 8b - 13$

2. The slope is about 150; this means the value of the antique painting increases \$150 every year after being found in the attic.

3. The y intercept is about (0, 11,000). This means the painting was worth \$11,000 when it was discovered.

4. a. $K = 308.15$ **b.** $A = 350$ **c.** $C = 5{,}111.25$

3-2 Applications

Name ______________________________

Unfortunately for the snow cone business we studied within the lesson, they're not the only game in town when it comes to frosty treats. They compete fiercely with Frosty the Cone Man, a nearby ice cream stand. The table describes the profit that Frosty can make based on the number of cones he sells. We'll be using this information throughout the Applications.

Cones sold	Profit
0	–$375
200	–$181
400	$13
600	$207
800	$401
1,000	$595

1. Use the grid to graph the points from our table. Put cones sold on the horizontal axis and profit on the vertical axis.

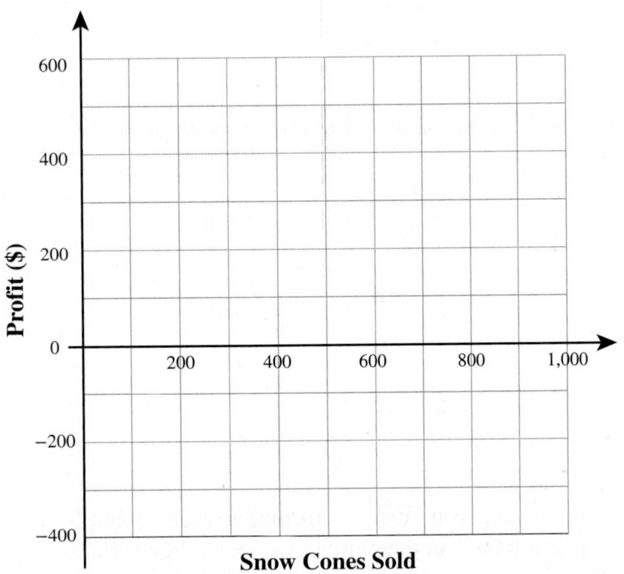

2. Based on the plotted points, does the graph appear to be a line? If so, draw it in.

3. Find the *y* intercept of your graph, and describe what it tells us about Frosty's business.

3-2 Applications

Name ______________________________

4. Estimate the x intercept based on your graph, and describe what it tells us about the business.

5. Find the slope of the line using two of the points from the table.

6. Notice that in calculating the slope, you subtracted two profits in the numerator, and two numbers of cones sold in the denominator. Based on this, what are the units for your slope? What does it tell us about the business?

7. Based on your answer to Question 6, how much would the profit go up if they sell 10 cones? Twenty cones?

8. How much would the profit go up if they sell x cones?

9. Now let's put the pieces together. Using your answers from Questions 3 and 8, write a formula in the form $P =$ ________ that describes the company's profit based on the number of cones they sell.

10. Remember that festival where the snow cone business sold 8,000 units? How much profit would Frosty the Cone Man make if he sells 10,500 cones at that festival?

Lesson 3-3 Prep Skills

SKILL 1: ADD OR SUBTRACT LIKE TERMS

Just in case you missed this in the last lesson. . . . In Lesson 1-2, we saw that quantities can only be added if they're like quantities, which means they're identical except possibly for the numeric part. For example, $8n$ and $12n$ are like quantities because they're both just n without the 8 and the 12, while $8n$ and $12n^2$ are not like quantities because n and n^2 aren't identical.

The individual pieces of an expression containing addition and subtractions are called **terms**. In this case, we use the phrase *like terms* rather than like quantities. Deciding which terms can be added or subtracted is an important part of simplifying algebraic expressions.

- $10k + 8 - 3k + 4$ $= 7k + 12$ — $10k$ and $-3k$ are like terms; 8 and 4 are like terms
- $4n - 7n^2 + 11n + 7 - 4n^2$ $= 15n - 11n^2 + 7$ — $4n$ and $11n$ are like terms; $-7n^2$ and $-4n^2$ are like terms

SKILL 2: EVALUATE EXPRESSIONS

Hey, here's another one we just covered in the last lesson. Good deal. When we use an algebraic expression to model some quantity, it's very common to calculate the value of that quantity by replacing an input variable with some number. This is what we called *evaluating an expression*. The only thing you need to think about is replacing every single occurrence of the variable(s) with the given number(s).

- For the future value formula $A = P(1 + rt)$, if we want to know the future value of a \$5,000 initial investment at 6% interest for 5 years, we're given values of $P = 5{,}000$, $r = 0.06$, and $t = 5$. So we replace each letter in the formula with the associated value:

  $A = \$5{,}000(1 + 0.06(5)) = \$6{,}500$

SKILL 3: WRITE A LINEAR EQUATION GIVEN A BEGINNING VALUE AND RATE OF CHANGE

This was one of the most important skills we practiced in Lesson 3-2. If some quantity begins at some value when the input is zero and has a *constant* rate of change, then we can write a linear equation that models that quantity. In every case, the equation will be of the form

Output = initial value + rate of change(input variable)

- If you start a job making \$11 per hour and get a raise of \$1.25 per hour every year, a formula for your hourly wage (W) in terms of years after you started is

  $W = 11 + 1.25t$

  Beginning value (11) Rate of change (1.25) Input variable (t)

PREP SKILLS QUESTIONS

1. Simplify each expression by combining like terms.

   a. $8t + 7 - 3t - 4$
   b. $14x + 12 - (7 - x)$
   c. $5(10 - 4x) - 5(2x + 2)$

2. Evaluate each expression for the given value of the variable(s).

   a. $V = 21{,}000 - 1{,}700t$; $t = 4$
   b. $C = 4.25 + 0.9x$; $x = 12$
   c. $P = 2l + 2w$; $l = 35$, $w = 50$

3. Marlene rented a new apartment for $1,400 a month, and was told that the rent increases by $20 per year if she decides to stay. Write an equation that describes the rent (R) in terms of years after she moves in (y).

4. A stock that has been tanking lately was at $42 per share a while ago, and has been losing value at the rate of $1.75 per week since then. Write an equation in the form $V =$ ______ that describes the value of the stock in terms of w, the number of weeks since it was at $42.

Lesson 3-3 All Things Being Equal

LEARNING OBJECTIVES

- ☐ 1. Explain what it means to solve an equation.
- ☐ 2. Demonstrate the procedures for solving a basic linear equation.
- ☐ 3. Solve a literal equation for a designated variable.

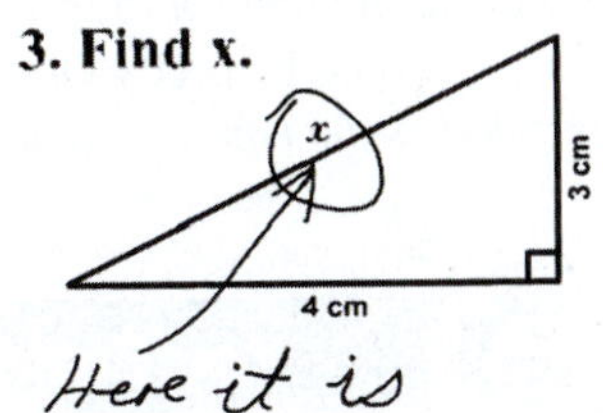

Thinking is like loving and dying. Each of us must do it for himself.
—Josiah Royce, American philosopher

What is an equation? If your answer is "a thing you solve to find x," you're (a) not alone, and (b) not correct. Of all the things that contribute to students struggling in algebra, probably the biggest is a simple lack of understanding of what the important words and procedures actually mean. As the famous graphic reproduced here illustrates, not understanding what the words mean can lead to some embarrassing answers. (By the way, if you don't think that's funny, you REALLY need this lesson.) In it, we'll review basic equation-solving techniques that you probably feel like you know. In fact, you might even feel like you're expert at them. Try not to focus so much on the *how* — this lesson is about understanding the *what* and *why* of solving equations.

0. What is your current view of what it means to solve an equation?

3-3 Class

Consider the following four statements:

- New York City is the capital of Guatemala.
- Over a million distinct species of animals have been cataloged and named.
- $8 = 5$
- $13 - 10 = 3$

The first and third statements are false, while the second and fourth are true. As you probably noticed, the connection between the last two statements is that each states that two things are equal. This simple exercise illustrates the two key ideas behind understanding equations.

Key Equation Ideas

1. An **equation** is simply a statement that two quantities are equal.
2. Like verbal statements, equations can be either true or false.

1. The statement below contains a variable quantity. Is it always true or always false? Explain.

The number of hours you'll spend on homework this week is 14.

An equation written to represent a statement like this is called a **conditional equation**, and is the type we use most commonly in areas of math (and other fields of study) that use algebra.

If we use the letter h to represent the number of hours you spend on homework, we can write our equation as $h = 14$. In this case, it should be pretty clear that if the value of h is 14, the equation is true, and otherwise it's false. So *14 is the only number that makes the equation true,* which brings us to some important definitions.

Key Equation Definitions

1. A number that makes an equation a true statement when substituted in for the variable is called a **solution** of an equation.
2. The set of ALL numbers that make an equation true is called **THE solution** of an equation.

When an equation looks like $h = 14$, the solution is pretty obvious. That's good. But when an equation looks like $-12y + 17 = 2y - 11$ . . . not so much. Our goal in solving equations is to develop a systematic approach to turning an equation whose solution isn't obvious into one with an obvious solution. But that only helps if we keep this next key idea in mind, which in some sense is the most important thing in this lesson:

The Process of Solving Equations

When changing the form of an equation in an attempt to find the solution, anything you do to the equation should not affect what makes the equation true or false.

2. The equation $5 = 5$ is obviously true, while the equation $5 = 10$ is false. Multiply BOTH SIDES of each equation by 2. What can you conclude?

Here's what you should have learned from Question 2: multiplying both sides of the equation by the same number (2 in this case) doesn't change when the equation is true and false, which means we can do that to a conditional equation *without changing the solution.* And ultimately, THAT is what this lesson is about.

3. We know the equation 1 ft = 12 in. is a true statement. So it's also true that 3 ft = 36 in. Explain why.

4. The equation 1 mi = 5,280 ft is true, while the equation 1 yd = 1 m is false. Multiply both sides of each equation by zero. Explain why multiplying both sides of a conditional equation by zero is a bad idea when trying to find the solution.

Math Note

When two equations have the same solution, we call them **equivalent equations**. So our goal in solving equations is to transform the original equation into an equivalent equation with an obvious solution.

5. Fill in the blank to complete a description of our first useful tool for solving equations, then use it to solve the equation, and check your answer.

Equation Solving Tool #1: Multiplication

We can ____________________ both sides of an equation by the same number or expression as long as that number or expression isn't equal to zero.

Example: Use your answer above to solve the equation $\frac{d}{360} = 0.35$.

Now check your answer by seeing if it makes the original equation a true statement.

6. Is the solution you found in the colored box the only solution to the equation? How can you tell?

Did You Get It

Try this problem to see if you understand the concepts we just studied. The answer can be found at the end of the Portfolio section.

1. Solve the equation $\frac{y}{8} = -12$.

7. We know that the equation 1 min = 60 sec is true. So is it also true that 1/2 min = 30 sec? What did we do to both sides to produce an equivalent equation?

8. Use the result of Question 7 to fill in the blank, completing a description of our next tool for solving equations. Then use it to solve the equation, and check your answer.

Equation Solving Tool #2: Division

We can __________ both sides of an equation by the same number or expression as long as that number or expression isn't equal to zero.

Example: Use your answer above to solve the equation $5x = 60$.

Now check your solution by seeing if it makes the original equation true.

An equation stating that two ratios are equal is called a **proportion**. The equation we solved in Question 5 can be written as a proportion if we write the percentage (35%) in fraction rather than decimal form:

$$\frac{d}{360} = \frac{35}{100}$$

A procedure called "cross multiplying" is often used to solve proportions: In this case, you'd multiply down one diagonal, giving you $100 \cdot d$, and multiply up the other diagonal, giving you $360 \cdot 35$. (This is shown in the diagram.) Setting those two results equal to each other, the resulting equation is $100d = 360 \cdot 35$, which you can solve using Tool #2. But why does this work?

$$\frac{d}{360} \times \frac{35}{100}$$

9. Use Tool #1 to multiply both sides of the original proportion by the number $100 \cdot 360$, and do any obvious reducing of fractions. What do you notice about the result?

10. Use cross-multiplying to solve the proportion $\frac{x}{12} = \frac{15}{36}$.

Did You Get It ?

2. Solve the equation $\frac{10}{y} = \frac{35}{81}$.

Now let's work on building our next equation-solving tool.

11. We know that 1 min = 60 sec. Is it okay to add the number 3 to both sides of the equation, resulting in 4 min = 63 sec? Why or why not?

12. Is it okay to add 3 seconds to both sides of the equation, resulting in 1 min 3 sec = 63 sec? Why or why not?

13. Use the result of Question 12 to fill in the blank, completing a description of our next tool for solving equations. Then use it to solve the equation, and check your answer.

Equation Solving Tool #3: Addition

We can ________ the same quantity or expression to both sides of an equation.

Example: Use your answer above to solve the equation $x - 142 = 70$.

Now check your solution by seeing if it makes the original equation true.

Did You Get It

3. Solve the equation $230 = 63 - x$.

14. Given the fact that 36 in. = 3 ft, subtract 5 in. from each side of the equation. Is the resulting equation still true?

15. Use the result of Question 14 to fill in the blank, completing a description of our last tool for solving equations. Then use it to solve the equation, and check your answer.

Equation Solving Tool #4: Subtraction

We can ________ the same quantity or expression from both sides of an equation.

Example: Use your answer above to solve the equation $n + 1{,}361 = 2{,}094$.

Now check your solution by seeing if it makes the original equation true.

The types of equations that we've been solving in this lesson are called **linear equations,** because we've seen that the graph of an equation with the input appearing only to the first power is a line. Here's a review of the steps we've discovered, along with some bonus material.

Summary of the Steps for Solving Linear Equations

Step 1: Simplify both sides of the equation, if necessary, by multiplying out any parentheses and combining any like terms.

Step 2: Use Tools #3 and/or #4 to rearrange the equation so that the term containing the variable is isolated on one side.

Step 3: Use Tool #1 or #2 to change the coefficient of the variable to 1.

Note: If the equation you're trying to solve contains any fractions, like the one in Question 9, the FIRST thing you should do is multiply both sides by a common denominator. This will eliminate any fractions.

Did You Get It

4. Solve the equation $4(x - 3) = 10 + x$.

3-3 Group

In Lesson 3-2, we were provided with a graph describing the profit made by a snow cone business in terms of the number of snow cones sold, which is reproduced here. We used the graph to estimate the net profit if 150 snow cones were sold, and also to estimate the number that we'd need to sell to make a profit of $50.

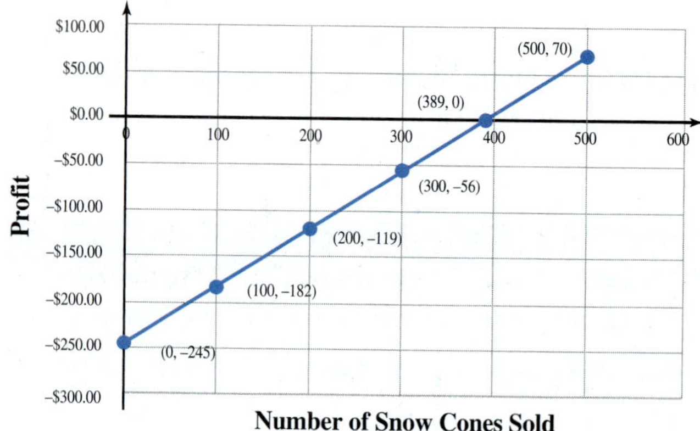

And using the graph was just fine, except for that pesky "estimate" thing. What if we want exact answers instead? In that case, we'd need to have an equation representing the profit, so it's a darn good thing that we found one in that lesson: $P = -245 + 0.63x$, where P is the profit, and x is the number of snow cones sold.

1. Given the fact that x represents the number of snow cones sold, what should we do to find the profit made from selling 150 snow cones? Describe what you'd do, then find that profit.

2. How does this compare to the estimate you found in the previous lesson?

3. Given that P represents the profit, what should we do to find the number of snow cones needed to make a profit of \$50? What happens when you do this? Why is this question in this lesson?

4. Now you should be able to solve an equation to find the number of snow cones needed to make a profit of \$50. If your solution requires rounding, think about what it represents before deciding on a final answer.

5. Check your answer by plugging it back into the equation you wrote. Does it make the equation exactly true? What can you conclude about the company making \$50?

6. Use the equation to find the profit if 700 snow cones are sold.

7. Use the equation to find the number of snow cones they'd need to sell to make a profit of \$400.

8. Look very carefully at the steps you did to solve the equations in Questions 4 and 7. Write those steps in words here.

9. Start with $P = -245 + 0.63x$. Perform the exact same steps you described in Question 8 on this equation to solve for x. Be prepared for the fact that you won't get a number this time as your solution: You'll get an expression involving P.

What you just did is solve a **literal equation.** This is an equation with more than one variable, so that when you solve for one of the variables, the result is an expression rather than a number. In this case, we were given an equation that told us P for a given value of x. We turned it into an equation that tells us x for a given value of P. Why would we do that? Let's see.

10. Suppose we were interested in using this spreadsheet to calculate the number of snow cone sales that would be necessary to make certain levels of profit. What formula could you type into cell B2 to calculate the number of snow cones needed to break even (that is, make a profit of \$0)? Your answer to Question 9 should be quite helpful.

	A	B
1	P	X
2	\$0	
3	\$250	
4	\$500	
5	\$1,000	

11. Use a calculator or spreadsheet to complete the table, then describe why solving the equation for x was helpful.

Earlier in the course, we worked with two equations for converting temperatures between Fahrenheit and Celsius:

$F = \frac{9}{5}C + 32$ and $C = \frac{5}{9}(F - 32)$

Let's start with the formula that calculates F in terms of C, and see if we can solve for C.

12. First, subtract ______ from both sides. Fill in the blank, then perform that step on the equation.

13. Now multiply both sides by ______, which is the reciprocal of the fraction in the equation.

14. What do you notice?

Solving a Literal Equation

1. Pretend that all variables other than the one you're solving for are just numbers, not letters.
2. Use normal equation-solving procedures to solve for the desired variable.

Did You Get It

5. Solve the equation $A = P + Prt$ for t.

3-3 Portfolio

Name ____

Check each box when you've completed the task. Remember that your instructor will want you to turn in the portfolio pages you create.

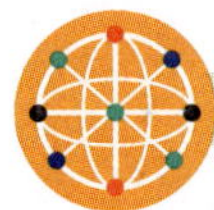

Technology

1. ☐ You'll need to complete the first few questions in the Applications before attacking this Tech assignment. One of the real advantages of being able to solve literal equations is that you can rearrange a formula to find the variable you're interested in calculating, rather than whatever variable happens to be isolated in the original formula. In the Applications, we'll revisit the cab problem from Lesson 3-1, and work with a formula that provides the number of miles that can be driven for a certain cost. Your job is to make a spreadsheet that calculates the trip length for every dollar amount from \$5 up to \$100, then draw a graph of your result using the connected scatter plot command. There's a template in the online resources for this lesson to help you get started.

Online Practice

1. ☐ Include any written work from the online assignment along with any notes or questions about this lesson's content.

Applications

1. ☐ Complete the Applications problems.

Reflections

Type a short answer to each question.

1. ☐ Explain in your own words what it means to solve an equation. If your answer doesn't use the word "true," it probably stinks.
2. ☐ Describe the four equation-solving tools we covered in this section, and what they all have in common.
3. ☐ What is a literal equation? What's the point of solving them?
4. ☐ Name one thing you learned or discovered in this lesson that you found particularly interesting.
5. ☐ What questions do you have about this lesson?

Looking Ahead

1. ☐ Complete the Prep Skills for Lesson 3-4.
2. ☐ Read the opening paragraph in Lesson 3-4 carefully and answer Question 0 in preparation for that lesson.

Answers to "Did You Get It?"

1. $y = -96$ **2.** $y = \frac{162}{7}$ **3.** $x = -167$ **4.** $x = \frac{22}{3}$ **5.** $t = \frac{A - P}{Pr}$

Answers to "Prep Skills"

1. a. $5t + 3$ **b.** $15x + 5$ **c.** $-30x + 40$

2. a. $V = 14{,}200$ **b.** $C = 15.05$ **c.** $P = 170$

3. $R = 1{,}400 + 20y$

4. $V = 42 - 1.75w$

3-3 Applications

Name ______________________________

In Lesson 3-1, we studied a cab ride by looking at the price for certain distances in table and graph form. We found that the initial cost of starting the trip was \$5.10, which means that you'd pay \$5.10 for zero miles traveled. We also found that the slope of the line was \$2.60, which means that you'd pay \$2.60 per mile. Using what we learned in Lesson 3-2, we can write a formula that describes the cost of a trip (C) in terms of miles traveled (m): $C = 5.1 + 2.6m$.

1. How much would it cost for a 17-mile trip to the airport?

2. If you have budgeted \$18 for a cab ride to tour the downtown area, how far can you go? Set up and solve an equation.

3. Here's a smaller version of the graph describing cab rides from Lesson 3-1. Does your answer to Question 2 match the information on the graph? Draw an arrow to the location on the graph that you used to decide.

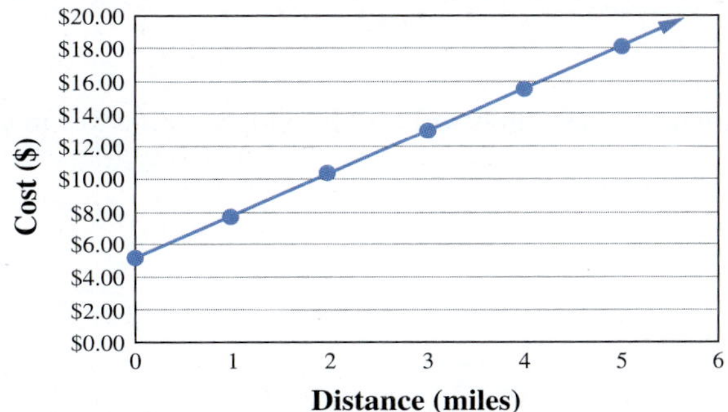

4. Solve the equation for m. What is the significance of the result?

3-3 Applications

Name ______________________________

5. Use your answer to Question 4 and a calculator or spreadsheet to fill in the table. Round to the nearest tenth of a mile.

Cost of trip	Miles travelled
$8	
$10	
$15	
$25	

In Lesson 3-1, we also studied how the value of a car depreciates after it's purchased. The original value of the car (zero years after purchased) was $15,000, and we found that the value decreased by $2,000 each year.

6. Write a formula in the form $V =$ ______ that describes the value of the car in terms of the number of years after it was purchased (t).

7. How much was the car worth 6 years after it was purchased?

8. When was the car worth half of what it was purchased for? Provide your answer in years and months.

9. Here's a smaller version of the graph describing the value of the car from Lesson 3-1. Does your answer to Question 2 match the information on the graph? Draw an arrow to the location on the graph that you used to decide.

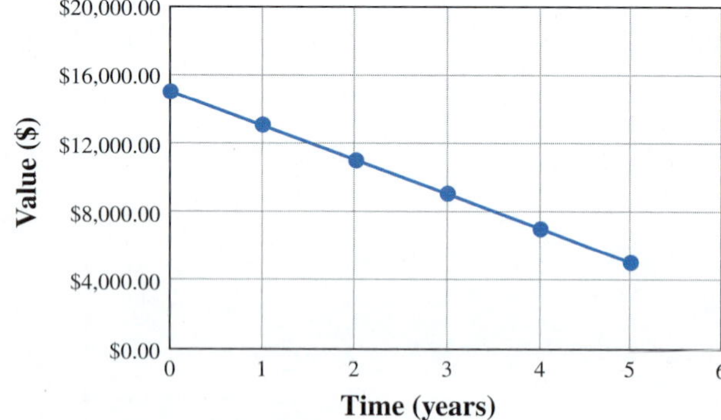

3-3 Applications

Name ______________________________

10. How long does it take the car to have no value?

11. Find an equation that provides the number of years since the car was purchased in terms of its value.

12. The formula $P = 2l + 2w$ calculates the perimeter of a rectangle (P) when you know the length (l) and width (w). Solve this literal equation for the width.

13. Let's examine the relationship between the perimeter of a rectangle and its area. If the perimeter is 200, fill in this table using your answer to Question 12 and a calculator or spreadsheet.

Perimeter	Length	Width	Area
200	10		
200	30		
200	40		
200	50		
200	80		

14. What conclusions can you draw about the relationship between perimeter and area for a rectangle?

Lesson 3-4 Prep Skills

SKILL 1: WORK WITH INEQUALITY NOTATION

Inequality notation in algebra is typically used to represent an interval of numbers. For example, the set of numbers that are more than 10 is an interval: We could represent that in inequality notation by $x > 10$. (Of course, the choice of specific letter is immaterial: $y > 10$ describes the same interval.) There are basically two skills you need to be comfortable with: interpreting the meaning of an expression in inequality notation, and the opposite, writing an expression in inequality notation when an interval is described.

- The set of all numbers less than or equal to −5 can be written as $x \leq -5$.
- The set of numbers that are between 6 and 12 can be written as $6 < x < 12$. This is called a **compound inequality**, and is always used to describe an interval between two specific numbers. You should NEVER write a compound inequality with both a $<$ and a $>$ symbol. To be on the safe side, always write the smallest number first, then $<$, then your choice of variable, then another $<$, then the largest number in the interval.
- The expression $t \geq 8$ represents the set of all real numbers that are 8 or greater.
- The expression $y < 4$ represents the set of all real numbers that are less than 4.
- The expression $-3 \leq x < 2$ represents the set of all real numbers between −3 and 2 including −3.

SKILL 2: RECOGNIZE TERMS THAT INDICATE AN INEQUALITY

The following are some of the terms commonly used to describe inequalities.

- Used to represent:

  $>$: greater than, more than

  $<$: less than, not as much as

  $\geq$: greater than or equal to, at least, no less than, a minimum of

  $\leq$: less than or equal to, at most, no more than, a maximum of

SKILL 3: SIMPLIFY ALGEBRAIC EXPRESSIONS

We've already reviewed adding and subtracting like terms in the Prep Skills for both Lessons 3-2 and 3-3. Sometimes we'll also need to use the distributive property (which was reviewed in the Prep Skills for Lesson 1-4). Combining the two, and using order of operations, we can develop a general strategy for simplifying algebraic expressions.

1. Use the distributive property to multiply out any parentheses.
2. Combine any like terms using addition or subtraction.

   - $2(x + 4) - 5x$ Distribute 2 through $(x + 4)$

     $= 2x + 8 - 5x$ Subtract $2x - 5x$

     $= -3x + 8$

   - $4y + 10 - (3y + 5)$ Distribute −1 through $(3y + 5)$

     $= 4y + 10 - 3y - 5$ Subtract $4y - 3y$; subtract $10 - 5$

     $= y + 5$

SKILL 4: SOLVE BASIC LINEAR EQUATIONS USING EQUATION-SOLVING TOOLS

You just completed an entire lesson based on solving linear equations, so consider the Prep Skills Questions here to be an opportunity to get a little more practice.

PREP SKILLS QUESTIONS

1. Write an inequality to represent each set of numbers described.

   a. All real numbers that are at least 17

   b. The numbers between −5 and 5, along with 5

   c. A group of test scores that are no more than 90

   d. The positive numbers that are a maximum of 32

   e. The variable quantity represented by y is at most $\frac{3}{2}$.

2. Write a verbal description of the interval described by each inequality.

   a. $t \leq 3$ b. $x > -10$ c. $300 < P < 500$ d. $0 \leq k < 9$

3. Simplify each expression as much as possible.

   a. $4(2t - 11) + 3t$

   b. $8y + 4 - 4(7 - 3y)$

   c. $5(2x + 1) - (5x + 3) + 2x + 10$

4. Find all solutions to each equation.

   a. $x + 2 = 5$

   b. $y - 3 = -10$

   c. $5x = 35$

   d. $\frac{P}{365} = 10{,}000$

Lesson 3-4 All Quantities Are Not Created Equal

LEARNING OBJECTIVES

- ☐ 1. Demonstrate the procedures for solving a linear inequality.
- ☐ 2. Solve application problems that involve linear inequalities.

©Stockbyte/Getty Images RF

We hold these truths to be self-evident: that all men are created equal . . ."
—The Declaration of Independence

The lesson quote above is one of the fundamental bedrocks that our nation was built upon. Society isn't perfect, and we can all work harder to live up to that ideal. But at the end of the day, we as a society believe that all people are equal in the eyes of the law. Notice that the Declaration of Independence didn't say anything about quantities, though. Equality certainly plays a key role in algebra, but what about inequality?

We've seen that solving linear equations is a very useful tool in solving problems from a wide variety of areas. But think about the following situation: You've finished school and are looking for a real job. As you nervously wait for your big interview, you decide not to be too demanding, but that the minimum compensation you're willing to accept in salary is \$40,000 per year. So "Acceptable salary = \$40,000" seems like an adequate description of your standards. But nobody has ever gone into an interview and said, "I want 40k and I won't accept a penny more!" Of course, you would be perfectly happy with any amount over \$40,000 as well. This is when using a statement like "Acceptable salary ≥ \$40,000" is much more reasonable. And this is just one of many modeling situations where inequality is a *good* thing.

0. Think of some of the scenarios we've modeled in this unit so far. Describe at least one where an inequality would be a more realistic model than an equation.

Remember the snow cone business from Lesson 3-2? Of course you do. That was like two days ago. In one of the questions in Lesson 3-3, we found the number of snow cones they'd need to sell in order to make a profit of exactly \$400. But wouldn't it make more sense to find the number they'd need to sell to make AT LEAST \$400?

A statement that one quantity is more or less than another is called an **inequality**. The inequality $2 < 5$ is the true statement that two is less than five. The inequality $x > 3$ is a statement that the variable quantity represented by x is more than 3. Just like with equations, **the solution to an inequality is the set of all numbers that make the inequality a true statement when substituted in for the variable**.

1. Name a number that makes $x > 3$ true.

2. Name a number that makes $x > 3$ false.

In many ways, solving inequalities is similar to solving equations. For example, the first step we identified in solving equations was to use the distributive property and combine like terms to simplify each side of the equation as much as possible. The same is true for inequalities: Surely most folks would prefer to work with a simpler inequality rather than a more complicated one.

After doing that, we'll use tools similar to the ones we learned for equations to rearrange inequalities, with the goal of isolating the variable we're solving for.

First, here's a summary of the tools we can use to solve equations:

#1. Multiply the same nonzero number or expression on both sides.
#2. Divide both sides by the same nonzero number or expression.
#3. Add the same number or expression to both sides.
#4. Subtract the same number or expression from both sides.

Do these tools work for solving inequalities too? Let's see. (Remember, by "work," we mean that we can do these procedures to an inequality without changing whether it's true or false.) We'll use less than (<) to explore new rules, but the rules will also apply to greater than (>), less than or equal to (≤), and greater than or equal to (≥). In Questions 1–4, fill in the comparison symbol that makes the inequality true after performing the operation.

1. Add 8 to both sides.

Left side	Comparison symbol	Right side
10	<	20

2. Add −8 to both sides.

Left side	Comparison symbol	Right side
10	<	20

3. Subtract 6 from both sides.

Left side	Comparison symbol	Right side
10	<	20

4. Subtract −6 from both sides. (Be careful!)

Left side	Comparison symbol	Right side
10	<	20

5. Fill in either "does" or "does not" to complete our first tool for solving inequalities. Then follow the remaining instructions in the box.

Inequality Solving Tool #1

Adding or subtracting the same number on both sides of an inequality ________________ change the direction of the comparison symbol.

Use inequality Tool #1 to solve each inequality. Then graph the solution set on the number line. Finally, choose any number in the solution set and plug it back into the inequality to check that it makes the inequality true. (Note: When graphing a set of numbers on a number line, include a filled-in circle at an endpoint that IS included in the set, and an empty circle if the point is NOT included.)

$x - 5 > 9$

Check:

$t + 11 \leq 5$

Check:

6. Write a verbal description of what the solution sets to the two inequalities in the colored box really tell us. The first has been done as an example.

Any value of x greater than 14 will cause $x - 5$ to be greater than 9.

Math Note

The inequality $3 > x$ says exactly the same thing as $x < 3$. In each case, the quantity that's less is on the closed end of the comparison symbol. But we'll almost always write $x < 3$, with the variable on the left.

Did You Get It

Try this problem to see if you understand the concepts we just studied. The answer can be found at the end of the Portfolio section.

1. Solve each inequality.
   a. $x - 32 < 11$
   b. $y + 6 \geq -2$

7. Multiply both sides by 2.

Left side	Comparison symbol	Right side
10	<	20

8. Multiply both sides by −2.

Left side	Comparison symbol	Right side
10	<	20

9. Divide both sides by 5.

Left side	Comparison symbol	Right side
10	<	20

10. Divide both sides by −5.

Left side	Comparison symbol	Right side
10	<	20

11. Write "stays the same" or "changes" in each blank to complete our second tool for solving inequalities. Then follow the remaining instructions in the box.

Inequality Solving Tool #2

If you multiply or divide both sides of an inequality by a positive number, the direction of the comparison symbol ________________. If you multiply or divide both sides by a negative number, the direction of the comparison symbol ________________.

Use inequality Tool #2 to solve each inequality. Then graph the solution set and choose one number from the solution set and check to see that it makes the inequality true.

$2x < 10$

Check:

$-3z \geq 12$

Check:

$\frac{1}{4}t > -3$

Check:

$-\frac{1}{3}b \leq 2$

Check:

Did You Get It

2. Solve each inequality.

a. $\frac{2}{3}x < -12$ b. $-4a \geq 20$

The types of inequalities that we've been solving in this lesson are called linear inequalities, because we've seen that the graph of an equation with the input appearing only to the first power is a line. Here's a review of the steps we've discovered, along with some bonus material.

Summary of the Steps for Solving Linear Inequalities

Step 1: Simplify both sides of the inequality, if necessary, by multiplying out any parentheses and combining any like terms.

Step 2: Use Tool #1 to rearrange the inequality so that the term containing the variable is isolated on one side.

Step 3: Use Tool #2 to change the coefficient of the variable to 1. Make sure that you change the direction of the inequality if you multiply or divide by a negative number.

Note: If the inequality you're trying to solve contains any fractions, the FIRST thing you should do is multiply both sides by a common denominator. This will eliminate any fractions.

Did You Get It

3. Solve the inequality $-2(x - 5) \geq x - 8 - 5x$.

3-4 Group

Now let's get back to our good friends in the snow cone business. We were able to model the company's profit using the equation $P = -245 + 0.63x$, where x is the number of snow cones sold. Let's say that the goal is to make a profit of at least \$100 in a day.

1. Write an inequality that states in symbols that the profit for this business will be at least \$100.

2. Solve this inequality.

3. Graph the solution on the number line.

4. Show that one value from the solution set checks.

5. What does the solution to the inequality tell us about the company?

6. Write a verbal description of what this inequality represents in the context of the snow cone business: $-245 + 0.63x < 0$.

7. Solve this inequality.

8. Graph the solution on the number line.

9. Show that one value from the solution set checks.

10. What exactly does the solution to the inequality tell us about the business?

Did You Get It

4. Find the number of snow cones the company will need to sell in order to make more than $600 in profit.

The temperature inside a room housing computer servers is important to proper function. A Google search for "server room temperature" yields the following (among other results): According to OpenXtra, server room temperatures should not dip below 50 degrees Fahrenheit, and should not exceed 82 degrees Fahrenheit. The optimal temperature range is between 68 and 71 degrees Fahrenheit.

11. Write an inequality to express the statement "server room temperatures should not dip below 50 degrees Fahrenheit" mathematically using the letter F to represent the temperature.

12. In Lesson 2-4, we learned that Fahrenheit and Celsius temperatures are related by the equation $F = \frac{9}{5}C + 32$. Use this to turn your inequality from Question 11 into an inequality with variable C.

13. Solve this inequality, and describe what the solution tells us.

14. Write an inequality to express the statement "and should not exceed 82 degrees Fahrenheit" mathematically using F.

15. Use the Celsius conversion formula to turn your inequality into one involving variable C.

16. Solve this inequality, and describe what the solution tells us.

17. Putting your answers to Questions 13 and 16 together, write a single compound inequality to describe the allowable server room temperatures.

18. Write a single compound inequality to describe the optimal server room temperatures in degrees Fahrenheit.

19. Use the conversion formula to rewrite your compound inequality from Question 18 with variable *C*. Then solve this inequality to determine the optimal server room temperatures in degrees Celsius. Does this match your answer to Question 17? (Note: To solve a compound inequality, use the same steps you've been using: Just do the same thing to all three sides.)

3-4 Portfolio

Name ______________________________

Check each box when you've completed the task. Remember that your instructor will want you to turn in the portfolio pages you create.

Online Practice

1. ☐ Include any written work from the online assignment along with any notes or questions about this lesson's content.

Applications

1. ☐ Complete the Applications problems.

Reflections

Type a short answer to each question.

1. ☐ Explain why inequalities are often a more realistic way to model situations than equations.
2. ☐ How does our procedure for solving linear inequalities differ from our procedure for solving linear equations?
3. ☐ What do the solution sets for inequalities look like? How do they compare to the solution sets for equations?
4. ☐ Name one thing you learned or discovered in this lesson that you found particularly interesting.
5. ☐ What questions do you have about this lesson?

Looking Ahead

1. ☐ Complete the Prep Skills for Lesson 3-5.
2. ☐ Read the opening paragraph in Lesson 3-5 carefully and answer Question 0 in preparation for that lesson.

Answers to "Did You Get It?"

1. **a.** $x < 43$ **b.** $y \geq -8$
2. **a.** $x < -9$ **b.** $a \leq -5$
3. $x \geq -9$
4. More than 1,341

Answers to "Prep Skills"

1. **a.** $x \geq 17$ **b.** $-5 < x \leq 5$ **c.** $0 \leq t \leq 90$ **d.** $0 < x \leq 32$ **e.** $y \leq \frac{3}{2}$
2. **a.** All real numbers that are at most 3
   b. All real numbers that are greater than –10
   c. All real numbers between 300 and 500
   d. All real numbers between 0 and 9, along with 0
3. **a.** $11t - 44$ **b.** $20y - 24$ **c.** $7x + 12$
4. **a.** $x = 3$ **b.** $y = -7$ **c.** $x = 7$ **d.** $P = 3{,}650{,}000$

3-4 Applications

Name ______________________________

Much of this assignment follows up on the work we did in the Applications for Lesson 3-3. In that lesson, we were solving equations to learn more about situations we had previously modeled. In this lesson, we'll use inequalities to learn a bit more about those situations.

In Lesson 3-1, we used the equation $C = 5.1 + 2.6m$ to model the cost of a cab ride (C) in terms of miles traveled (m).

1. What does the inequality $5.1 + 2.6m \leq 20$ represent in this scenario?

2. Solve the inequality and graph the solution on a number line.

3. Write a sentence describing what your solution tells us.

4. Explain why an inequality is a more natural choice than an equation for this situation.

5. Suppose that you're running some errands for your boss, and she says, "I imagine you'll need to spend at least 30 bucks on cab fare. Don't spend more than 50, though." Write an inequality that matches these conditions. (Think about whether to use $<$ or $\leq$.)

6. Solve the inequality you wrote in Question 4.

7. Describe what information is provided by your solution.

3-4 Applications

Name ____

8. Suppose that you have a maximum of $50 to spend on cab fare, and you need to make a round trip, both to and from a certain location. Which of the following inequalities best describes this situation?

   a. $5.1 + 2.6x \leq 2 \cdot 50$

   b. $10.2 + 2.6x \leq 50$

   c. $2(5.1 + 2.6x) \leq 50$

   d. $\frac{5.1 + 2.6x}{2} \leq 50$

9. Solve the inequality you chose and write an interpretation of the solution.

10. In Lesson 3-1, we also studied the depreciation in the value of a car, which was modeled by the equation $V = 15{,}000 - 2{,}000t$. The owner of the vehicle decides he'll hang onto it at least until it's worth less than $2,000. Write and solve an inequality that describes how long he'll keep the vehicle, and write a sentence that describes your solution.

Lesson 3-5 Prep Skills

SKILL 1: TRANSLATE STATEMENTS INTO SYMBOLS

In the last few lessons, we've seen that when using algebra to model data or solve problems, it's important to recognize what can change (or vary) in a situation, and represent those quantities (called variables) with letters. Once you've done that, the key skill becomes writing verbal statements regarding those variable quantities in symbols.

- If you work a part-time job making $9 per hour, the number of hours you work per week might vary. If you represent the hours using h, then $9h$ describes your pay for working h hours.
- If a rectangular parking lot is being designed to be 120 feet long, but the width hasn't been decided, then the area is unknown. We could use w for the width and A for the area, then the equation $A = 120w$ describes the area.
- If you plan to study a certain number of hours this week, then 3 hours more than that next week because you have a test coming up, you can use N for the number of hours this week, and $N + 3$ for the number of hours next week.

SKILL 2: SOLVE LINEAR EQUATIONS

This has been the main skill we've been working on for the past two lessons, so you know where to go for reference on this skill. Above all else, remember the basics:

- Simplify both sides of an equation separately before trying to solve.
- When using the equation-solving tools to isolate the variable, make sure you do the same thing to BOTH SIDES of the equation.

SKILL 3: SOLVE FORMULAS FOR ONE OF THE VARIABLES

This was the third objective in Lesson 3-3, where we called it solving literal equations. This just means solving an equation with more than one variable for one of those variables. The procedure is exactly the same as the one used for solving equations with one variable: You just have to pretend that every letter *other than the one you're solving for* is just a number.

- $A = lw$ — To solve this formula for l, divide both sides by w.

  $l = \frac{A}{w}$

Why would anyone want to do this? Convenience. In its original form, the formula above tells you area when you know length and width. After solving for l, it's a formula that tells you the length if you know the area and the width. That's how we'll use this skill in the forthcoming lesson.

- $A = P(1 + rt)$ — To solve this formula for t, first simplify the right side.

  $A = P + Prt$ — Subtract P from both sides.

  $A - P = Prt$ — Divide both sides by Pr.

  $t = \frac{A - P}{Pr}$

SKILL 4: USE DIMENSIONAL ANALYSIS

We initially learned dimensional analysis as a means for converting units of measure, but it can actually be used for many types of calculations.

- How long would it take you to drive 130 miles if you average 48 miles per hour?

$$130\text{ mi} \times \frac{1\text{ hr}}{48\text{ mi}} \approx 2.7\text{ hr}$$

As always, the key is the units: We started with a number of miles and wanted a time, so we used a fraction that had miles in the denominator and hours in the numerator.

PREP SKILLS QUESTIONS

1. An account starts out with an initial deposit of \$400, and has \$30 added to it each week. Write an expression using variable w to describe the total deposits after w weeks.

2. The length of a rectangular climbing wall is 8 feet more than the height. Write two expressions, one describing the length and one the height. Make sure you describe what any letters you use represent.

3. Lindor truffles (my personal weakness) are on sale for \$12 per pound. Write an equation that describes the cost of buying truffles if you plan to spend \$20.

4. Solve each equation.

   a. $3x + 5 = 20$ b. $\frac{y-5}{3} = 8$ c. $4 = -2\,(5 - 2t)$

5. Solve each equation for the requested variable.

   a. $3x + y = k$ for y b. $\frac{y-n}{3} = W$ for n c. $z = -2(y - 2t)$ for t

6. Use dimensional analysis to solve each problem.

   a. Marlene makes \$14 per hour. How many hours would she have to work to make \$100?

   b. A groundskeeping crew can mow 12,000 square yards in an hour. How long would it take them to mow a 125,000-square-foot field?

Lesson 3-5 What's Your Problem?

©Brand X Pictures/PunchStock RF

LEARNING OBJECTIVES

- ☐ 1. Solve application problems using numerical calculations.
- ☐ 2. Solve application problems using linear equations.

Avoid problems, and you'll never be the one who overcame them.
—Richard Bach

It just occurred to me that snakes and word problems in math have a lot in common. Most people are afraid of them, and that fear is usually irrational. In each case, education is the key to overcoming your fear. While you may be able to avoid snakes, you can't avoid problems—they're an inevitable part of life. And there's a good reason that math problems are called "problems"! The skills and thought processes you practice when working your way through word problems will come in handy when dealing with other types of problems outside of the hallowed halls of academia. So in this lesson, we'll focus on problem-solving techniques.

0. What's the first thing you think of when you hear the phrase "word problem" in math class? Be honest.

3-5 Class

We were first exposed to Polya's procedure for solving problems in Lesson 2-5. This lesson is all about solving problems using the algebra that we've been practicing, so it seems like a good time to review the steps. If you want to review them in more detail, refer to the colored box at the beginning of Lesson 2-5.

A Review of Polya's Four-Step Procedure for Solving Problems

Step 1: Understand the problem.

Step 2: Devise a plan to solve the problem.

Step 3: Carry out your plan to solve the problem.

Step 4: Check your answer.

Now let's practice using Polya's procedure.

Here's the first problem we'll tackle: Jack and Diane are going to start saving for retirement. They deposit $1,000 in an account to start out. Then they plan to deposit $50 each month. How long will it take until they've deposited $4,500?

Step 1: Understand the problem.

1. What information is provided in the statement that you're likely to need? Write it all down.

2. What is it that the problem is asking you to find?

Step 2: Devise a plan

3. First let's plan to do this using arithmetic. How much is added each month? How much needs to be added total to get to $4,500?

Step 3: Execute the plan

4. Perform an arithmetic operation based on your answers to Question 3 to solve the problem.

Step 4: Check your answer

5. What would the total deposits be after 70 months? Perform a calculation to check.

Now let's redo steps 2 and 3, this time using algebra.

Step 2: Devise a plan

6. Use letter m to represent the quantity in this problem that can change, and describe what that quantity is. (With a variable, we can write an equation to describe the given information.)

Step 3: Execute the plan

7. If $50 is being deposited each month, how much will be deposited after m months? Use this to write an equation that describes the problem. Then solve the equation and write a solution to the problem.

When you develop a strategy to solve a specific problem, then adapt that strategy to a variety of related problems, that's called **generalizing** your approach. In this case, the essence of the problem is that we had a starting value of some quantity ($1,000 in this case) and were adding $50 to it at regular intervals.

8. Generalize this problem to develop a framework that would allow someone to input any starting amount and any monthly amount, and be able to quickly determine how long it would take to reach a specific savings goal. (Hint: Look very carefully at the equation you wrote in Question 7, and identify EXACTLY what every number or letter in your equation represents.)

Did You Get It

Try this problem to see if you understand the concepts we just studied. The answer can be found at the end of the Portfolio section.

1. A manufacturer of custom drinking mugs has a fixed setup cost of $1,000 per order plus a variable cost of $1.50 per mug. How many mugs were produced when the total cost was $2,800?

A useful reminder before moving on: The two most important elements of understanding the problem are the two things you should ALWAYS do when attacking a word problem:

Key First Steps in Attacking Word Problems

1. Write down any relevant information AS YOU COME TO IT, rather than reading the whole problem all at once. If you're not sure if something is relevant or not, write it down and decide later.
2. Identify AND WRITE DOWN exactly what it is the problem is asking you to find. After you've done that, carefully read the problem again to make sure you understand the context.

3-5 Group

In this activity, you'll be working on four problems of different types. We'll help you to follow Polya's general guidelines on the first. After that, you're on your own. I know you won't let me down.

1. Suppose that you have four people in your family, and you want to hang a picture of each family member. Aww, that's sweet. Anyhow, the wall you choose is 60 inches wide, and each of the picture frames is 8 inches wide. To make everything look just right, you want to make the wall space between the edges of each frame identical, and put that same amount of space between the ends of the wall and the edge of the nearest frame. Each frame has a single hook in the center of its back. Where should you put the four nails needed to hang the pictures?

Step 1: Understand the problem.

a. Write down the relevant information provided by the problem.

b. What exactly are you being asked to find?

Step 2: Devise a plan.

When a problem describes something physical that can be drawn, a diagram is usually a good idea.

c. Draw a diagram based on the description of the problem including the relevant information you wrote down.

Based on your diagram, you should be able to note how much space will be covered by frames, and how many blank spaces there will be. Once you know that, you can find the total blank space and divide by the number of spaces to find how far from the edges of the wall the edge of each frame will be.

Step 3: Carry out the plan.

d. How much of the 60 inches will be covered by picture frames? How many empty spaces are there? What is it we're trying to find? Give that quantity a variable name.

e. Note that the total 60 inches of the wall can be described by space covered by frames plus space in between frames. Use this idea to write and solve an equation that finds the space between frames. Then use that to describe the locations of all four nails.

Step 4: Check your answer.

f. Use your diagram to decide if the spots you found for the nails will work according to the statement of the problem.

2. Now let's think about generalizing our approach. Here's a verbal description of the equation you should have written in part e:

(# of pictures)(Width of each picture) + (# of pictures + 1)(x) = Total width

where x is the space between pictures. Choose variable names for all quantities in this problem and write a general equation as described above.

Math Note

What you're doing in Question 3 is a GREAT way to test out a general solution: See if it works on a specific problem that you already know the answer to.

3. Use the numbers from the problem we already solved to verify that your formula provides the same solution we already found.

Courtesy of Dave Sobecki

4. Hey, great news! Your family now has five members since you decided to adopt a puppy. Puppies are awesome! Now you have five pictures to evenly space on that wall. Use your answer to Question 3 to find where to put the nails, and describe why this is better than just starting over from scratch.

Did You Get It

2. In the interest of thwarting the efforts of nosy neighbors, a homeowner plans to plant three 10-foot-wide bushes along a 48-foot fence. If she wants consistent spacing between the bushes, and between the bushes and the end of the fence, where should she dig the holes?

Now you're on your own. Remember that the point of this activity is to think about a problem-solving procedure that can be used on any problem, so focus on our systematic procedure rather than just trying to figure out an answer.

©AF archive/Alamy

5. The riding mower in the classic movie *Forrest Gump* was a 1962 Snapper HiVac with a 28″ cut that averaged about 4 miles per hour in regular use. A standard football field is 120 yards long and 160 feet wide. How long would it take Forrest to mow a football field with this fine machine?

6. Can you generalize what you did to solve Problem 5? How long would it take Forrest to mow two football fields if he upgraded to a 54-inch John Deere ZTrak that can mow at 9 miles per hour? (Okay, so the timing doesn't work, but indulge us. Anything can happen in the movies!)

7. What are we not accounting for in our calculations in Problem 6? How can you make your answer more accurate?

Did You Get It

3. A field cultivator is the tool that is pulled behind a tractor to work ground in a field. How long would it take to work a 320–acre field (1 mi long and 0.5 mi wide) with a field cultivator that is 60 ft wide if the tractor pulling it goes 7.5 mi/hr? (Don't worry about turnaround time.)

8. Last problem: A crop-dusting service is hired to spray a chemical pesticide over a field of broccoli, which I personally think tastes like feet, but some people seem to like. The plane has a 350-gallon tank that at the moment contains 80 gallons of a 0.5% solution, which means that 0.5% of the liquid in the tank is actually pesticide: The rest is water. The goal is to find how many gallons of a pre-mixed 1% solution would need to be added to increase the concentration in the tank to 0.8%.

a. How many gallons of pesticide are currently in the tank?

b. If we add x gallons of the 1% solution, how many gallons total will be in the tank?

c. If we're adding x gallons of a 1% solution, how many gallons of pesticide are we adding? How many gallons of pesticide will we have total?

d. Now set up an equation that has two different expressions to describe the new amount of pesticide in the tank. One expression uses your answer to Question 8b, along with the fact that the new solution is 0.8% pesticide. The other expression is your answer to Question 8c. Then solve to find the number of gallons that would be needed to make the mixture 0.8% pesticide.

e. How many gallons of the 1% mix would we need to raise the concentration of the tank to 0.9%? Don't start over from scratch! Just adapt your work from part c.

3-5 Portfolio

Name ______________________________

Check each box when you've completed the task. Remember that your instructor will want you to turn in the portfolio pages you create.

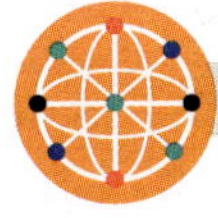

Technology

1. ☐ **a.** Trial-and-error isn't the greatest strategy for solving every problem, but it does come in handy from time to time. And technology can help shoulder the burden of repeated calculations. Suppose that Jack and Diane from Class Questions 1–7 start with $2,000 and invest $175 every month. Set up a spreadsheet with number of months in column A and the value of the account in column B. Use a formula that you can fill down each column after putting in the original values. Then find how long it will take for the retirement account to reach $10,000. A template to help you get started can be found in the online resources for this lesson.

b. On a second tab of your spreadsheet, use the generalized formula developed in Class Question 8 to build a calculator that will tell you the time required to reach an investment goal, and use it to solve the problem in part a without trial-and-error.

Online Practice

1. ☐ Include any written work from the online assignment along with any notes or questions about this lesson's content.

Applications

1. ☐ Complete the Applications problems.

Reflections

Type a short answer to each question.

1. ☐ Write an honest assessment of your experience with word problems in math: the good, the bad, and the ugly. What do you think is most responsible for these experiences, either good or bad?

2. ☐ What are some advantages to using algebra instead of arithmetic to solve problems?

3. ☐ What does it mean to generalize a solution to a specific problem?

4. ☐ Name one thing you learned or discovered in this lesson that you found particularly interesting.

5. ☐ Do you have any questions about this lesson?

Looking Ahead

1. ☐ Do an Internet search for "exchange rates" and find the value of one U.S. dollar in Russian rubles today. You'll need this number at the beginning of Lesson 3-6.

2. ☐ Complete the Prep Skills for Lesson 3-6.

3. ☐ Read the opening paragraph in Lesson 3-6 carefully and answer Question 0 in preparation for that lesson.

Answers to "Did You Get It?"

1. 1,200 mugs
2. The first goes 9.5 ft from the end of the fence, the next goes 14.5 ft away, then the last is another 14.5 ft past the second.
3. About 5.87 hours, or 5 hours 52 minutes.

Answers to "Prep Skills"

1. $400 + 30w$
2. The length is $h + 8$, where h is the height of the wall; the height is $l - 8$, where l is the length of the wall.
3. $20 = 12p$, where p is the number of pounds purchased.
4. **a.** $x = 5$ **b.** $y = 29$ **c.** $t = \frac{7}{2}$
5. **a.** $y = k - 3x$ **b.** $n = y - 3W$ **c.** $t = \frac{z + 2y}{4}$
6. **a.** About 7.1 hours **b.** About 1.16 hours or 1 hour and 9.6 minutes

3-5 Applications

Name ______________________________

For the first question, we'll provide some suggestions on how to proceed. Of course, if you can think of a different way that works for you, that's not just okay, it's GREAT.

On the way to the airport a few years ago, my wife's car hit a huge hole in a construction area, damaging the front right wheel. Huge. I think there may have been alligators living in it. The temporary tire forced us to decrease our average speed by 15 miles per hour. If the 39-mile drive usually takes us 45 minutes, how much time should we have budgeted for the drive home from the airport?

1. Write down all the information that seems relevant to a solution. (Hint: The alligator thing isn't relevant.)

2. What is the usual speed described by the problem?

3. Now you should have enough to get you started. Good luck. We're all counting on you.

4. Want to read about something wild? Of course you do. Do a search for "Barkley Marathons." This is a race so grueling that often nobody actually finishes. The race consists of five loops of 20 miles through the mountains of Tennessee. For those who make it to the last lap, the racers go in alternate directions around the course, which of course means they'll meet in the middle somewhere. Let's say that only two racers make it to the last lap, and it's close: In fact, both start the last loop around the course at the same time. If one averages 2 miles per hour and the other 1.92 miles per hour, how many hours would it be until they meet? (Hint: Use a letter to represent the time until they meet. Then write expressions for the distance traveled by each racer. What do their distances have to add up to?)

3-5 Applications

Name ______________________________

5. A landscape architect is planning a new nature area in the middle of an urban campus. She wants the length to be twice the width, and wants to put a 3-foot-high retaining wall around the perimeter. There will be 309 total feet of wall installed. How wide will this area be?

6. Adapt the procedure you did in Question 5 to solve the problem if the length is three times the width instead of two times.

7. A pool is being built in a new student rec center at Falcon Community College. The pool is designed to be a 60 ft by 26 ft rectangle, and the deck around the pool is going to be lined with slate tiles that are 1-ft squares. How many tiles are needed? (This is not quite as easy as it seems at first . . .)

3-5 Applications

Name ____________________

8. Generalize your result from Question 7 to solve any problem like this where 1-foot square tiles are bordering a rectangular area with length l feet and width w feet.

A landscaping company sells two types of grass seed mixes that are popular in its area: a premium mix that sells for \$1.25 per pound and a standard mix that sells for \$ 0.95 per pound.

9. If the store manager dumps a 20-lb bag of each mix into a large barrel, what is the total value of the grass seed in the barrel?

10. At what price per pound should the company sell this new mix to produce the same income as selling the two mixes separately?

11. If the store manager started with 20 lb of the premium mix in a large barrel, how many pounds of the standard mix should be added to create a mix that could be sold for \$1.00 per pound?

3-5 Applications

Name ____________________

12. How many pounds of grass seed must the barrel hold in order to contain all the mixture created in Question 11?

13. How many pounds of the standard mix should be put in a barrel with 20 lb of the premium mix to create a mix worth \$1.10 per pound?

Lesson 3-6 Prep Skills

SKILL 1: WORK WITH PROPORTIONS

We were introduced to proportions in Lesson 3-3. A proportion is simply a statement that two ratios (fractions) are equal. Like any equation, proportions can be true or false. Here's a true proportion:

$$\frac{4}{12}=\frac{1}{3}$$

Proportions often arise in practical situations. For example, if you make \$12.50 for 1 hour of work, you'd make \$25 for 2 hours: This can be described by a proportion:

$$\frac{\$12.50}{1\text{ hr}}=\frac{\$25}{2\text{ hr}}$$

When a proportion contains a variable, then solving it (like any equation) means finding all values for the variable that make it a true statement. The main tool we have for solving proportions is cross-multiplying. When you multiply the numerators from each fraction by the denominators from the other, the two products that result are also equal.

- Solve the proportion: $\frac{25.00}{2}=\frac{x}{7.5}$ Cross-multiply

  $2x = 187.50$ Divide both sides by 2

  $x = 93.75$

Did you notice the connection to the previous proportion? The left side of this equation is the ratio of dollars earned (\$25) to time worked (2 hours). So that must be what the ratio on the right side of the equation represents as well. It's comparing $$x$ to 7.5 hours, so our solution tells us how much you'd make for working 7.5 hours. This is exactly how we'll use proportions in the upcoming lesson.

SKILL 2: SOLVE LINEAR AND LITERAL EQUATIONS

Yep, here it is again. Are you starting to get the impression that solving equations is an important skill?

SKILL 3: WRITE CONVERSION FACTORS FROM AN EQUIVALENCE

When using dimensional analysis to convert units, you need conversion factors to multiply by. The thing that makes conversion factors work is that they're really just different ways to say 1: As such, the numerator and denominator have to be equal to each other. When we're told that two quantities are equal even if they're measured differently, we can turn that into two different conversion factors.

- The amount of snowfall that corresponds to 1 inch of rain depends on temperature and some other factors, but a general rule of thumb used by meteorologists is that 10 inches of snow corresponds to 1 inch of rain. We can write two conversion factors from this equivalence:

$$\frac{10\text{ in. snow}}{1\text{ in. rain}} \quad\text{and}\quad \frac{1\text{ in. rain}}{10\text{ in. snow}}$$

Each of these is equal to 1 so we could multiply by either in a calculation without changing the actual size of the quantity.

SKILL 4: USE DIMENSIONAL ANALYSIS

Another important skill that keeps coming up, this was reviewed in the Prep Skills for Lessons 3-1 and 3-5.

PREP SKILLS QUESTIONS

1. Write a proportion that describes this situation: 8 of 10 people surveyed liked a certain new soft drink, so if 200 people are surveyed we'd expect about 160 to like it.

2. Solve each proportion.

   a. $\frac{y}{84} = \frac{5}{7}$ b. $\frac{4}{9} = \frac{x+2}{20}$

3. Solve each equation.

   a. $34 = k \cdot 5$ b. If $y = 220$ and $x = 40$, find k in the equation $y = kx$.

4. Solve for n: $B = nw$.

5. On February 21, 2017, one ounce of gold was worth $1,283.50. Write two conversion factors based on this information.

6. Use one of the conversion factors you wrote to find the value of 10.5 ounces of gold, and use the other to find how much gold could be purchased for $30,000.

Lesson 3-6 Big Mac Exchange Rates

LEARNING OBJECTIVES

- ☐ 1. Identify situations where direct variation occurs.
- ☐ 2. Write an appropriate direct variation equation for a situation.
- ☐ 3. Solve an application problem that involves direct variation.

©Ingram Publishing/SuperStock RF

Decide what you want, decide what you are willing to exchange for it. Establish your priorities and go to work.

—H.L. Hunt

One of the most confusing things about traveling abroad is different currency. Since you're so familiar with dollars, you instantly have an idea of how cheap or expensive something is when the price is in dollars. What if I told you that a Big Mac costs about 75 rubles in Russia and 43 kroner in Norway? Those certainly sound like awful prices, but unless we know what a ruble or a kroner is worth, we have no idea how much or how little that actually is. In this lesson, we'll use exchange rates between currencies to study the topic of variation, which can be used to model the connection between many useful quantities.

0. Have you ever heard the phrase "is directly proportional to"? What do you think it means?

3-6 Class

An **exchange rate** is a number that describes how much of one currency you can trade for another currency. For example, if the U.S. exchange rate for Canadian currency is 1.2, it means that you could trade one U.S. dollar for $1.20 Canadian. When travelers talk about how expensive or cheap a certain country is, it's often a reflection of the exchange rate. The Big Mac costs mentioned earlier? The average cost in the U.S. in July 2016 was $5.06. In Russia it was just $2.15, and in Norway the cost was almost $6.

1. Using the exchange rate you found online in the Looking Ahead portion of Lesson 3-5, write the exchange rate as an equation: $1 US = ________ rubles.

2. Write the rate above as a conversion factor with U.S. dollars in the denominator.

3. Complete the table that shows the relationship between the number of U.S. dollars and Russian rubles that can be exchanged for those dollars.

U.S. Dollars	Russian Rubles
0	
10	
20	
30	
40	
50	

4. Draw a graph based on your table. Make sure to mark a scale on each axis, and include a verbal label that shows the quantity represented on each axis.

5. Find the slope of the line you graphed. What does it represent?

6. Write an equation using x and y that calculates the number of rubles (y) that you can trade x American dollars for. Which variable is the input and which is the output?

7. Use your equation to find the current cost of a Big Mac in Russia, using \$2.15 as the equivalent in American dollars.

8. Use your graphing calculator to verify the table and graph from Questions 3 and 4.

9. How many rubles do you think you'd get if you exchanged zero U.S. dollars?

10. If you doubled the number of dollars you are exchanging, what do you think that would do to the number of rubles you'd get in exchange? Find an example in the table to confirm your answer.

11. If you tripled the number of dollars you are exchanging, what do you think that would do to the number of rubles you'd get in exchange? Find an example in the table to confirm your answer.

12. If you cut the number of dollars you are exchanging in half, what do you think that would do to the number of rubles you'd get in exchange? Find an example in the table to confirm your answer.

Here are some conclusions we can draw about exchange rates:

- Zero dollars will get you zero rubles.
- If the amount of dollars goes up, the number of rubles goes up by the same factor.
- If the amount of dollars goes down, the number of rubles goes down by the same factor.

When this happens, it means one variable quantity is found simply by multiplying a second variable quantity by a constant. We say these two quantities **vary directly** or that they are **directly proportional**. We call this type of relationship **direct variation**. Algebraically, the thing that shows us that two quantities x and y vary directly is an equation of the form $y = kx$, where k is some real number. (So if your equation from Question 6 doesn't look like that, you might think about changing it.)

Did You Get It

Try this problem to see if you understand the concepts we just studied. The answer can be found at the end of the Portfolio section.

1. As of February 2017, 1 U.S. dollar equals 0.81 British pounds. Write an equation to give the number of British pounds (y) you would receive in exchange for x U.S. dollars. Then use the equation to find the number of British pounds you would receive if you exchanged $50.

Direct Variation

The following are various ways of illustrating what it means to say that two variable quantities vary directly.

Verbally

The quantity y varies directly as the quantity x, and the constant of variation is k.

Algebraically

$y = kx$

Example:
$y = 3x$

Numerically

x	$y = 3x$
1	3
2	6
3	9
4	12
5	15

Graphically

3-6 Group

1. Earlier in this unit, we put a lot of energy into modeling the cost of taking a cab in Vegas. There was an initial charge of $5.10, plus $2.60 per mile. Is this an example of direct variation? Explain.

Now let's talk about a painful experience that we almost all go through at some point: buying gasoline.

2. Do you think the amount spent at a gas pump and the number of gallons of gasoline purchased are quantities that vary directly? Explain.

3. Let's say you stop in at the local gas station just to buy an energy drink, and don't pump any gas. How much would you pay for zero gallons of gas?

If 5 gallons of gas costs \$11.10:

4. How much would you expect to pay if you double the amount purchased to 10 gallons?

5. How much if you cut the amount in half to 2.5 gallons?

Now the more interesting question: How much would you expect to pay for 15.2 gallons of gasoline? That's not quite as straightforward. First we'll attack this problem using proportions. Then we'll see if maybe it's more efficient to use variation.

6. In Lesson 3-3, we learned that an equation stating that two ratios are equal is called a proportion. Solve this problem by setting up a proportion. On one side, the ratio should compare the 5 gallons purchased to the cost of that purchase. On the other side, the ratio should compare 15.2 gallons of gas to the cost (which is what you're trying to find).

Did You Get It

2. On my campus last year, I was involved in a survey of first-year students. We got 140 responses, and 105 students indicated that they had declared a major at that point. At the time, there were about 1,600 first-year students on campus. How many would you guess had declared a major?

7. Now let's see how we can solve the gasoline problem using direct variation, then see why that might be a good idea. First, write a direct variation equation describing the cost C of buying G gallons of gas.

8. We know that 5 gallons costs \$11.10, so plug those values into your variation equation. Then solve the resulting equation for the constant of variation.

9. Now you can rewrite your variation equation so that the only unknowns in it are C and G. Do it!

10. What does the constant of variation you found describe about buying gas at this station?

11. How much would you expect to pay for 15.2 gallons?

12. Which method of finding the cost of 15.2 gallons makes it easier to complete this table? Pick one method, explain why, and fill in the table.

Gallons	Cost
0	
3	
6	
9	
12	
15	

13. What would it cost you to put 11.55 gallons in your car from this pump?

14. If you spent $40 on gasoline at this pump, how many gallons did you get?

Solving Problems with Direct Variation

1. Identify the two quantities that vary directly.
2. Note which is the input and which is the output, and write an initial equation of the form $y = kx$.
3. Use a given input and output to find the constant of variation k.
4. Rewrite the equation in the form $y = kx$ where you've replaced k with a number. (It might help to write units on y, k, and x and to think about dimensional analysis.)
5. Use your equation to solve the problem.

15. If you get paid hourly at your job, the total amount of money you earn varies directly with the number of hours you work. Let's say you worked 14 hours last week and made \$163.80. Write a variation equation describing your total pay P in terms of hours worked h. Then use it to find how much you'd make for working 20 and 25 hours.

Did You Get It

3. When taking a multiple choice test, the number of points that you get varies directly with the number of questions you get right. If you got 12 questions right and scored 75 points, write a variation equation to describe the test, then use your equation to find the score for someone who got 14 questions right.

16. Solve the general direct variation equation $y = kx$ for k.

17. What does this tell you about the ratio (fraction) involving x and y? (Hint: Does k vary in a direct variation equation?)

Some advantages of using a direct variation equation rather than a proportion:

1. Proportions answer questions one at a time, but you can use a direct variation equation to answer multiple questions very quickly.
2. With the table feature on a graphing calculator, or a spreadsheet, you can speed up the process even more when you have a direct variation equation.
3. With proportions, you never actually see the conversion factor (constant of variation). Look at the gasoline problem: The variation equation shows us easily that gas was \$2.22 per gallon.

The best gift I got for my birthday last year was my very own Mini Me. Not a living one, which would have been the best gift that *anyone* ever got for their birthday, but a bobblehead made to look like me. (Don't laugh at the shorts. Custom clothes would have cost extra.) When scale models of real objects are made, the scale is often given as a ratio. In this case, the ratio of my Mini Me's size to the real thing (me) is 1/13.

17. My Mini Me is 5.82 inches tall, that handsome little devil. How tall am I in feet and inches? Either set up and solve a proportion, or use an equation of variation.

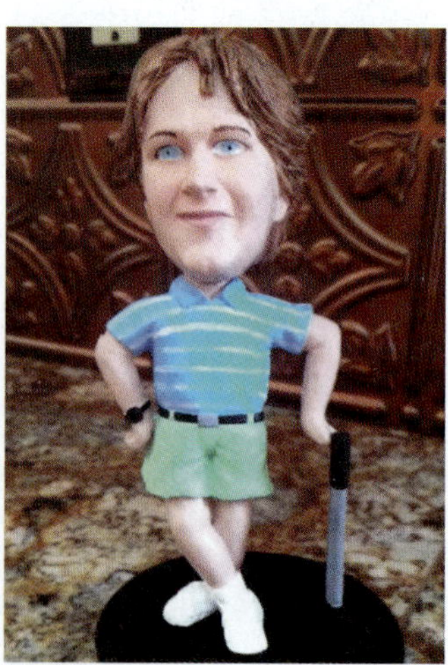

Courtesy of Dave Sobecki

18. The actual Mini Me from the *Austin Powers* movies is 32 inches tall. If he had a bobblehead made at the same scale as mine, how tall would it be?

19. The sales commission earned by a saleswoman varies directly with her gross sales. In one quarter, her commission was $7,200 on $180,000. What was her sales commission rate?

20. Use the sales commission rate to write an equation relating the commission earned to the amount of gross sales for that quarter. Make sure you describe what any letters you use represent.

21. Use your equation to complete the table of values.

	A	B
1	Gross Sales ($)	Commission ($)
2	$40,000	
3	$80,000	
4	$120,000	
5	$160,000	
6	$200,000	
7	$240,000	
8	$280,000	
9	$320,000	
10	$360,000	

22. Earlier, we observed that as the number of rubles goes up, so does the number of dollars. And we also saw that as the amount of gas you buy goes up, so does the cost. Is that always the case? If the constant of variation is negative, what is the effect on y if x increases? Explain. Studying the generic equation that describes direct variation will help.

In Questions 23–25, decide if the graph illustrates a situation where y varies directly as x, and explain your reasoning.

23.

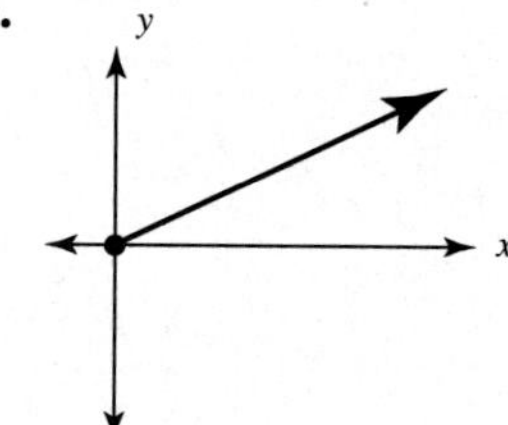

24.

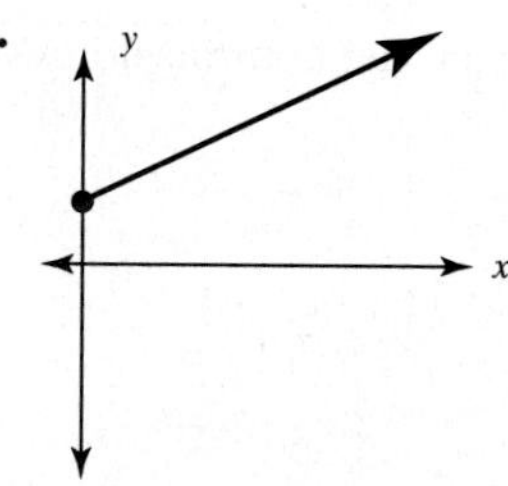

25.

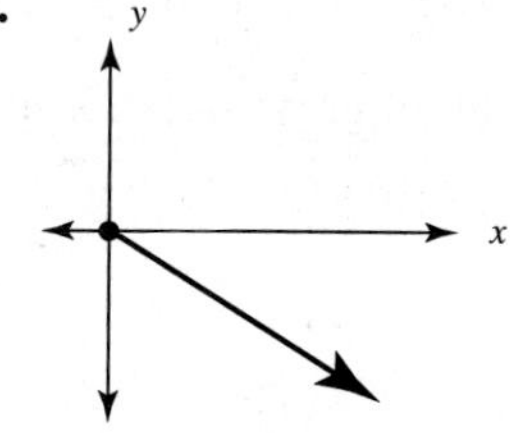

3-6 Portfolio

Name ______________________________

Check each box when you've completed the task. Remember that your instructor will want you to turn in the portfolio pages you create.

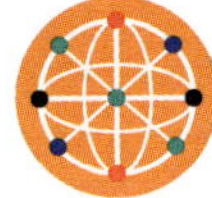

Technology

1. ☐ As you know, when you copy and paste a formula in Excel, it changes the cell reference accordingly: If the original formula is =A1/B1 in cell C1, copying and pasting in C2 changes the formula to =A2/B2. But sometimes we want to have all copied calculations use the same cell in a formula. In that case putting a dollar sign in front of the row and column will "lock" the cell reference and keep it from changing when we paste. This is useful in building an Excel table to convert currency.

   Look up the exchange rate for converting U.S. dollars to the currency for a country you would like to visit. Type that rate in cell E1. Then create a table and graph similar to the ones in the Class portion of this lesson. Make sure you include the name of the country and the website where you got your exchange rate. A template to help you get started can be found in the online resources for this lesson.

Online Practice

1. ☐ Include any written work from the online assignment along with any notes or questions about this lesson's content.

Applications

1. ☐ Complete the Applications problems.

Reflections

Type short answer to each question.

1. ☐ How would you decide whether or not two quantities vary directly?
2. ☐ What are some advantages to writing a direct variation equation rather than solving a problem using a proportion?
3. ☐ Name one thing you learned or discovered in this lesson that you found particularly interesting.
4. ☐ What questions do you have about this lesson?

Looking Ahead

1. ☐ Complete the Prep Skills for Lesson 3-7.
2. ☐ Read the opening paragraph in Lesson 3-7 carefully and answer Question 0 in preparation for that lesson.

Answers to "Did You Get It?"

1. $y = 0.81x$; 40.5 pounds **2.** $y = 6.25x$; 87.5 points **3.** 1,200 students

Answers to "Prep Skills"

1. $\frac{8 \text{ like it}}{10 \text{ people}} = \frac{160 \text{ like it}}{200 \text{ people}}$

2. **a.** $y = 60$

b. $x = \frac{62}{9}$

3. **a.** $k = \frac{34}{5}$

b. $k = 5.5$

4. $n = \frac{B}{w}$

5. $\frac{1 \text{ oz gold}}{\$1{,}283.50}; \frac{\$1{,}283.50}{1 \text{ oz gold}}$

6. 10.5 ounces of gold is worth $13,476.75; $30,000 will purchase about 23.4 ounces of gold.

3-6 Applications

Name ______________________________

1. Use the Internet to find and write the current exchange rate for converting the U.S. dollar to the Chinese yuan.

2. Complete the table using the current exchange rate.

U.S. Dollar	Chinese Yuan
0	0
20	
40	
60	
80	
100	

3. Write an equation that will convert U.S. dollars to Chinese yuan.

4. Which is the independent variable in your equation? Which is the dependent variable? Describe each using the terms input and output.

5. The average cost in China of an imported Volkswagen is the equivalent of $24,000. How much is that in yuan?

At various times over the past ten years, economists in Western countries have felt that China was manipulating the value of its currency, keeping it artificially low in order to give the country an export advantage. The point is that if the yuan is valued at less than it's actually worth, imported goods will cost more in yuan than they should, and that will make imports less attractive to Chinese buyers, leading to an export advantage.

Some economists feel that the yuan is valued at 20% less than it should be compared to the U.S. dollar.

6. If that's accurate, one dollar should be worth less yuan than the actual exchange rate. If the yuan is undervalued by 20%, then one dollar should be worth 20% less yuan. Find what the exchange rate should be.

7. How much would that affect the price of a Volkswagen in China?

8. Forensic scientists often examine scaled-up crime scene photos to search for evidence that might be hard to identify at regular size. Studying bite marks on a photo of a victim's leg at 500% magnification, a technician measures the distance between two punctures from the bite to be 165 mm. What is the actual distance between the punctures? $\left(\text{Hint: The ratio of the photo size to actual size is } \frac{5}{1}\right)$.

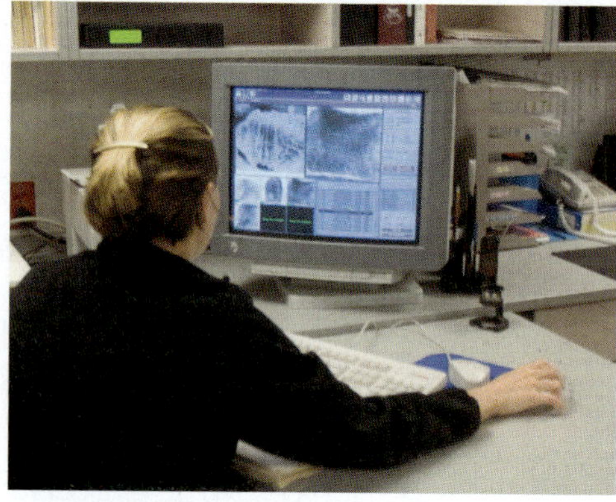

©Mikael Karlsson/Alamy

Here's a look at a map of the east side of Dubuque, Iowa. It shows a driving path from a casino on the river to a golf course on the other side. According to the maps app on my phone, the driving distance is 3.0 miles. We're most interested in the graphic at the bottom left, which indicates that 1 inch on the map (measure if you don't believe me) corresponds to 2,000 feet in real life.

9. Write a ratio that compares size on the map to size in real life.

10. Points A and B on the map are 4.06 inches apart. How far apart are the casino and golf course in miles? Use a proportion to decide.

11. Write an equation of variation that relates the distance on the map m to the distance in real life R. Then use information you know to solve for the constant of variation and update your equation.

12. If we looked at a satellite photo at the same scale as this map, do you think we could see any boats on the river? Answer yes or no and explain your reasoning.

13. Let's say that a 50-foot boat was out on the river. That's a pretty good-sized boat: The typical recreational motorboat is in the 18- to 24-foot range. Use your variation equation to find how big that boat would be on the satellite photo. Do you think we'd be able to see it?

Lesson 3-7 Prep Skills

SKILL 1: TRANSLATE STATEMENTS INTO SYMBOLS

This is a skill that we've been using throughout Unit 3, and you're probably getting better at it than you ever thought you would. More practice certainly doesn't hurt though.

- One conservative estimate for the global rise in sea levels over the next 100 years is 0.22 inches per year. If that happens, sea levels will rise $0.22y$ inches in y years.
- If a new TV show got a lot of hype but turns out to stink, and it loses 700,000 viewers each week after the premiere, then $-700{,}000w$ represents the total change in viewership after w weeks.
- If 12.5 million people watched the first episode, then the number of viewers after w weeks looks like $12{,}500{,}000 - 700{,}000w$.

SKILL 2: DRAW A GRAPH FROM A TABLE OF VALUES

We've practiced scatter plots several times—this is when we match up inputs and outputs from a table as points, then plot those points on a grid with a choice of scale that displays all the points distinctly in the space provided. You can also connect those points with lines or curves to illustrate trends in data.

- This table displays the number of viewers for the TV show described in Skill 1 over the first five weeks of its (presumably short) life. The data are illustrated in graphic form next to the table.

Weeks after premiere	Viewers
0	12,500,000
2	11,100,000
4	9,700,000
6	8,300,000
8	6,900,000
10	5,500,00

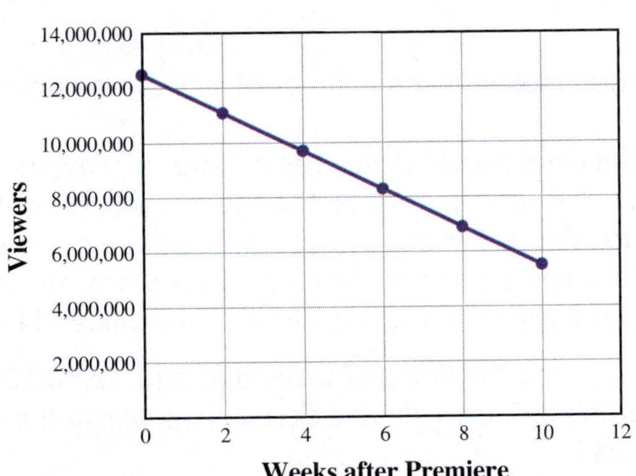

SKILL 3: SOLVE LINEAR INEQUALITIES

This was the key topic in Lesson 3-4. The key thing to know is that solving a linear inequality uses the same steps as solving a linear equation with one important exception: If at any point you multiply or divide both sides by a negative number, you have to reverse the direction of the inequality sign. That is, $<$ becomes $>$ and vice versa.

- $y + 13 \leq 5$ Subtract 13 from both sides
  $y \leq -8$

- $-6t > 54$ Divide both sides by -6; change $>$ to $<$
  $t < \dfrac{54}{-6}$
  $t < -9$

- $12 + 3x \geq 80$ Subtract 12 from both sides
  $3x \geq 68$ Divide both sides by 3; inequality sign stays the same
  $x \geq \dfrac{68}{3}$

SKILL 4: COMPUTE SLOPE

In this unit, we've been working with slope in terms of a known constant rate at which some quantity changes, which is great. But remember that we can also compute slope when we have two points on a graph. Rather than think of it as a formula, like you may have learned it in an algebra class, just think of it as the change in *y* value divided by the change in *x* value.

- The TV show we've been working with had 12,500,000 viewers for its premiere, and 9,700,000 four weeks after its premiere. The slope of the graph connecting the two points corresponding to this information is

$$m = \frac{\text{Change in viewers}}{\text{Change in weeks}} = \frac{9{,}700{,}000 - 12{,}500{,}000}{4 - 0} = \frac{-2{,}800{,}000}{4} = -700{,}000$$

SKILL 5: INTERPRET SLOPE AND *y* INTERCEPT

Speaking of slope, when we find the slope of a line, as we just saw, we're finding a comparison between the change in output and the change in input. This measures the rate at which the quantity described by the output is changing.

- In this case, the −700,000 indicates the number of viewers is decreasing at the rate of 700,000 viewers per week.

  Notice the units: Since the output describes number of viewers and the input describes number of weeks, when we divide change in output by change in input, we get units of viewers per week. This is often the easiest way to interpret what slope is telling us in a given situation.

  The *y* intercept of a graph is the point that has first coordinate zero. The first coordinate of any point represents an input, so the *y* intercept always tells us the value of the output when the input is zero.

- In the TV show example, since the first point we have is (0, 12,500,000), this means that there were 12,500,000 viewers for the premiere. (In this case, we're calling the premiere time zero, because it's zero weeks after the premiere.)

PREP SKILLS QUESTIONS

1. Translate each statement into symbols.
   a. According to a report on CNNMoney in 2015, 869,000 new pieces of computer malware were released per day in that year. Write an expression for the number released in *x* days.
   b. If you owe your parents \$600 and pay them \$50 per month, write an expression describing how much you owe after *m* months.

2. The table describes the amount of money earned by a cotton candy vendor at a ballpark based on how many bags she sells. Draw a graph of the data.

Bags	0	20	40	60	80
Money earned	0	\$8	\$16	\$24	\$32

3. Find the slope of your graph, and describe what it means.

4. What is the *y* intercept of your graph? What does it mean?

Lesson 3-7 The Effects of Alcohol

LEARNING OBJECTIVES

- ☐ 1. Write an equation of a line given a description of the relationship.
- ☐ 2. Write an equation of a line that models data from a table.
- ☐ 3. Write an equation of a line from a graph of the line.
- ☐ 4. Graph a line by plotting points.

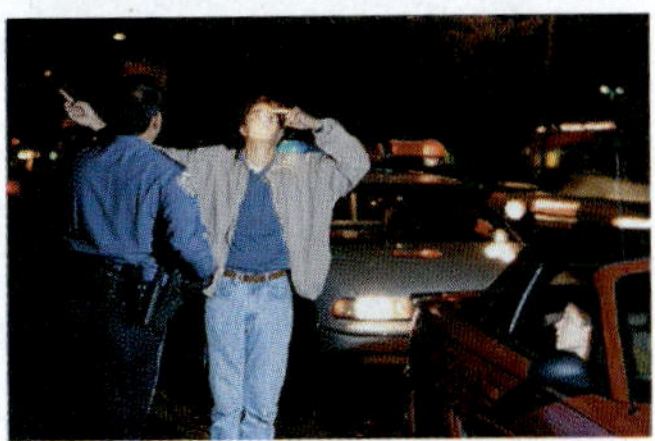

©Doug Menuez/Photodisc Green/Getty Images RF

As long as algebra is taught in school, there will be prayer in school.
— Cokie Roberts

It's no secret that alcohol consumption by college students is common, and that overindulgence can have many negative effects, in both the short and long term. And if anything, the problem is getting more serious. At my school alone, 21 students were hospitalized for alcohol poisoning during ONE recent weekend. When someone makes the choice to drink, they're affected as long as the alcohol remains in their blood stream. There are many theories on how to hasten the sobering up process: black coffee, ice cold showers, greasy food, drinking lots of water. . . . The truth is that none of these have any effect whatsoever on blood alcohol concentration (BAC). You can try every mythical treatment known to man, but at the end of the day the only thing that will bring down BAC is time. So far, we've studied a wide variety of quantities that can be modeled using linear equations. In this lesson, we'll use the time required to sober up, and other quantities, to study writing linear equations that model situations in greater depth. But first, let's talk about something much more positive: pizza!

0. When a person drinks, how do you think alcohol leaves the body over time? At a steady rate, or some other way?

3-7 Class

A student org you're involved in is planning a year-ending pizza party to celebrate another successful year on campus. As chair of the planning committee, you've wisely worked a deal with a local pizza place to provide large pizzas at \$9 each. The budget will allow at most \$175 to cover the food, and of course you'd like to know how many pizzas you can get and stay under budget. We'll attack the problem in several different ways.

1. Use a numerical calculation. Include all details, and don't forget to write the number of pizzas you can buy.

2. Write an equation that represents the total cost C of ordering p pizzas. What does this tell you about how the cost varies when you change the number of pizzas ordered?

3. Complete the table, then use it to estimate the number of pizzas you can buy.

Pizzas Bought	Total Cost
0	
5	
10	
15	
20	
25	
30	

4. Use the information from the table to draw a graph with the number of pizzas on the horizontal axis and the total cost on the vertical axis. Make sure you label the scale on each axis. Use your graph to estimate the number of pizzas that can be bought for $150. Try to ignore the answer you found in Question 1, and use only the graph.

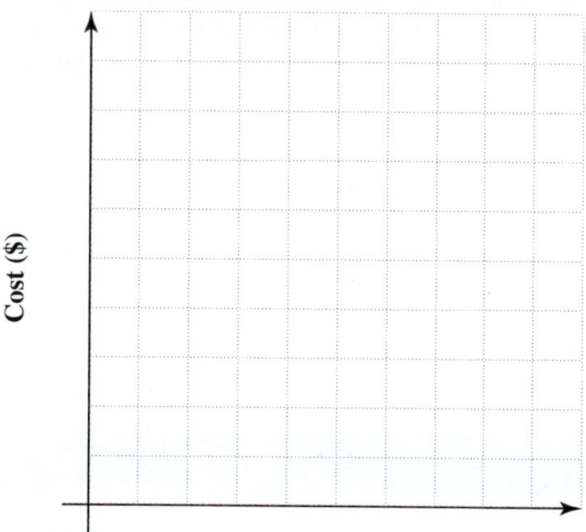

5. Since we want to spend AT MOST $175, this is a good place to use an inequality. Write an inequality that can be used to find the number of pizzas you can buy for at most $175. Then solve that inequality and use the result to decide on the number of pizzas you can buy.

Did You Get It

Try these problems to see if you understand the concepts we just studied. The answers can be found at the end of the Portfolio section.

1. If hot dogs cost $1.50 at Ball Park Frank night at the ballpark, write an equation that describes the cost C of buying h hot dogs.
2. Write and solve an inequality that will tell you how many hot dogs you can buy for your family if you want to spend less than $20.

6. What's the slope of the line you drew in Question 4? What does it represent?

7. What's the y intercept of the line? What does it represent?

Nice job so far, but did we forget to mention the $10 delivery charge? Sorry. You'll have to rework the problem, keeping in mind that we want to know how many pizzas we can get for at most $175 at $9 each, plus a flat fee of $10 for delivery.

8. Write an equation that describes the cost C of buying p pizzas with the delivery charge included.

9. Complete the table, then use it to estimate the number of pizzas you can buy.

Pizzas Bought	Total Cost
0	
5	
10	
15	
20	
25	
30	

10. Use the information from the table to draw a graph with the number of pizzas on the x axis and the total cost on the y axis. Don't forget to label a scale on each axis. Use your graph to estimate the solution to the problem.

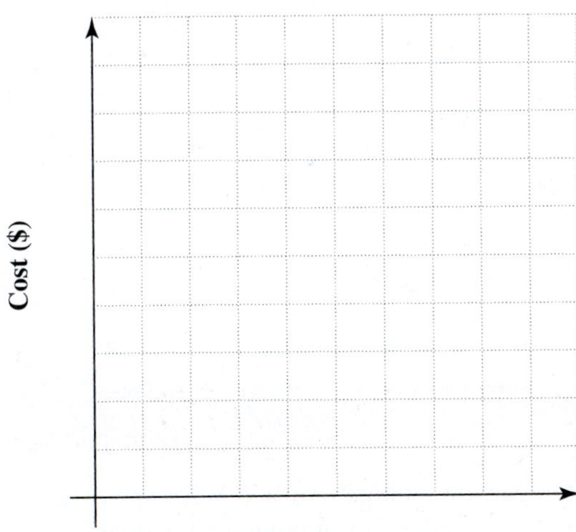

11. Write and solve an inequality that can be used to find the number of pizzas you can buy with the delivery charge added. **Write the solution to the problem in a full sentence.**

12. In this situation, would you say that the cost of ordering pizza varies directly with the number of pizzas? Explain.

13. What is the slope of the line you drew? What does it represent?

14. What is the y intercept of the line you drew? What does it represent?

Did You Get It

3. Refer to Did You Get It 2. You can't have the family choke down 13 hot dogs with nothing to drink. What is wrong with you? If you need to spend $18 on drinks, write a new equation that describes the cost of buying h hot dogs.
4. What are the slope and y intercept of your equation? What do they tell us about the situation?

15. Now here's the key idea: Look back at your answers to Questions 8, 13, and 14. The slope was __________, the y intercept was __________, and the equation was ____________________.

This fits the form of many of the linear models we've studied previously, where the equation was of this form:

$$y = \text{starting value of the quantity} + \text{constant rate of change} \cdot x$$

(Remember, "starting value" means the output associated with input zero.) In algebra, we call this the **slope-intercept form** of a line, because the two constants in the equation are the slope (rate of change) and the y intercept (starting value). The slope-intercept form is usually written as $y = mx + b$ but can also be written as $y = b + mx$ to match the order we've been using.

Writing the Equation of a Line

When a line has slope m and the second coordinate of its y intercept is the number b, then the equation of that line is $y = mx + b$.

16. A small business had just 12 employees when they opened their doors, but have been adding employees at the rate of 4 per month since then. Write a linear equation that describes the number of employees in terms of the number of months after the company was founded.

17. What are the slope and y intercept of this line? What do they tell us about the company?

18. Use this information to draw a graph of your equation. Be sure to label a scale and the axes.

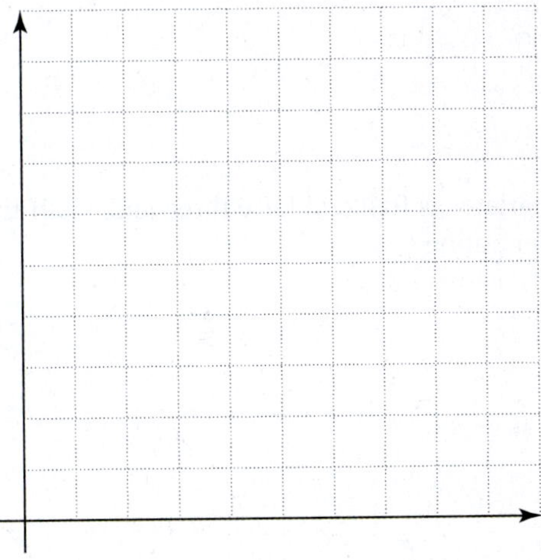

Graphing an Equation in Slope-Intercept Form

Step 1: Plot the *y* intercept; its first coordinate is always zero, and its height is the constant term in the equation.

Step 2: Find at least one other point on the graph. You can either plug an input into the equation, or use the slope. For example, if the slope is 2/3, that indicates a vertical change of 2 units for every 3 units of horizontal change. So you can find another point by moving two units above and three units to the right of the *y* intercept. Remember, the slope is the coefficient of the variable in the equation.

Step 3: Draw a line through the points you plotted. If they don't all line up, then you must have plotted one or more of them wrong.

Did You Get It

5. Use the procedure in the colored box to graph the line

$y=\frac{7}{3}x-5.$

3-7 Group

Now we turn our attention to studying the effects of alcohol. Blood alcohol concentration is a percentage of alcohol in the blood stream. The legal limit for operating a motor vehicle in most states is 0.08: This means that 0.08% of a person's blood is alcohol. The table here is borrowed from WebMD. It shows the ways that alcohol can affect a person at different BAC levels.

The Effects of Drinking Alcohol	
Estimated blood alcohol concentration (BAC) %	**Observable effects**
0.02	Relaxation, slight body warmth
0.05	Sedation, slowed reaction time
0.10	Slurred speech, poor coordination, slowed thinking
0.20	Trouble walking, double vision, nausea, vomiting
0.30	May pass out, tremors, memory loss, cool body temperature
0.40	Trouble breathing, coma, possible death
0.50 and greater	Death

1. After being arrested for driving under the influence, a well-known starlet was found to have three and a half times the legal limit of alcohol in her blood. What was her blood alcohol concentration?

2. Explain exactly what your answer to Question 1 means.

3. Describe some of the effects the arresting officers may have been able to observe at the time of the arrest.

John Q. Bonehead decides it would be a good idea to drive when he shouldn't. Fortunately, he gets pulled over before anything tragic happens, and a BAC test finds a level of 0.16 at 2 A.M. Unlike most other drugs, studies have shown that alcohol leaves the blood stream at a constant rate. The table shows his BAC every hour after the arrest.

Hours after 2 A.M.	BAC
0	0.16
1	0.145
2	0.13
3	0.115
4	0.1

4. What is the slope of the line describing John's BAC? What does it tell us?

5. What is the y intercept of the line describing John's BAC? What does it tell us?

6. Write the equation describing John's BAC, with x representing hours after 2 A.M., and y representing BAC.

7. When using an equation to model a situation, it's important to recall what specific information the equation provides. Restate your answer to Question 6 in the form of a statement like this: "If x represents __________, then y = __________ represents __________." (Don't physically fill in the blanks: write a full sentence below.)

8. We've learned about several features of graphing calculators that allow us to study the information provided by an equation: substituting in values for variables, making a table, and making a graph come to mind. Using these features, discuss whether or not the equation you wrote tells the story that it's supposed to. More detail is always better!

9. What would Mr. Bonehead's blood alcohol concentration be at 3:30 A.M.? (Remember, eating greasy food won't change it!)

10. How many hours would it take for his BAC to drop below the legal driving limit of 0.08? (He can take a cold shower if he likes, but it won't help.) Set up and solve an inequality to answer this question, please.

Did You Get It

6. Set up and solve an inequality that will tell you what times John's BAC is more than half of the legal limit.

11. What information would you need in order to draw the graph of your equation? Be specific.

12. Graph the line on the coordinate system provided. Make sure that you choose and label an appropriate scale on each axis.

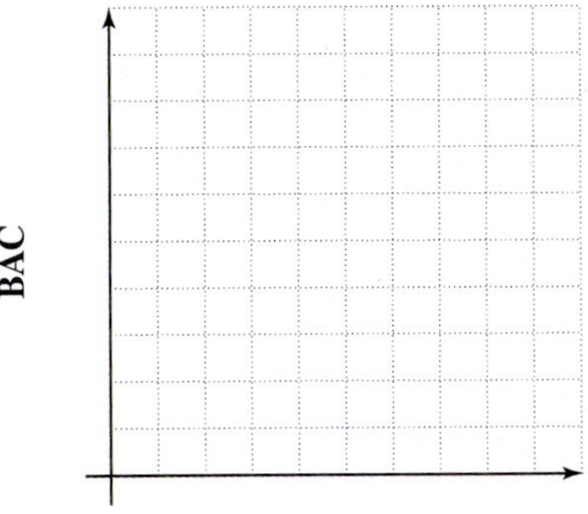

13. What is the x intercept? What does it mean?

The Connection Between Slope and Direction

- When the slope of a line is positive, it means that the quantity modeled by the line is increasing. This in turn means that the graph will slant upward from left to right.
- When the slope of a line is negative, it means that the quantity modeled by the line is decreasing. This in turn means that the graph will slant downward from left to right.

14. Sketch an example of a line with a positive slope. Label this line on the graph below.

15. Sketch an example of a line with a negative slope. Label this line on the graph below.

16. Can you describe what a line with a slope of 0 would look like? Sketch an example and label this line on the graph below.

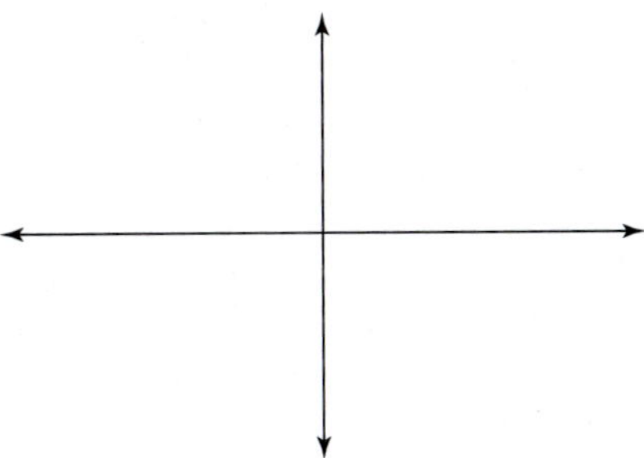

3-7 Portfolio

Name ______________________________

Check each box when you've completed the task. Remember that your instructor will want you to turn in the portfolio pages you create.

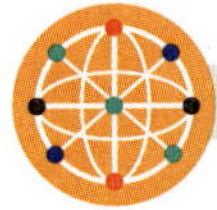

Technology

1. ☐ Use Excel to create a table and graph for both of the pizza party problems in the Group portion of this lesson. Make sure that you use a formula for calculating the costs. For each formula, explain how that formula relates to the equation that describes the cost of the pizzas. Type your explanation in the text box which is included in the template that you can find in the online resources for this lesson.

Online Practice

1. ☐ Include any written work from the online assignment along with any notes or questions about this lesson's content.

Applications

1. ☐ Complete the Applications problems.

Reflections

Type a short answer to each question.

1. ☐ Describe the connection between the graph of a line and how we use the line's equation to solve problems like the pizza party problem. How would you find the solution on a graph?
2. ☐ Explain how to find the equation of a line when you know the slope and the y intercept.
3. ☐ Describe what you learned about the time required to sober up from the Group portion of this lesson.
4. ☐ Explain how you would graph a line if you had the equation in $y = mx + b$ form.
5. ☐ Name one thing you learned or discovered in this lesson that you found particularly interesting.
6. ☐ What questions do you have about this lesson?

Looking Ahead

1. ☐ Complete the Prep Skills for Lesson 3-8.
2. ☐ Read the opening paragraph in Lesson 3-8 carefully and answer Question 0 in preparation for that lesson.

Answers to "Did You Get It?"

1. $C = 1.50h$
2. $1.5h < 20$; you can buy at most 13 hot dogs.
3. $C = 18 + 1.5h$
4. The slope is 1.5: This is the cost per hot dog. The y intercept is (0, 18); that's the cost if you buy no hot dogs, which is the drinks alone.
5.

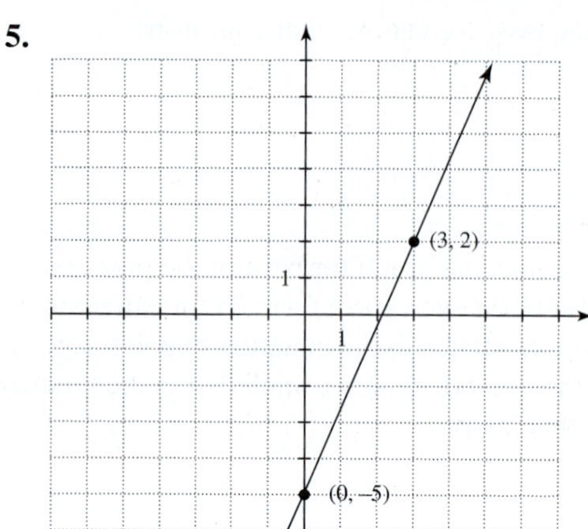

6. $0.16 - 0.015x > 0.04$; when less than 8 hours have passed, his BAC is still more than half the legal limit.

Answers to "Prep Skills"

1. **a.** $869{,}000x$ **b.** $600 - 50m$
2.

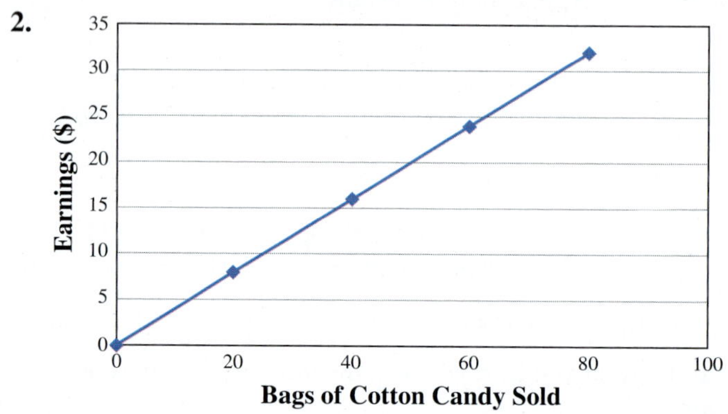

3. 0.4; The vendor earns \$0.40 for every bag of cotton candy she sells.
4. (0, 0); If she doesn't sell any cotton candy, she earns no money.

3-7 Applications

Name ____

If you're interested in weight loss, there certainly isn't a shortage of information, diet plans, supplements, and flat-out gimmicks available to you. But the simple truth is this: Gaining or losing weight ultimately comes down to one thing. If you consume more calories than you burn, you'll gain weight, and if you consume less calories than you burn, you'll lose weight. In this activity we'll study some aspects of weight change.

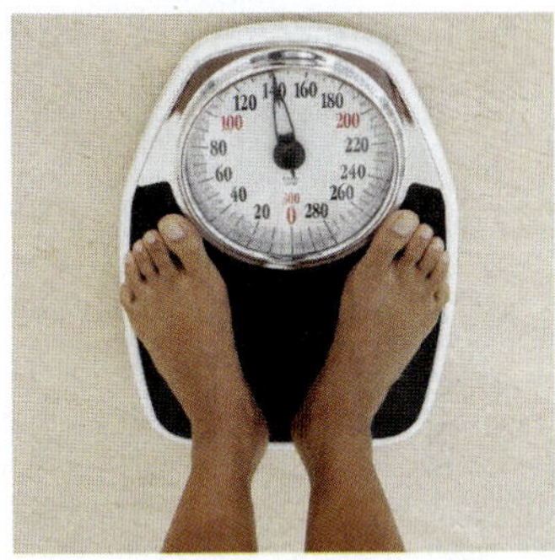

1. Bubba weighed 170 pounds when he started college, and never really thought about his weight much. Then a funny thing happened: He got super-busy, moderately poor, and started eating a lot of cheap fast food. He also started sleeping less and not drinking enough water. The result? He started gaining weight at the rate of 0.4 pounds per month. No biggie right? We shall see. Write a linear equation describing Bubba's weight (y) in terms of months after he started college (x).

2. Restate your answer to Question 1 in the form of a statement like this: "If x represents ________, then $y =$ ________ represents ________." (Don't physically fill in the blanks: write a full sentence below.)

3. If Bubba continues this modest-sounding weight gain, how much will he weigh after three years of college?

4. How long will it take our buddy Bubba to pass the big two-zero-zero at this rate? Use an inequality, please.

3-7 Applications

Name ______________________________

5. Draw the graph of your equation, including (as usual) an appropriately chosen scale on each axis.

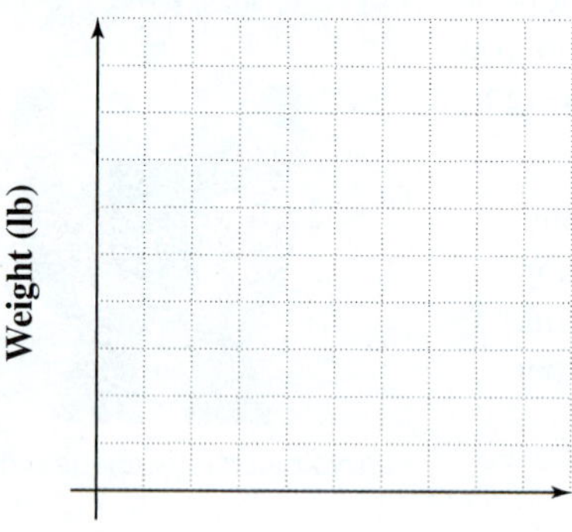

6. Find the slope of the line and explain what it means.

7. Find the y intercept and explain what it means.

Now let's study what it takes to lose weight. In Lesson 2-5, we saw that an average person can lose one pound if he burns 3,500 calories more than he consumes. Let's say that after reaching the 200-pound plateau, Bubba has had enough and decides to work on losing some weight. The table shows his weight if a certain number of calories are burned.

8. Find the slope of a line based on the data in the table. Use weight for y coordinates and thousands of calories burned as x coordinates. Recall that slope is change in output divided by change in input.

Calories burned (thousands)	Weight
0	200
14	196
28	192
42	188
56	184

9. Write a linear equation that describes Bubba's weight (y) in terms of thousands of calories burned (x). Your answer to Question 8 will help.

3-7 Applications

Name ______________________________

10. Restate your answer to Question 9 in the form of a statement like this: "If x represents __________, then y = __________ represents __________."

11. Use your equation to find Bubba's weight if he burns 50,000 calories above what he consumes.

12. Use your equation to find how many extra calories Bubba would have to burn to get back to the weight at which he started college.

13. According to the Mayo Clinic website, a man of Bubba's size would burn about 755 calories when running for an hour. How many hours of running will Bubba need to put in to get back to his original weight?

Math Note

Interested in overcoming the famous "freshman fifteen"? A Google search for that phrase yields some very informative and useful results.

3-7 Applications

Name ______________________________

14. Based on this entire group activity, what would have been Bubba's best approach when starting college?

A movie download service has a $14 monthly fee for membership; it then costs $3 to rent each movie.

15. If we model the cost (y) of renting a certain number of movies in a month (x), what is the slope of the line? What does it mean?

16. What is the y intercept of the line? What does it mean?

17. Use slope-intercept form to write an equation that models the cost in terms of movies rented.

Lesson 3-8 Prep Skills

SKILL 1: FIND AND INTERPRET SLOPE

We've worked with slope quite a lot in Unit 3. In some sense, knowing how to compute AND interpret the slope of a line are very closely related skills. If you think of slope as some abstract formula, it's hard to interpret what it means. But if you think of slope as change in output divided by change in input it's a lot easier to recognize that the slope measures the rate at which the output changes compared to a one-unit change in the the input.

This is especially simple to recognize when you consider the units attached to both the numerator and denominator of a slope calculation.

- A podcast has 423,000 downloads of its first episode, and 510,000 for its fifth. If we connect two points on a graph representing that data, the slope is

$$\frac{(510{,}000 - 423{,}000)\text{ downloads}}{(5-1)\text{ episodes}} = \frac{87{,}000\text{ downloads}}{4\text{ episodes}} = 21{,}750\,\frac{\text{downloads}}{\text{episode}}$$

The units (downloads per episode) are really helpful in interpreting what this means: The number of downloads went up by 21,750 downloads per episode.

SKILL 2: GRAPH A LINE

The nice thing about drawing the graph of a line is that it doesn't change direction. This means all we really need is two points on the graph, since you can then draw a straight line that connects them. On the other hand, unless you never, ever make mistakes, it's a pretty good idea to find at least three points when drawing a line: If the three points you found don't line up, then you know at least one of them must be wrong.

Of course, this relies on knowing in advance that the graph is a line. And how do we do that? There are two main ways. First, if the situation described involves a quantity changing at a constant rate, then the graph is going to be a line. Why? See Skill 1 above! Constant rate of change means constant slope, and constant slope means that the graph never changes direction — that's a line. Second, if you're graphing an equation with two variables, and both variables appear only in the numerator and only to the first power, then the graph will be a line.

- To graph $y = 15x - 12$, we first recognize from the form of the equation that the graph is a line. Then we find three points by substituting three values in for x:

x	y
0	−12
−5	−87
5	63

Now we set up a coordinate system, labeling the axes so that all of the points will fit on the graph. Then we plot the points, connect with a line, and include arrows on each end because there's nothing that says the line should have a beginning or an end.

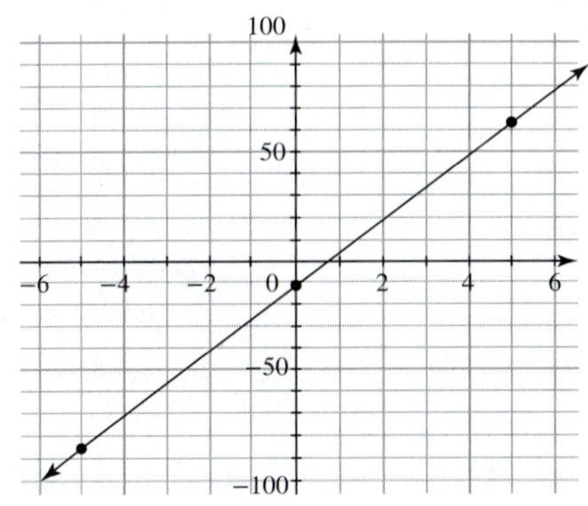

SKILL 3: SOLVE LINEAR EQUATIONS AND INEQUALITIES

There's not much we can say here that hasn't already been said like nine times, but a couple extra practice problems certainly won't hurt you any. Heck, it might even help.

PREP SKILLS QUESTIONS

1. After spending only 10 minutes on his first homework assignment, Ben slowly begins to realize that he needs to take his math class a little more seriously, and starts to increase his study time. By the sixth assignment, he's spending 45 minutes on homework. Find the slope of a line that describes the amount of time Ben is spending on homework, and describe what it means.
2. Draw the graph of the equation $y = -20x + 45$.
3. Find all solutions to each equation or inequality.

   a. $5x + 12 = 100$ b. $3(2 - y) = y - 10$ c. $50 + 12.5x \geq 99$

Lesson 3-8 Party Planning

LEARNING OBJECTIVES

- ☐ 1. Find the y intercept and equation of a line given two points.
- ☐ 2. Find the equation of a line using point-slope form.
- ☐ 3. Convert between forms of a linear equation.

©Ingram Publishing RF

Spring is nature's way of saying "Let's Party!"
– Robin Williams

Everyone loves a good party, with the possible exception of the person in charge of planning it. A backyard barbecue with a handful of friends is not a big deal, I suppose, although they can be surprisingly expensive depending on who you invite. But the big shindigs—weddings, New Year's Eve, charity balls—those take a TON of planning. And a lot of money. Anybody can throw a great bash with an unlimited budget, but for most folks the most important aspect of planning is keeping track of spending based on a target amount that's been budgeted. And that takes math skills! In this lesson we'll take our modeling to the next level: the party level.

0. What aspects of a big party would require math skills? Write down some ideas.

3-8 Class

Congratulations! Your boss thinks highly enough of your organizational skills to put you in charge of a gala to benefit a local no-kill shelter. (Whatever you do, don't screw it up.) You find an appropriate banquet hall and are told that hosting 150 people will cost $6,500. If you up the number of guests to 200, the bill jumps to $8,500.

1. If you make the reasonable assumption that the relationship between the number of people and cost is linear, do we have enough information to find a model for the cost? (By model, we mean an equation that provides the cost for a certain number of people.) Discuss.

2. Which quantity should be the independent variable, and which should be the dependent variable? Explain.

3. Using your answer to Question 2, write two points based on the cost information provided. Then use those points to find the slope of a line.

4. What does your slope tell us about the cost?

Now the big question: Can we find the equation of the line? If we knew the y intercept, we'd be golden, but we don't. So we'll need to be clever.

5. Using x for your independent variable and y for your dependent variable, write a generic equation for cost using the slope-intercept form. Then substitute in the slope you found and information from one of the points. This will give you an equation that you can solve to find the y intercept of the line.

6. What does your y intercept mean? Explain.

7. Write the linear model in the form $y = mx + b$.

Finding the Equation of a Line Through Two Points Using Slope-Intercept Form

1. Find the slope in the usual way: Divide the difference of the y coordinates by the difference of the x coordinates.
2. Substitute the slope in for m, and the x and y coordinates of one of the points in for x and y in the equation $y = mx + b$. This results in an equation you can solve to find b.
3. Now that you know both m and b, you can use $y = mx + b$ to write the equation of the line.

8. Graph the line you found, and then the two points you were originally given. If the line doesn't go through the points, what does that mean?

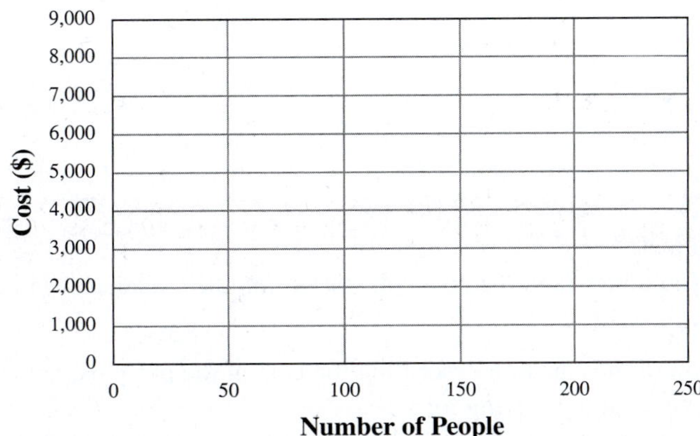

Now we're going to use our new favorite word: generalize.

9. We know that the slope of the line connecting ANY two points on the line is ___________.

10. Call one point on the graph (x, y). Find the slope of the line between that generic point and the point (150, 6,500).

11. Multiply both sides of the equation you wrote by the denominator to clear fractions.

What you have now is called the **point-slope** form of a line. Since that equation has to be true for every point on the line, it acts as a formula for finding the equation when you know the slope and one point on the line. How is this different from slope-intercept form? Glad you asked. Slope-intercept form requires that you know a *specific point* on the line, namely the *y* intercept. Point-slope works with ANY point on the line, making it more flexible.

Point-Slope Form

$$y - y_0 = m(x - x_0)$$

x and y are the variables; m is the slope; (x_0, y_0) is any point on the graph.

Example: A line through (2, 7) with a slope of −3 has equation $y - 7 = -3(x - 2)$.

12. Solve the equation from Question 11 for y. What do you notice?

Did You Get It

Try these problems to see if you understand the concepts we just studied. The answers can be found at the end of the Portfolio section.

1. Write the point-slope form of the equation describing the cost of the party using the point (200, 8,500). Then show that you get the same result by solving for y.
2. Find the equation of a line through (−3, 8) with slope 5. Write your answer in slope-intercept form.

3-8 Group

A small business has been selling 200 digital locking cookie jars per week, which has led to them losing $150 each week (a net profit of −150). According to their business plan, their net profit will increase by $12 for each unit sold.

1. What evidence do you see that would lead you to conclude that this is a linear relationship between net profit and units sold?

2. Identify the independent and dependent variables.

3. What's the slope? How do you know that?

4. Write the equation in slope-intercept form. (If you use point-slope form, convert it to slope-intercept form.)

5. What is the *y* intercept? What does it mean?

6. How many cookie jars do they need to sell in order to break even? What is the significance of this point on the graph? Think about what the profit would be if the business breaks even.

7. Fill in the table of values, then use it to draw a graph of your equation. Label the scale and axes as appropriate.

x	y
0	
200	
400	
600	

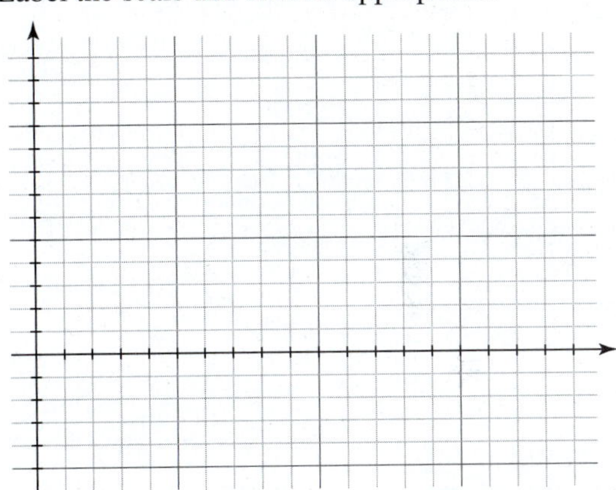

8. Find the number of units needed to make a profit of at least $500 per week.

A custom shirt shop charges a setup fee plus an additional fixed amount per shirt. An order for 20 shirts costs $188, and an order for 25 shirts costs $228.

9. What evidence do you see that would lead you to conclude that this is a linear relationship between the number of shirts ordered and the total cost of the order?

10. Identify the independent and dependent variables.

11. What is the slope? What does it mean?

12. Find an equation that will allow you to calculate the cost of ordering any number of shirts.

13. What is the y intercept? What does it mean?

14. Graph your equation. Label the scale and axes as appropriate.

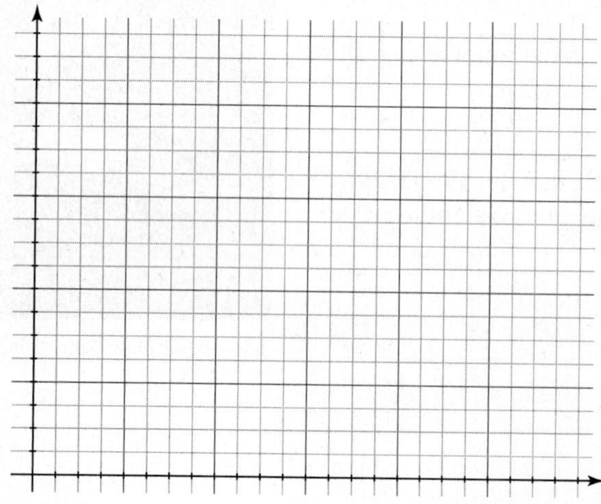

15. How much would 35 shirts cost?

16. How many shirts could you order if you wanted to spend at most $500?

Did You Get It

3. Find the slope-intercept equation of the line that goes through the points (8, 150) and (18, 400).

3-8 Portfolio

Name ____________________

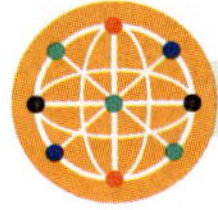

Technology

1. □ We now know how to find the equation of a line through two points. But what if you have data that are kind of linear, but not exactly? The world isn't always as orderly as we'd like, so many quantities that can be modeled effectively by a linear equation aren't exactly linear. In that case, we'd like to find the line that comes closest to matching the data. This is actually the topic of the next lesson, but in this tech assignment we'd like you to start kicking the tires, so to speak. Follow the instructions on the tech template, located in online resources for the course, and see if you can find the line that comes closest to modeling some sort-of-linear data.

Online Practice

1. □ Include any written work from the online assignment along with notes or questions about this lession's content.

Applications

1. □ Complete the Applications problems.

Reflections

1. □ When finding the equation of a line, do you prefer using point-slope or slope-intercept form? Why? Does it depend on the situation?
2. □ Once you have the equation for a linear model, what's the advantage of writing it in slope-intercept form?
3. □ Describe the process you prefer for finding the equation of a line when you know two points on that line.
4. □ Name one thing you learned or discovered in this lesson that you found particularly interesting.
5. □ What questions do you have about this lesson?

Looking Ahead

1. □ Make sure you complete the Technology assignment! It's a really important part of preparing for the next lesson.
2. □ Complete the Prep Skills for Lesson 3-9.
3. □ Read the opening paragraph in Lesson 3-9 carefully and answer Question 0 in preparation for that lesson.

Answers to "Did You Get It?"

1. $y - 8{,}500 = 40(x - 20)$; this also simplifies to $y = 40x + 500$
2. $y = 5x + 23$
3. $y = 25x - 50$

Answers to "Prep Skills"

1. 7; The amount of time Ben spends on homework increases by 7 minutes per assignment.
2. 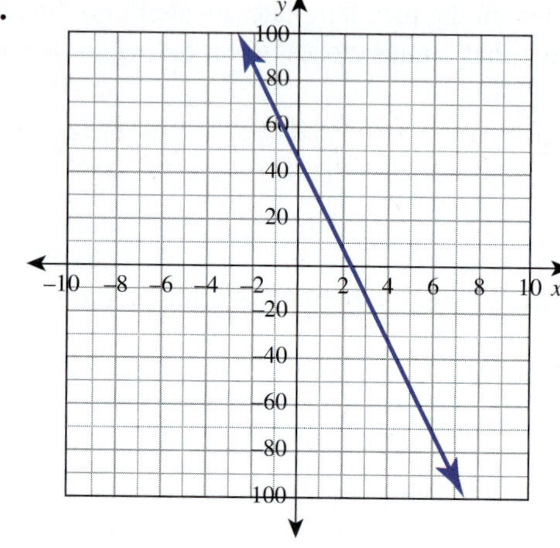

3. **a.** $x = \dfrac{88}{5}$ **b.** $y = 4$ **c.** $x \geq 3.92$

3-8 Applications

Name ______________________________

Tires have tread on them to channel away water and improve grip on both wet and dry surfaces. The height of the treads on a tire is referred to as tread thickness. The table shows the tread thickness of a particular set of truck tires based on the number of miles on those tires.

Miles (thousands)	Tread thickness (mm)
15	7.2
35	4.8

1. As you might expect, this table shows that the more miles you put on a set of tires, the less tread there is left on those tires. That's why tires don't last forever. Do you think it's reasonable to predict that the relationship between tread thickness and miles driven is linear? Why or why not?

2. Let's continue on the assumption that tread thickness and miles driven are linearly related. Find the rate at which the thickness is changing compared to thousand miles driven by comparing change in thickness to number of miles driven in thousands. What does this number describe about a line that can be used to model tread thickness?

3. Why was it smart to write miles in thousands, rather than the actual number of miles?

4. Use either one of the ordered pairs in the table to write the point-slope form of the equation that describes the relationship between the tread thickness and the number of thousands of miles the tire has been used.

3-8 Applications

Name ______________________________

5. Write the slope-intercept form of the equation that describes the relationship between the tread thickness and the number of thousands of miles the tire has been used. You can either use the previous question or create a new equation from the data in the table.

6. Graph the equation and the two points from the table. If they don't live on the line, that would be bad, right? Label the scale and axes as appropriate.

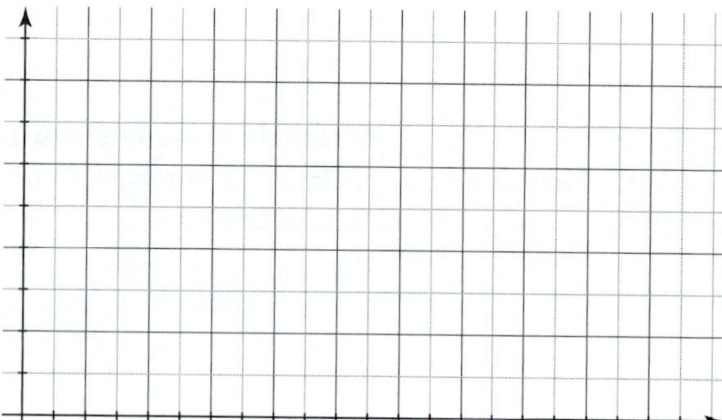

7. Use your equation to predict the tread thickness when the tires have been used for 50,000 miles.

3-8 Applications

Name ______________________________

8. According to the U.S. Department of Transportation, the minimum safe tread thickness is $\frac{1}{16}$ in. Convert this thickness to millimeters, then decide how many more miles the owner of these tires can drive before he risks life and limb driving on unsafe tires. (1 in. is about 25.4 mm. Assume that the current number of miles is the later data in the table.)

9. That particular model of tire has tread thickness of 9.2 mm when brand new. Does that match what your model predicts? What can you conclude?

Lesson 3-9 Prep Skills

SKILL 1: DRAW A SCATTER PLOT

This skill was introduced in Lesson 1-5 and reviewed in the Prep Skills for Lesson 3-1. Remember that the first coordinates go along the horizontal axis and the second coordinates go along the vertical axis. Always focus on the scale you choose for the axes: Identify the biggest value on each axis and use that as a guide to help you choose. The horizontal axis should extend a little bit beyond the largest first coordinate, and the vertical axis should extend a little bit beyond the largest second coordinate.

SKILL 2: FIND THE EQUATION OF A LINE THROUGH TWO POINTS

Of course, we just studied this skill in the last lesson, so hopefully a quick refresher will suffice. In order to find the equation of any line, you need to know the slope, so that's the first step. If you have two points on the line, you can just use

$$\frac{\text{Difference of } y \text{ coordinates}}{\text{Difference of } x \text{ coordinates}}$$

to find the slope, and you're off and running.

The next thing you need is a point on the line. In this situation, you'll have two to choose from. Either is fine, but be smart about it: If one point has a zero coordinate, always choose that one. In general, choose the point that has the simplest coordinates. Avoid fractions and negative numbers if you can, as those make you more likely to make an arithmetic mistake.

Once you have the slope, a known x coordinate, and a known y coordinate, you can plug directly into the formula

$$y - \text{known } y \text{ coordinate} = \text{Slope}(x - \text{known } x \text{ coordinate})$$

and there's your equation. We'll usually get a simpler equation by multiplying out parentheses on the right, then solving for y.

- Find the equation of the line through (2, 6) and (−1, 20).

$$m = \frac{20-6}{-1-2} = \frac{14}{-3} = -\frac{14}{3}$$

I would pick (2, 6) since both coordinates are positive.

$$y - 6 = -\frac{14}{3}(x-2)$$

$$y - 6 = -\frac{14}{3}x + \frac{28}{3}$$

$$y = -\frac{14}{3}x + \frac{28}{3} + 6$$

$$y = -\frac{14}{3}x + \frac{28}{3} + \frac{18}{3}$$

$$y = -\frac{14}{3}x + \frac{46}{3}$$

PREP SKILLS QUESTIONS

1. Draw a scatter plot for the data here, which relate the test scores of 15 math students to their homework average.

HW average (x)	40	95	77	100	22	85	88	81	93	100	95	0	50	78	90
Test score (y)	51	88	62	91	15	83	94	77	89	97	98	32	67	72	67

2. Find the equation of the line through each pair of points.

   a. (0, 9) and (12, 57)

   b. (−3, −4) and (−4, 8)

   c. (2, 5) and (−3, 8)

Lesson 3-9 The Great Tech Battle

LEARNING OBJECTIVES

- ☐ 1. Determine whether two variables have a linear relationship.
- ☐ 2. Calculate the line of best fit for a set of data using a spreadsheet.
- ☐ 3. Calculate the line of best fit for a set of data using a calculator.
- ☐ 4. Interpret the correlation coefficient for a data set.

One machine can do the work of fifty ordinary men. No machine can do the work of one extraordinary man.

— Elbert Hubbard

That quote says an awful lot about the dangers of too much reliance on technology, a peril with the potential to affect every modern student. The most interesting thing about it, though, is that it was written in 1903! When used correctly in math, technology allows us to focus more on understanding and interpretation, and less on computation. That's great. But when you try to use any technology as a substitute for thinking, nothing good is likely to happen. In this lesson, we'll learn how technology can help us to decide when a given data set is likely to be modeled well with a linear equation. Better still, calculators and spreadsheets will be able to find the best linear model for such situations, freeing us to focus on interpreting the model and using it to further study the data.

0. Without using technology, what are some ways you could tell if a data set might be modeled well with a linear equation?

©DonNichols/Getty Images RF

3-9 Class

In Lesson 3-7, we examined the relationship between calories burned and weight loss. While we needed to study data to model the exact nature of the relationship, you were probably able to figure out that the two quantities are related in some way. There are many quantities that *seem* like they might be related, but it's not clear that they really are, or if so, how. So it would be nice to have a way to study relationships (if they even exist) between data sets. Fortunately, the lovely folks who program spreadsheets and graphing calculators have done most of the heavy lifting for us. Still, in our continuing quest to not rely on technology more than our brains, it's a good idea to think about when quantities are likely to be related in some way.

1. Decide if you think the two quantities are likely to be related numerically, and explain your answer.

a. Dollars spent by a political candidate and the number of votes he gets

b. An adult test subject's height and IQ

c. A person's height and shoe size

d. The age of a car and its resale value

2. Is there a relationship between the total fat in a fast-food burger and the total calories? The table here lists calorie and fat information for a standard hamburger at a variety of fast-food restaurants. Begin your study of this question by drawing a scatter plot based on the table.

Fast Food Restaurant	Total Fat (g)	Calories
McDonald's	9	250
Burger King	12	290
Wendy's	8	230
Hardee's	12	310
Carl's Jr	17	470
Sonic	15	310
White Castle	7	140
Dairy Queen	14	350
Jack in the Box	14	310
In-N-Out Burger	19	390

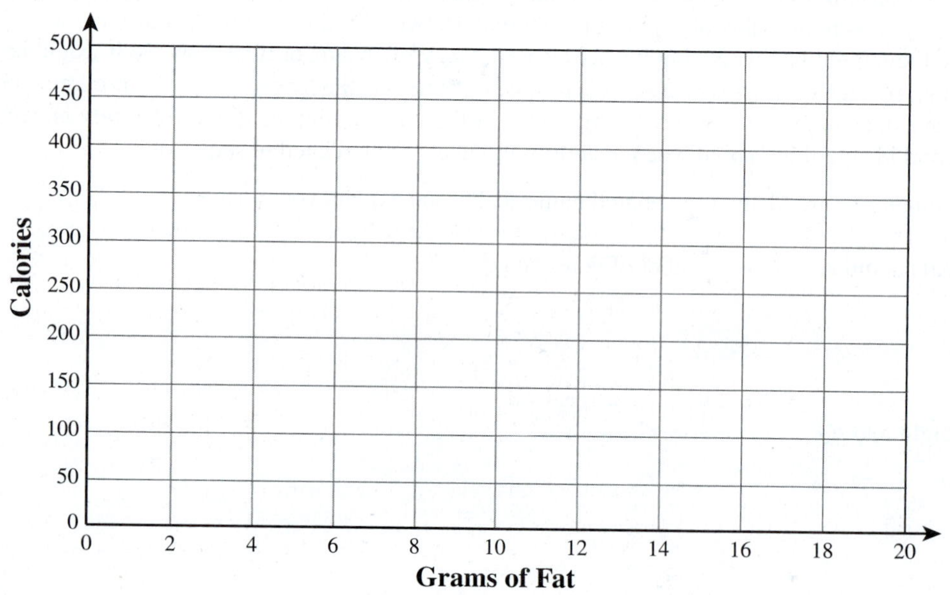

3. The points on your scatter plot don't form a straight line, but there should be a linear trend you can observe. Use a straightedge to draw a line that you think looks like the best fit for your scatter plot. Note that it's possible that the line that fits the data best doesn't go through any of the points.

4. Identify two points on the line of best fit that you drew on your scatter plot, and use them to find an equation for the line.

Using Technology: Finding the Line of Best Fit with a Graphing Calculator

To find a line of best fit for two data sets that appear to have an approximate linear relationship:

1. Press STAT ENTER to get to the list editor.
2. Enter the data set you want to use as inputs (*x* values) under **LI**.
3. Use the right arrow key ▶ to access **L2**. Then enter the data set you want to use as outputs (*y* values). When entering data from a table, make sure you enter the second list in the same order as the first.
4. Turn on **Plot1** by pressing 2nd Y= 1. Set up the screen as shown below.
5. Press Y= and if the **Y** = screen isn't blank, move the cursor over any entered equations and press CLEAR.
6. Press ZOOM 9 which is the **Zoom Stat** option; this automatically sets a graphing window that displays all of the plotted points.
7. Press STAT followed by the right arrow key to access the **STAT CALC** menu, and choose the **LinReg** (for linear regression) option, which is choice 4. Then press ENTER to calculate the line of best fit. Depending on which version you have, it might be necessary to use the arrows to move down to select **Calculate** before pressing enter.
8. Press Y= and enter the equation of the line of best fit, then press GRAPH to display the scatter plot along with the line of best fit.

```
Plot1 Plot2 Plot3
On Off
Type:
Xlist:L1
Ylist:L2
Mark: ▫ + ·
```

See the Lesson 3-9-1 video in class resources for further instruction.

5. Use the procedure in the Using Technology box to find the equation of the line of best fit using a graphing calculator. Round to one decimal place.

6. Use your answer to Question 5 to predict the number of calories in a Whataburger, which has 25 grams of fat. How does the answer compare to the number of calories predicted by your own line of best fit in Question 4? The Whataburger actually has 590 calories. Which line gave a more accurate prediction?

Did You Get It

Try this problem to see if you understand the concepts we just studied. The answer can be found at the end of the Portfolio section.

1. The table shows the gold-medal lengths for the men's Olympic long jump for selected years from 1900 to 2000. Using years after 1900 as input and length as output, find the line of best fit for this data. Round to three decimal places. Then use it to predict the length of the winning jump in 1984.

Year	x	Length (meters)
1900	0	7.18
1912	12	7.60
1920	20	7.15
1932	32	7.64
1948	48	7.82
1960	60	8.12
1972	72	8.24
1980	80	8.54
1996	96	8.50
2000	100	8.55

7. A classic roast beef sandwich from Arby's has 360 calories and 14 grams of fat. What does your line of best fit predict for calories? Is this expected? Explain.

When finding the line of best fit using a graphing calculator, the last line in the display should be a value for a quantity labeled *r*, which is known as the **correlation coefficient**. If your calculator is not displaying *r*, press 2nd 0, then use the down arrow to scroll down and choose "DiagnosticOn" and try the calculation again. This number measures how well the line seems to fit the data. The closer *r* is to 1 (positive slope) or −1 (negative slope), the more accurately the data you entered can be modeled using a linear equation. Here are some examples, based on data from two of my classes last semester:

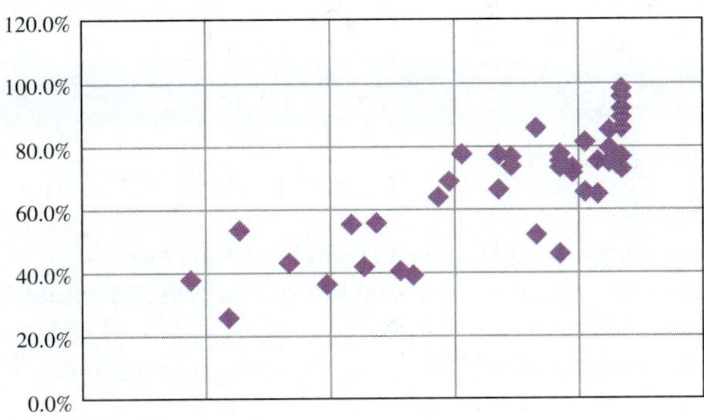

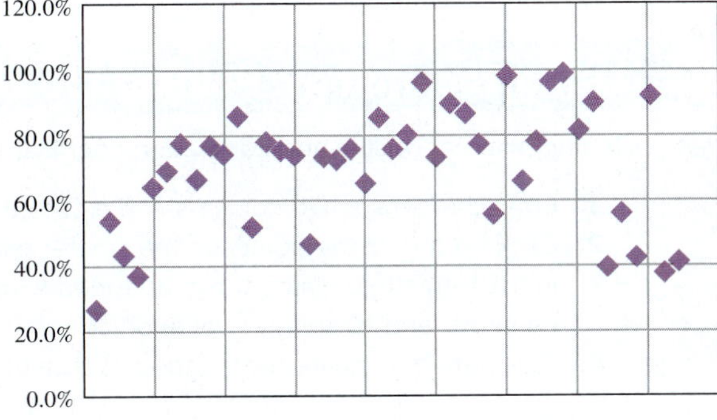

8. One of these scatter plots has a correlation coefficient of 0.19; the other has a correlation coefficient of 0.82. Discuss which you think is which, and why.

9. One of the scatter plots has each student's homework score for the semester on the *x* axis and the corresponding overall percentage in the course (from 0 to 100) on the *y* axis. The other again has overall percentage on the *y* axis, but along the *x* axis is the student's position alphabetically in the class. (The student whose name comes first alphabetically is 1, the next alphabetically is 2, and so on.) Which plot is which? Again, discuss your reasons.

10. What is the value of the correlation coefficient *r* for the fast-food sandwich plot? Round to two decimal places.

Did You Get It

2. Find the correlation coefficient for the line of best fit you found in Did You Get It 1. Does this indicate a strong linear relationship?

Using Technology: Finding the Line of Best Fit with a Spreadsheet

To find a line of best fit and correlation coefficient on a spreadsheet:

1. Enter the data in two columns and create a scatter plot.
2. Click on one of the points on the scatter plot, then choose "Add Trendline" from the **Chart** menu.
3. In the formatting dialog box that appears, click "Options," then click the checkboxes for "Display equation on chart" and "Display r^2 value on chart."
4. Calculate the square root of the r^2 value that appears on the chart. This is the correlation coefficient.

See the Lesson 3-9-2 video in class resources for further instruction.

11. Based on all of the work done in this activity, how strong do you think the relationship is between the number of grams of fat in a fast-food hamburger and the number of calories in it? Make sure you justify your answer.

An Important Note on Lines of Best Fit

Hopefully, you noticed right from the beginning of our study of burgers that you can't draw a straight line that goes exactly through all of the points. That in turn means the predictions made by our equation for any sandwich are unlikely to be perfectly accurate. This will almost always be the case when finding a line of best fit for a data set with more than two values. There's a good reason we call it "the line of best fit," not "the line of perfect fit"! The equations we're finding are models for data that allow us to make approximations and predictions; they're not exact representations of data.

12. Look back at the original table of fat and calories near the beginning of this lesson. Use the calculator-generated line of best fit to predict the number of calories in the Hardee's burger. Then find the difference between the actual number of calories and your predicted value.

When using a line of best fit to model data, we use the term *residual* to refer to the difference between predicted values obtained from a model and actual values. You just computed the residual for the Hardee's burger. Remember: the output we get from the equation of the line of best fit is the predicted value.

The Residual of a Data Point

When a data set is modeled by a line of best fit, the difference between the actual value from that data set and the value predicted by the equation is called the **residual** for that value.

Residual = Actual value − predicted value

The sign of the residual tells you if the actual value was higher (positive residual) or lower (negative residual) than the predicted value.

13. Find the residual for the White Castle burger.

Did You Get It

3. Continuing our study of Olympic men's long jump champions from Did You Get It 1, find the residuals for the winning long jumps in 1920 and 1980.

3-9 Group

Correlation Lab: How Much Candy Can You Grab?

Supplies needed:

- A large bowl filled with Hershey's miniatures candy (or some other relatively small, WRAPPED candies). One bowl per group is nice, but somewhat costly.
- A ruler for each group

(Thanks to our friends from Cincinnati State Technical and Community College for inspiring this activity.)

1. What factors would be likely to go into how many candy bars you can grab out of a bowl? List some.

In this activity we're going to see how closely hand size correlates to the number of candy bars you can grab. First you'll need to decide on some parameters to make sure that everyone's data are comparable.

2. Decide which hand everyone will use. Right? Left? Dominant? Nondominant? Describe why you made the choice you did.

3. How should you measure hand size? And what units will you use? Describe your choice.

At this point, you should have a short class discussion to decide on choice of hand and how to measure for the entire class so that results are consistent. Write the choices made here for reference.

4. Now it's time to begin the activity. For each person in your group, measure the size of his or her hand, then have each person reach into the bowl and grab as many candy bars as possible IN ONE TRY. Count the number, put the candies back in the bowl, then record the data in the table. Finally, share your data with your instructor or other groups so that data for the entire class are included. (Note: In this question, you're only using the first two columns of the table. The other columns will be filled in later.)

Hand size (x)	# Candy bars (y)	Predicted #	Residual

5. Draw a scatter plot of the data on the grid supplied.

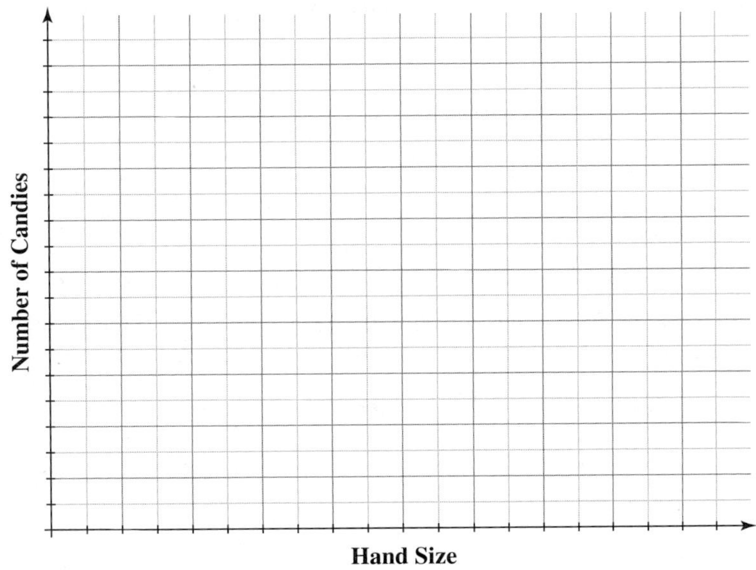

6. Use a calculator or spreadsheet to draw a scatter plot for the data, then find and write the equation of the line of best fit.

7. Add the line to your hand graph.

8. Find the correlation coefficient for the data. How closely are the two quantities related?

9. Use the line of best fit to predict the number of candy bars that your instructor should be able to grab. Then have your instructor grab some candy. How accurate was the prediction?

10. Use your equation and either the table feature on a graphing calculator or a spreadsheet to calculate the predicted number of candy bars for everyone in your chart based on hand size. Fill that data into the table in the Predicted # column.

11. Compute the residual for everyone in the class, and add that to the table as well.

12. Find the sum of all residuals. What can you conclude?

13. Write any conclusions you can draw about the connection between hand size and the number of candy bars you can grab, then start eating those candies (like you haven't already).

3-9 Portfolio

Name ______________________________

Check each box when you've completed the task. Remember that your instructor will want you to turn in the portfolio pages you create.

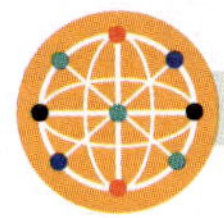

Technology

1. ☐ Enter the Olympic men's long jump data from Did You Get It 1 in the first two columns of a spreadsheet, then create a scatter plot and find the line of best fit. In the third column, use the equation of best fit to calculate the predicted winning jump for each year in the table, then in the fourth column compute the residual for each winning jump. Finally, use the SUM command to add all residuals.

2. ☐ One of the years conspicuously absent from the long jump data was 1968. American Bob Beamon jumped 8.90 meters (over 29 feet) that year to beat the previous world record by almost 2 feet—in a sport where new records usually pass the old by an inch or two. Many people consider this the single most remarkable athletic feat in human history—Beamon himself literally fell over from shock when he realized how far he'd jumped. Rework Question 1, but this time include that 1968 jump in the data set. How much does it affect the line of best fit? Discuss.

Online Practice

1. ☐ Include any written work from the online assignment along with any notes or questions about this lesson's content.

Applications

1. ☐ Complete the Applications problems.

Reflections

Type a short answer to each question.

1. ☐ What is a line of best fit for two sets of data?

2. ☐ Does the line of best fit provide useful information about every pair of data sets? Why or why not? Your answer should probably mention the correlation coefficient.

3. ☐ Does a line of best fit have to go through a lot of the data points? Explain.

4. ☐ Name one thing you learned or discovered in this lesson that you found particularly interesting.

5. ☐ What questions do you have about this lesson?

Looking Ahead

1. ☐ Complete the Prep Skills for Lesson 3-10.

2. ☐ Read the opening paragraph in Lesson 3-10 carefully then answer Question 0 in preparation for that lesson.

Answers to "Did You Get It?"

1. $y = 0.015x + 7.179$; 8.44 m
2. 0.96; this does indicate a strong linear relationship.
3. 1920: –0.329; 1980: 0.161

Answers to "Prep Skills"

1.

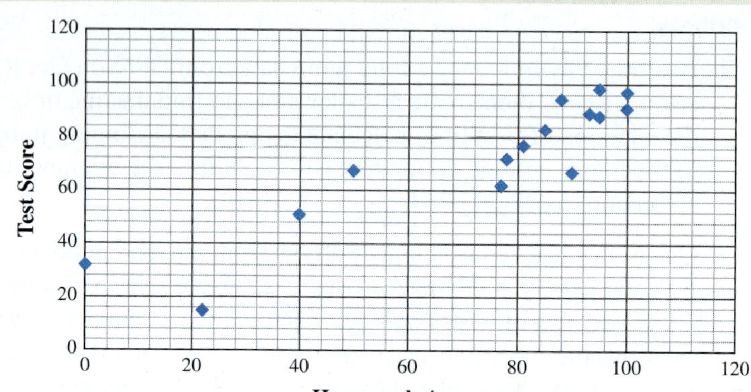

2. **a.** $y = 4x + 9$ **b.** $y = -12x - 40$ **c.** $y = -\frac{3}{5}x + \frac{31}{5}$

3-9 Applications

Name ______________________________

This table shows the average annual cost for tuition, room, and board at all colleges in the United States between the 2000 and 2014 school years.

1. Draw a scatter plot on the axes provided. The x values are given in the table; use values from the Cost column as y values. Do you think a linear equation will model the data accurately?

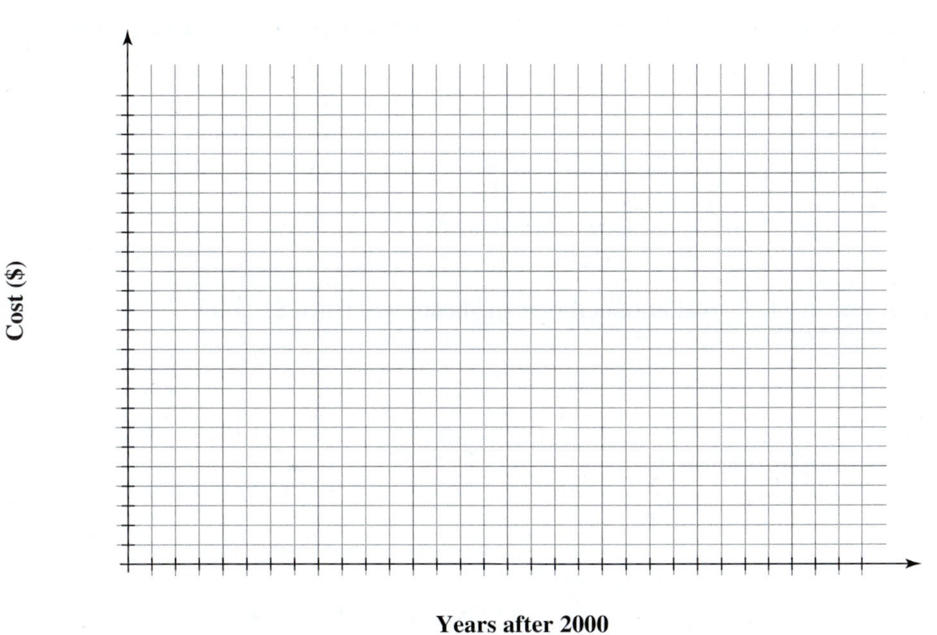

School Year	x	Cost ($)
2000–01	0	$10,820
2001–02	1	$11,380
2002–03	2	$12,014
2003–04	3	$12,953
2004–05	4	$13,793
2005–06	5	$14,634
2006–07	6	$15,483
2007–08	7	$16,231
2008–09	8	$17,092
2009–10	9	$17,650
2010–11	10	$18,475
2011–12	11	$19,401
2012–13	12	$20,233
2013–14	13	$20,995
2014–15	14	$21,728

2. Use a graphing calculator or spreadsheet to create a scatter plot and find the line of best fit for the data, then add the line to your scatter plot.

3. What is the slope of the line? What does it mean?

3-9 Applications

Name ______________________________

4. Use your equation to predict total tuition, room, and board costs for the 2019–2020 school year.

5. Use your equation to predict total tuition, room, and board for the 1990–1991 school year. (Hint: What x value would correspond to 10 years BEFORE 2000?)

6. Use your equation to predict the school year in which total tuition, room, and board will reach $30,000.

7. What is the correlation coefficient for the data? Are you surprised based on your answer to Question 1? Explain.

8. Find the residuals for the 2002–2003 and 2008–2009 school years. What do your results mean?

Lesson 3-10 Prep Skills

SKILL 1: COMPUTE A UNIT PRICE

Unit pricing is used very commonly in grocery stores as a way for shoppers to compare prices. If one box of coffee k-cups has 12 cups for \$8.99 and another has 16 cups for \$12.89, it's not at all obvious which is the better deal. So the tag in the store (if you look closely enough) probably tells you the unit price, which is the cost of one unit of a given item.

- In this case, for the first box: $\dfrac{\$8.99}{12 \text{ cups}} \approx \$0.749$ per cup

  For the second: $\dfrac{\$12.89}{16 \text{ cups}} \approx \$0.806$ per cup

Now we can see that the first box is the better price: Each cup is about 75 cents, compared to about 81 cents for the bigger box.

SKILL 2: SUBSTITUTE AN EXPRESSION INTO AN EXPRESSION

By now you should be very comfortable evaluating expressions, which typically involves replacing the variable with a number. For example, to evaluate the expression $120 + 9x$ for $x = 4$, we'd get $120 + 9(4) = 156$. Sometimes it's useful to evaluate expressions not just for numbers, but for other expressions as well.

- Substitute $15 - 2y$ for x into the expression $120 + 9x$:

  $120 + 9(15 - 2y) = 120 + 135 - 18y = 255 - 18y$

- If $y = 100 - x$, rewrite the equation $4x - 2y = 425$ with only variable x.

  Since we're told that $100 - x$ is the same as y, we can consider $100 - x$ to simply be a new name for y. In that case, we can replace y in that equation with $100 - x$, to get $4x - 2(100 - x) = 425$. The left side could then be simplified by multiplying out parentheses and combining like terms, to get $6x - 200 = 425$.

SKILL 3: WRITE A LINEAR EQUATION FROM A DESCRIPTION

This specific skill was reviewed in Lesson 3-3 Prep Skills, but we've practiced writing algebraic expressions based on given information throughout this entire unit. When it comes to writing equations, one of the key ideas is that an equation comes from two different ways to express the same quantity.

- Let's say you make \$12.75 per hour at a part-time job and you're interested in setting up an equation to calculate how many hours you'd need to work to make \$200. You could use letter h to represent the number of hours worked, in which case $12.75h$ is the amount you'd make for h hours. The situation states that we want that amount to be \$200, so in essence both $12.75h$ and 200 are ways to describe the amount of money earned. So it makes perfect sense to write the equation $12.75h = 200$.

This can also be done with situations that have more than one variable quantity. If you had a second part-time job that pays \$10 per hour, and you represented the number of hours worked at that job with variable k, then the total amount you earn is now $12.75h + 10k$, so if you're still interested in earning \$200, the equation would be $12.75h + 10k = 200$.

PREP SKILLS QUESTIONS

1. Which has the cheaper unit price: a bag of 40 Reese's cups for \$6.99, or a package of 16 for \$3.79?

2. Which has the cheaper unit price: 4 pounds of ground chuck for \$8.19, or 24 ounces of ground chuck for \$3.49?

3. Substitute $18 + 3y$ for x into the expression $15 - x$.

4. Substitute $3x - 1$ for y into the expression $2y - 25$.

5. If $y = 14 - x$, rewrite the equation $5x - 10y = 85$ with x as its only variable.

6. For the unit price of Reese's cups in the bag of 40, write an equation that describes how many of those Reese's cups you could buy for \$25.

7. A semi-private golf course has an initiation fee of \$800 which allows players to play golf as much as they like. But they do have to pay a \$12 cart fee each time they play. Write an equation that describes how many times a member could play for \$1,100.

8. Players can choose to walk the course, in which case they only have to pay a \$2 fee that goes toward course maintenance each time they play. Write a new equation describing the number of times a player could play in a riding cart or walking for \$1,100.

Lesson 3-10 All Systems Go

LEARNING OBJECTIVES

- ☐ 1. Solve an application problem involving a system of equations.
- ☐ 2. Illustrate the solution to a system of equations using a table.
- ☐ 3. Illustrate the solution to a system of equations using a graph.

©REB Images/Blend Images LLC RF

In the long run, we shape our lives, and we shape ourselves. The process never ends until we die. And the choices we make are ultimately our own responsibility.
— Eleanor Roosevelt

If a problem comes knocking at your door, you can hide behind the sofa and hope it goes away, or you can answer the door and see an opportunity. Many successful people will tell you that this mindset is what separates the winners from the pretenders in life. It's about taking personal responsibility for your own success, and seeing every obstacle as one more chance to achieve success. Throughout this unit, we've learned skills and strategies that can be used to write equations describing situations, and use those equations to solve problems.

But the world is a complicated place, and there are many very real situations where more than one quantity varies, which means we'll need more than one variable. In that case, we'll also need more than one equation that models the situation. Fortunately, we'll have a system in place for dealing with these situations: a system of equations, in fact.

0. Think of all of the equations we've solved in Unit 3. Can you think of any situations where we solved equations that had more than one variable in them?

3-10 Class

1. Summarize Polya's problem-solving strategy from Lesson 2-5.

We do have some experience in solving equations that have multiple variables in them: We called that solving literal equations. The idea was to solve a formula for one of the variables to turn it into a more convenient form. But we never got a number as a solution when we did that: The result was always another formula with variables in it. This isn't ideal when solving specific problems: If I ask you how big the area of your campus is and you say *"l times w,"* I'm probably going to look at you funny and slowly back away.

When we're studying a situation with two variable quantities, we may be able to set up a **system of equations**. This is two or more related equations that share common variables. Let's start off with a situation that might lead to a system of equations.

The shipping manager at a small business sent out a shipment of three coffee mugs and six promotional packets on Monday, and the total shipping weight was 42 ounces. On Wednesday, she sent six more mugs and three more packets, and this time the total weight was 48 ounces. Now the boss wants to know the shipping weight for each packet, and for each mug. Our goal is to help the shipping manager out.

2. In this problem, there are two variable quantities that we don't know: the shipping weight of a packet and the shipping weight of a mug. If we use x to represent the weight of one mug and y to represent the weight of one packet, write an expression that describes the combined weights of three mugs and six packets.

3. Use your answer from Question 2 to write an equation that describes the weight of the first shipment.

If we repeat what you did in Question 3 for the second shipment, we get our first look at a system of equations:

$3x + 6y = 42$

$6x + 3y = 48$

4. Solve each equation for y.

5. Make a table of values for each equation.

x	1st Equation: $y =$	2nd Equation: $y =$
2		
4		
6		
8		
10		

6. Graph both equations on the same grid. Make sure you label which is which.

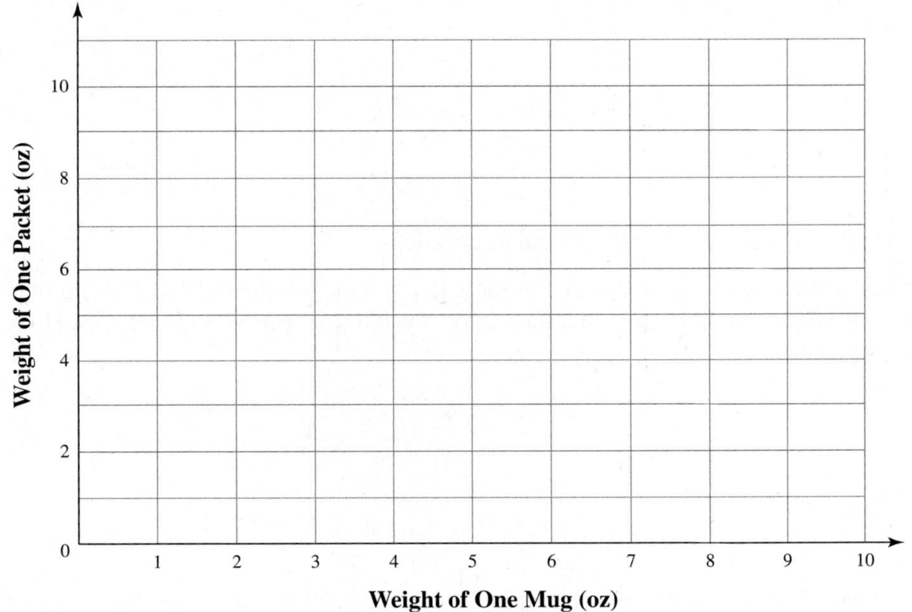

7. Where does it look like the two lines intersect?

8. Plug the coordinates of that point back into each of the ORIGINAL equations. What do you notice?

Congratulations! You just solved a system of equations! The point (6, 4) provides a value for each variable ($x = 6$ and $y = 4$) that makes both equations true, and that's what a solution is.

A **solution to a system** of two equations is a pair of numbers, one for each variable, that make both equations true when substituted in. **Solving a system** means finding all pairs of numbers that are a solution.

9. What does the solution we found tell us about the shipping weights?

10. How could you have found the solution to the system from the table you made without graphing?

Our next goal will be to develop an algebraic method for solving a system of equations.

11. Write the result you got earlier when you solved the second equation for y. If you look at it the right way, this provides another name for y: the right side of the equation. Substitute that for y in the ORIGINAL FIRST EQUATION. How many variables are in the resulting equation?

12. Solve the resulting equation for x, then find the value of y that corresponds to it. (You'll need to decide for yourself how to do that.) Did you end up with the same solution you got from graphing?

The method that we used to solve this system is called substitution, because what makes it work is substituting a new name for one of the variables that leaves an equation with only one variable.

Solving a System of Equations Using Substitution

Step 1: Solve one of the equations for one of the variables. You can choose either equation and either variable, but it's almost always worth thinking about a choice that makes it less likely you'll end up with fractions.

Step 2: Substitute the result of Step 1 for the variable you solved for in the OTHER equation. This results in an equation with only one variable.

Step 3: Solve that equation, and you'll have a value for one of the variables.

Step 4: Substitute the value from Step 3 back into the result of Step 1. This will give you a value for the other variable.

Step 5: Write the solution of the system. You can write it either as a point, like (6, 4), or as two equations, like $x = 6$, $y = 4$.

Did You Get It ?

Try this problem to see if you understand the concepts we just studied. The answer can be found at the end of the Portfolio section.

1. Solve the system of equations.

$4x + 3y = -2$
$y - 4x = -6$

Interestingly, this technique can also be used to solve problems that could be solved by writing a single equation, like the next one.

While traveling on business, Eldrick bought a \$50 prepaid international cell phone to call his girlfriend Lindsey. Calls cost \$0.07 per minute. How many minutes has he used if the display shows \$32.15 remaining?

To answer this question we could think about two different equations. One states that the value of the card V is \$32.15.

13. Write that equation here:

Next we can write another equation that states that the value V started at \$50 and decreased by \$0.07 for each minute used x.

14. Write that equation here:

15. Complete the table.

x	1st equation: $V =$	2nd equation: $V =$
0		
100		
200		
300		
400		
500		

16. Graph both equations on the grid.

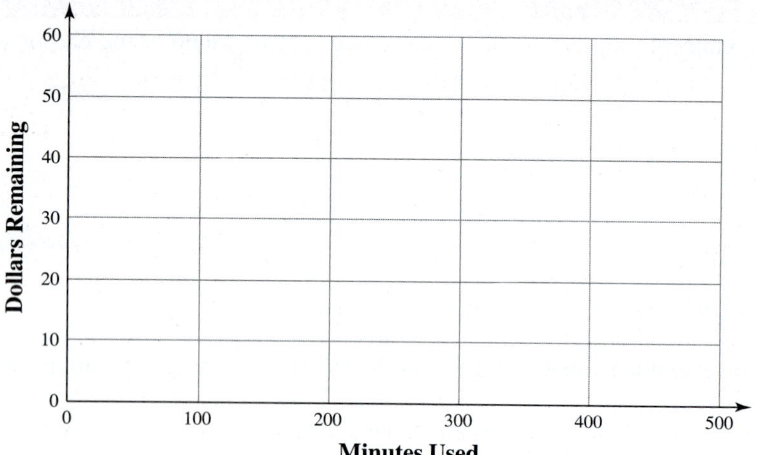

Math Note

Notice that the equation $V = 32.15$ is a horizontal line. This makes sense because it can be written in slope-intercept form as $V = 0x + 32.15$, which shows that it has slope zero.

17. Use your graph to estimate the solution to the system of equations. How confident are you of your answer? Why?

18. Now solve the system using the substitution method. Did you get the same solution? Use this result to solve the original problem.

19. How would you determine the second coordinate of the solution?

Did You Get It

2. A taxi company charges a fixed initial charge of \$4 plus an additional \$1.80 per mile. How long was the taxi ride if the total cost was \$14.62? Write an equation showing the constant total cost. Write a second equation showing the total cost based on the fixed initial cost and the cost per mile. Solve this system of equations and answer the question.

3-10 Group

1. The members of an intramural softball team decide to get custom t-shirts made. There are two screening shops in town that they can choose from. Wave Graphics charges a setup fee of $22, and then each shirt is $7. The Shirt Shack doesn't charge a setup fee, but each shirt is $9. If for some bizarre reason the team decided to only order three shirts, which shop would be the cheaper choice?

2. The goal is to find the number of shirts for which the two shops would charge the same.

   a. Write an equation that describes the cost C of buying x shirts from the Shirt Shack.

   b. Write an equation that describes the cost C of buying x shirts from Wave Graphics.

   c. Now you have a system of equations! Good job. Let's examine the costs associated with each shop by making a table of values.

Number of shirts (x)	Cost at Shirt Shack (C)	Cost at Wave Graphics (C)
3		
6		
9		
12		
15		

   d. Based on the table, estimate the number of shirts that would make the total cost the same at each shop.

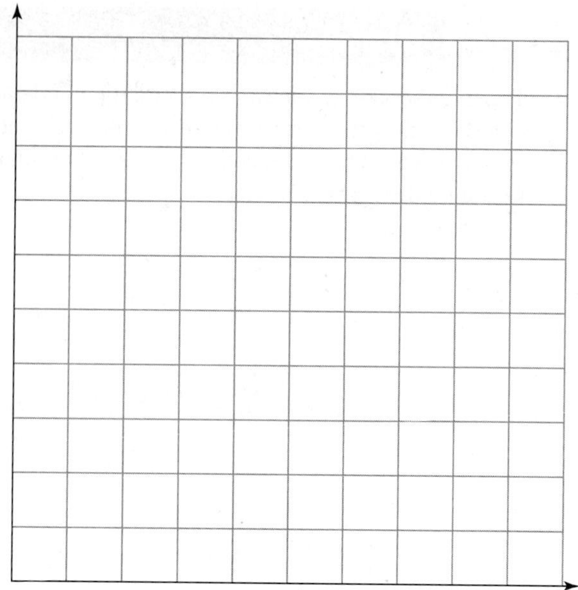

e. Graph the lines corresponding to each of the equations you wrote in parts a and b on the same coordinate system. Make sure you choose a scale so that the point where the two lines cross is visible.

f. What quantity does the height of every point on each line represent?

g. The point where the graphs cross is where the two heights are the same. What does this represent in the problem?

h. Based on your graph, estimate the number of shirts that will lead to the same cost at each shop.

i. Solve the system of equations algebraically to find the exact solution.

j. What will the total cost be for the number of shirts found in part i?

Now that we've practiced the process of setting up and solving a system of equations to solve a problem, we'll close the lesson by giving you a problem to solve, and providing just a couple of suggestions to get you started.

3. The Shirt Shack operates on a daily fixed cost plus a variable cost that depends on the number of shirts screened in one day. The total cost for screening 260 shirts on Friday was $1,015. The total cost for screening 380 shirts on Saturday was $1,345. What is the fixed daily cost? What is the cost to screen each shirt?

a. We're asked to find two things in this problem. Assign a variable to each.

b. Now use your variables to write and solve a system of equations, then write your solution to the problem in the form of a sentence or two.

Did You Get It

3. The Sugar Buzz has two popular kinds of candy. The owner is trying to make a mixture of 100 pounds of these candies to sell at $3 per pound. If the gummy crickets are priced at $2.50 per pound and the sour cubes are $3.75 per pound, how many pounds of each should be mixed in order to produce the mixture he's shooting for?

3-10 Portfolio

Name ______________________________

Check each box when you've completed the task. Remember that your instructor will want you to turn in the portfolio pages you create.

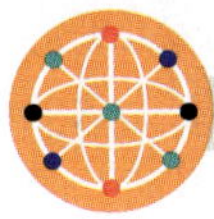

Technology

1. ☐ This refers to the problem in Questions 1–7 of the Applications section of this lesson. Use your equations for the cost of buying diapers with the Amazon Family membership and the cost without the membership to complete the table shown in an Excel spreadsheet. Then graph the system of equations. Explain how the graph shows when it is worth joining Amazon Family.

	A	B	C
1	Boxes of diapers bought	Cost with membership	Cost without membership
2	0		
3	2		
4	4		
5	6		
6	8		
7	10		
8	12		

Online Practice

1. ☐ Include any written work from the online assignment along with any notes or questions about this lesson's content.

Applications

1. ☐ Complete the Applications problems.

Reflections

Type a short answer to each question.

1. ☐ What is a system of equations? What kinds of situations are likely to be modeled by a system of equations?

2. ☐ How can you solve a system of equations using graphing?

3. ☐ If it's possible to solve a system of equations by graphing, why is it often better to solve algebraically?

4. ☐ Name one thing you learned or discovered in this lesson that you found particularly interesting.

5. ☐ What questions do you have about this lesson?

Looking Ahead

1. ☐ Complete the Prep Skills for Lesson 4-1.
2. ☐ Read the opening paragraph in Lesson 4-1 carefully and answer Question 0 in preparation for that lesson.

Answers to "Did You Get It?"

1. $x = 1, y = -2$
2. $C = 14.62$, $C = 4 + 1.80x$; the ride was 5.9 miles long.
3. 60 pounds of gummy crickets, 40 pounds of sour cubes

Answers to "Prep Skills"

1. A bag of 40 Reese's cups for \$6.99 is the better deal at \$0.17 per cup compared to \$0.24 per cup.
2. Four pounds at \$8.19 is the better deal at \$2.05 per pound compared to \$2.33 per pound.
3. $-3 - 3y$
4. $6x - 27$
5. $15x - 140 = 85$
6. $0.17c = 25$, where c represents the number of Reese's cups purchased.
7. $800 + 12c = 1{,}100$, where c represents the number of rounds of golf played with a cart.
8. $800 + 12c + 2w = 1{,}100$, where c represents the number of rounds of golf played with a cart and w represents the number of rounds played walking.

3-10 Applications

Name ______________________________

Parents with young babies buy a lot of diapers (to say the very least), so getting the best deal is pretty important. Comparing prices can be harder than you would think, though — retailers don't all sell diapers in the same size boxes. A recent check of prices on Pampers Swaddlers in Size 1 revealed the following:

Amazon.com: 234 diapers for $46.99
Meijer: 160 diapers for $34.99

1. Find the unit price for each box of diapers and use it to decide which is the best deal. (There's free shipping, so don't worry about shipping or tax.)

Amazon offers an interesting deal: If you pay $99 per year to join their Amazon Family club, you get 20% off all diaper purchases.

2. What would the $46.99 box of diapers cost with that discount?

3. Write an equation that provides the total cost C of buying x boxes of diapers in one year with the Amazon Family membership.

4. Write another equation that provides the total cost C of buying x boxes of diapers in one year from Amazon without being a member of Amazon Family.

3-10 Applications

Name ______________________________

5. Complete the table of values for each equation.

x	1st equation: $C =$	2nd Equation: $C =$
0		
3		
6		
9		
12		
15		

6. The equations you wrote in Questions 3 and 4 form a system of equations. Solve that system using substitution and use your result to find how many boxes of diapers you'd need to buy in a year for the two costs to be the same. What would the cost be?

7. Under what circumstances would you want to join Amazon Family to buy diapers? (Well, obviously having a baby would be one of those circumstances. What other ones?)

8. After her first two exams in psychology, a student has a mean score of 90 and a range of 12. What were her scores on the two exams?

a. Define two variables corresponding to two unknowns in the problem.

3-10 Applications

Name ______________________________

b. Write an equation stating that the mean of the two exam scores was 90, and one stating that the range was 12.

c. Solve the system of equations you wrote using your favorite method, then write a sentence answering the question about exam scores.

9. A hospital needs 80 liters of a 12% solution of disinfectant. This solution is to be prepared from a 33% solution and a 5% solution. How many liters of each should be mixed to obtain this 12% solution? There are different ways to solve this problem, but using a system of equations is the simplest way.

Unit 3 Language and Symbolism Review

Carefully read through the list of terminology we've used in Unit 3. Consider circling the terms you aren't familiar with and looking them up. Then test your understanding by using the list to fill in the appropriate blank in each sentence.

building up
conditional equation
correlation coefficient
direct variation
equation
equivalent
equivalent equations
exchange rate
generalizing
inequality
line of best fit
linear equation
literal equation
$m = \frac{y_2 - y_1}{x_2 - x_1}$
proportion
rate of change
slope
solution
solution to a system
solving a system
system of equations
terms
x intercept
$y = mx + b$
y intercept
$y = kx$
$y - y_0 = m(x - x_0)$

1. Rewriting fractions so that they have a bigger numerator and denominator is called ______________ a fraction.
2. A ______________ is a rate that compares the change in one quantity to the change in another.
3. Two quantities are ______________ if they have the same value.
4. The point where any graph crosses the y axis is called the ______________ for the graph.
5. When applied to the graph of a line, the ______________ of the line describes how steep the line is.
6. The formula for the slope of a line is ______________.
7. The individual pieces of an expression containing addition and subtraction are called ______________.
8. A point where a graph crosses the x axis is called an ______________ of the graph.
9. An ______________ is simply a statement that two quantities are equal.
10. A ______________ can be either true or false, depending on what value we choose for the variable quantity.
11. A number that makes an equation a true statement when substituted in for the variable is called a ______________ of an equation.
12. When two equations have the same solution, we call them ______________.
13. An equation stating that two ratios are equal is called a ______________.
14. An equation with the input appearing only to the first power has a graph that is a line and is called a ______________.
15. A ______________ is an equation with more than one variable, so that when you solve for one of the variables, the result is an expression rather than a number.
16. A statement that one quantity is more or less than another is called an ______________.
17. When you develop a strategy to solve a specific problem, then adapt that strategy to a variety of related problems, that's called ______________ your approach.
18. An ______________ is a number that describes how much of one currency you can trade for another currency.
19. It's called a ______________ relationship when one variable quantity is found simply by multiplying a second variable quantity by a constant.
20. The algebraic equation that describes the relationship when two quantities vary directly looks like ______________.

21. The slope-intercept form of the equation of a line is ________________.

22. The point-slope form of the equation of a line is ________________.

23. When two data sets appear to have an approximately linear relationship, ________________ is the line that best represents the data on a scatter plot.

24. The ________________ is a number that measures how well the line of best fit seems to fit the data.

25. A ________________ is two or more related equations that share common variables.

26. A ________________ of two equations is a pair of numbers, one for each variable, that make both equations true when substituted in.

27. ________________ of two equations means finding all pairs of numbers that are a solution.

Unit 3 Technology Review

This is a short review of the technology skills we've used in Unit 3. In each case, rate your confidence level by checking one of the boxes, If you feel like you're struggling with these skills, consult the online resources for extra practice.

				1. Graphing an equation with a graphing calculator (Lesson 3-2)
				2. Finding the line of best fit with a graphing calculator (Lesson 3-9)
				3. Finding the line of best fit with a spreadsheet (Lesson 3-9)

1. Use a graphing calculator to graph the equation $P = 0.25x - 50$. Use a scale that displays the x values from 0 to 500 with a distance of 50 units between each tick mark and displays the y values from -100 to 100 with a distance of 10 units between each tick mark.

2. Use a graphing calculator to create a scatter plot and determine the equation of the line of best fit for the data in the table. Also include the correlation coefficient.

x	y
2	4
5	7
7	8
9	11
12	15
14	19

3. Use a spreadsheet to create a scatter plot and determine the equation of the line of best fit for the data in the table. Also include the correlation coefficient.

x	y
10	18
15	16
19	12
26	9
29	6
34	2

Unit 3 Learning Objective Review

This is a short review of the learning objectives we've covered in Unit 3. In each case, rate your confidence level by checking one of the boxes. If you feel like you're struggling with these skills, consult the lesson referenced next to the objective and see the online resources for extra practice.

1. Interpret a rate of change. (Lesson 3-1)
2. Predict a future value from a rate of change. (Lesson 3-1)
3. Calculate a rate of change. (Lesson 3-1)
4. Find the intercepts of a line. (Lesson 3-1)
5. Interpret the meaning of the intercepts of a line. (Lesson 3-1)
6. Write expressions based on given information. (Lesson 3-2)
7. Interpret algebraic expressions in context. (Lesson 3-2)
8. Evaluate and simplify expressions. (Lesson 3-2)
9. Explain what it means to solve an equation. (Lesson 3-3)
10. Demonstrate the procedures for solving a basic linear equation. (Lesson 3-3)
11. Solve a literal equation for a designated variable. (Lesson 3-3)
12. Demonstrate the procedures for solving a linear inequality. (Lesson 3-4)
13. Solve application problems that involve linear inequalities. (Lesson 3-4)
14. Solve application problems using numerical calculations. (Lesson 3-5)
15. Solve application problems using linear equations. (Lesson 3-5)
16. Identify situations where direct variation occurs. (Lesson 3-6)
17. Write an appropriate direct variation equation for a situation. (Lesson 3-6)
18. Solve an application problem that involves direct variation. (Lesson 3-6)
19. Write an equation of a line given a description of the relationship. (Lesson 3-7)
20. Write an equation of a line that models data from a table. (Lesson 3-7)
21. Write an equation of a line from a graph of the line. (Lesson 3-7)
22. Graph a line by plotting points. (Lesson 3-7)
23. Find the y intercept and equation of a line given two points. (Lesson 3-8)
24. Find the equation of a line using point-slope form. (Lesson 3-8)
25. Convert between forms of a linear equation. (Lesson 3-8)
26. Determine whether two variables have a linear relationship. (Lesson 3-9)
27. Calculate the line of best fit for a set of data using a spreadsheet. (Lesson 3-9)
28. Calculate the line of best fit for a set of data using a calculator. (Lesson 3-9)
29. Interpret the correlation coefficient for a data set. (Lesson 3-9)
30. Solve an application problem involving a system of equations. (Lesson 3-10)
31. Illustrate the solution to a system of equations using a table. (Lesson 3-10)
32. Illustrate the solution to a system of equations using a graph. (Lesson 3-10)

After you've evaluated your confidence level with each objective and gone back to review the objectives you weren't sure about, use the problem set as an additional review.

1. The following two graphs show the distance traveled in miles based on time in hours for two different cars.

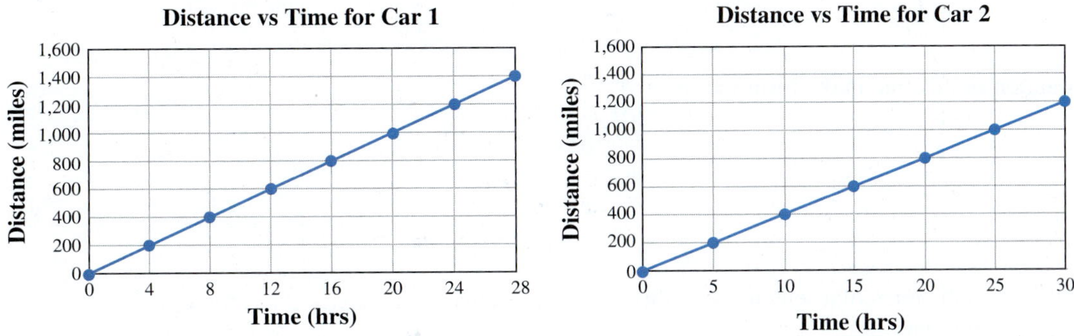

Which statement is more accurate? Circle your choice and show good mathematical evidence to support your choice.

Choose:

A. Car 1 is moving at a faster speed than car 2.
B. Car 2 is moving at a faster speed than car 1.
C. Both cars are moving at the same speed.
D. It cannot be determined from these graphs.

Explain:

In this graph the input x is the number of units produced by a machine in a factory. The output y is the profit made by the sale of these units when they are produced.

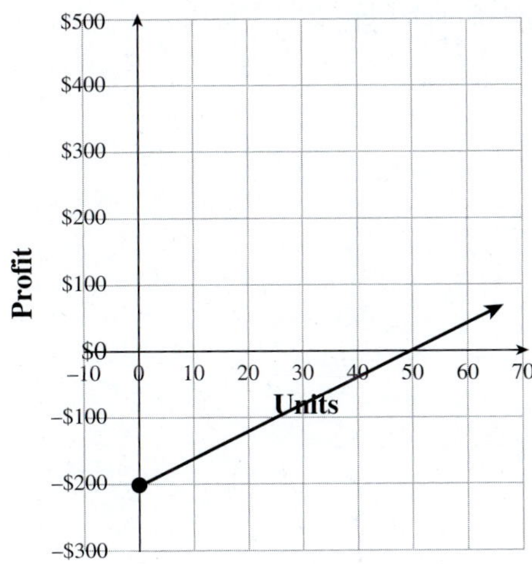

2. Determine the x intercept and interpret its meaning.

3. Determine the y intercept and interpret its meaning.

4. Determine the slope of this line and explain what it means.

5. Determine the equation of this line in the form $y = mx + b$.

6. Describe your equation in the form of a sentence like this: "If x represents _______________, then $y =$ _______________ represents _______________."

7. Use your equation to determine the profit if 60 units are produced and sold.

8. Use your equation to determine the number of units that must be produced and sold in order to generate a profit of $100.

9. The following equation has been solved. Provide an explanation of what was done in each step.

$$-2.5x + 70 = 65$$

$$-2.5x = -15$$

$$x = 6$$

Solve each equation in Questions 10–13.

10. $3x = 12$

11. $t + 4 = 28$

12. $b - 250 = 200$

13. $\frac{y}{2} = 11$

In Questions 14 and 15, solve each equation for a.

14. $P = a + b + 2c$

15. $y = 2ab - c$

16. The following inequality has been solved. Provide an explanation of what was done in each step, and insert the correct inequality symbol between the left side and the right side of the inequality in each step.

$-2x - 13 < 9$

$-2x \quad 22$

$x \quad -11$

17. Is -15 a solution to the inequality in Question 16? Is 0? Is 15?

18. Graph the solution to Question 16 on a number line.

Solve each inequality in Questions 19–22.

19. $\frac{a}{4} \leq 20$

20. $z - 4 > 10$

21. $-3b \leq 60$

22. $x + 5 < 11$

For Questions 23–25, refer to the following: A marketing firm has been hired to make custom foam fingers. They charge a setup fee of $40 and then charge an additional $5 per foam finger.

23. Write an equation that describes the total cost for an order of x foam fingers.

24. Write the algebraic inequality needed to indicate that the cost for an order of foam fingers would be at most $150. Solve this inequality.

25. Explain the significance of the answer you found for Question 24.

26. In a 200-gallon tank, a farmer has 120 gallons of an insecticide and water solution that is 4% insecticide and 96% water. If he adds 30 gallons of pure water to this tank, what is the percentage of the new solution that is insecticide? What is the percentage of the new solution that is water?

27. Turns out the farmer in Question 26 wanted to get the concentration lower. If he started with the 120 gallons of 4% insecticide mixture in the 200-gallon tank, how many gallons of water would he need to add to get a mixture that is 3% insecticide? Set up and solve an equation to answer this question. Be sure to clearly define any variable used.

For Questions 28–32, assume $20 U.S. can be exchanged for 18.16 Euro.

28. How many Euros do you think you would receive in exchange for $0 U.S.?

29. Would you say that $y = mx + b$ or $y = kx$ would be an appropriate form for an equation that gives the number of Euros based on the number of U.S. dollars? Explain your reasoning.

30. Determine the equation that gives the number of Euros based on the number of U.S. dollars.

31. Use your equation and a calculator or spreadsheet to complete the table.

U.S. dollars	Euros
20	18.16
40	
60	
80	
100	

32. Use your equation to determine the number of Euros that you would receive in exchange for $500 U.S.

33. An appliance repair shop charges $50 for a house call. They charge an additional $30 per hour for labor. Write the equation of a line in slope-intercept form to model this situation. Explain the meaning of your independent and dependent variables.

34. Graph the equation you wrote in Question 33.

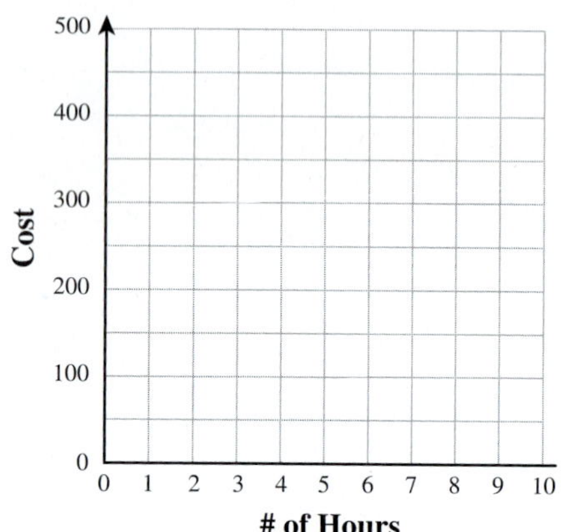

35. An airplane starts at an altitude of 36,000 ft and begins descending at a constant rate. Write the equation of a line in slope-intercept to model this situation. Explain the meaning of your independent and dependent variables.

Minutes	Altitude
0	36,000
5	28,500
10	21,000
15	13,500

36. Determine an equation of the line in slope-intercept form that passes through (−4, 9) with a slope of −2.

37. Determine an equation of the line in point-slope form that passes through $(-2, 4)$ and $(-1, -3)$. Then write your equation in slope-intercept form.

Consider the following as you work Questions 38–43. A math teacher asked his students to keep track of the number of hours they spent doing homework for a particular unit of a math course. After the unit exam, the grade obtained by each student was recorded next to the amount of time that student spent doing homework. The results are displayed in the table.

Study Time	Exam Score
22	94
18	85
13	77
14	88
17	89
12	91
15	61
8	73
16	84
16	90
5	55
20	99

38. Determine the equation of the line of best fit for this data. Round all coefficients to 4 decimal places.

39. Determine the meaning of the slope of this equation.

40. Determine the meaning of the y intercept of this line.

41. Determine the value of the correlation coefficient r. What does this tell you about how closely the quantities are related?

42. Based on this equation, predict the exam score (rounded to the nearest whole number) if a student spends 18 hours doing homework for this unit.

43. Compute the residual for the point where the study time was 18 hours, and explain what it means.

44. A jet has a fuel tank containing 40,000 gallons of jet fuel when it takes off on an international flight at midnight. If the plane consumes 3,000 gallons of fuel per hour, determine the number of hours into the flight when the jet's fuel tank contains 4,000 gallons of jet fuel. Set up and solve an equation to answer this question.

45. Use part of the equation you wrote in Question 44 to create a graph and a table. Use the graph and the table to illustrate the solution to Question 44. Indicate where you would find the solution on the graph and in the table.

Gallons Remaining

Hours

Hours	Gallons Remaining

Paul works for an organization that prints a color newsletter. The organization is trying to decide whether to send the newsletter out for printing, or buy a color copier of its own. The local copy shop will print the newsletters for $0.35 per page. If the organization buys its own printer for $2,200, then it can print the newsletters itself for $0.13 per page.

46. Write equations for the cost C of each option.

47. Solve the system of equations you wrote in Question 46. Explain the significance of the solution for this system.

Unit 4
Living in a Nonlinear World

Outline

Lesson 4-1 Prep Skills

SKILL 1: ADD OR SUBTRACT LIKE TERMS

In Lesson 1-2, we saw that quantities can only be added if they're like quantities, which means they're identical except possibly for the numeric part, which we call the **coefficient**. For example, $8n$ and $12n$ are like quantities because they're both just n without the 8 and the 12, while $8n$ and $12n^2$ are not like quantities because n and n^2 aren't identical.

When an expression has several pieces that are added or subtracted, the individual pieces are called **terms.** In this case, we use the phrase *like terms* rather than like quantities. Deciding which terms can be added or subtracted is an important part of simplifying algebraic expressions.

- $10k + 8 - 3k + 4$ $\quad$ $10k$ and $-3k$ are like terms; 8 and 4 are like terms
  $= 7k + 12$
- $4n - 7n^2 + 11n + 7 - 4n^2$ $\quad$ $4n$ and $11n$ are like terms; $-7n^2$ and $-4n^2$ are like terms
  $= 15n - 11n^2 + 7$

SKILL 2: DIVIDE TWO TERMS BY ONE TERM

When performing a division like $\frac{4+12}{2}$, begin by combining like terms in the numerator, then divide. But when the numerator has terms that can't be combined, as in $\frac{4n+12}{2}$, that won't work.

Instead, to simplify the expression, we'll essentially do the OPPOSITE of adding fractions. Look at the fraction as the *answer* to an "adding fractions" problem: There are two terms added in the numerator, with a single denominator. As an addition problem, this would look like $\frac{4n}{2} + \frac{12}{2} = \frac{4n+12}{2}$. To perform the division, we'll just do that process in reverse, breaking the original fraction into two separate fractions, each with the common denominator.

- $\frac{4n+12}{2} = \frac{4n}{2} + \frac{12}{2}$ $\quad$ Now perform the two easier divisions
  $= 2n + 6$

PREP SKILLS QUESTIONS

1. Simplify each expression by combining like terms.

   a. $12x + 7 - 13 - 4x$ $\quad$ b. $4m^2 - 3m + 5 + 7m - 10m^2$ $\quad$ c. $4y + 7y^3 + 9$

2. Perform each division.

   a. $\frac{20x+16}{4}$ $\quad$ b. $\frac{-12n-96}{6}$

Lesson 4-1 Oh Yeah? Prove It!

LEARNING OBJECTIVES

- ☐ 1. Apply inductive reasoning to make a conjecture.
- ☐ 2. Disprove a conjecture by finding a counterexample.
- ☐ 3. Apply deductive reasoning to solve a problem.

©Pixtal/age Fotostock RF

When you have eliminated the impossible, whatever remains, however improbable, must be the truth.
– Sherlock Holmes

Here's a hypothetical situation: You're in an English literature course, and the prof reads a passage from a Shakespearean sonnet, then says "Clearly, Shakespeare was referring to the forthcoming rise of reality television when he wrote that passage." Would you (a) mumble "Wow, no kidding" and write his claim down in your notebook, or (b) respectfully disagree and challenge him to back up his claim? If you understand what getting an education is all about, you'll choose option b. Most students would. But for some reason, those same students will dutifully scribble down every formula and example a math professor puts on the board without ever questioning where it all comes from or why it makes sense. Today, we fight back! In this lesson we'll study two different types of reasoning used in math. Ultimately, what it's about is being able to think in a logical, orderly way. And what skill could possibly be more useful than that?

0. What do you think the word "reasoning" means?

4-1 Group

©MCT/Getty Images

Each day that the United States Supreme Court is in session, the nine justices perform the traditional "conference handshake" that began during the late 19th century under Chief Justice Melville W. Fuller. Each justice shakes hands with all of the others to indicate that their differences of opinion won't prevent them from focusing on justice and a fair judicial process.

Here's a question to ponder: Just how many total handshakes are necessary? Will there be any time left over for things like doing their actual job? It's not such an easy question. So let's see if we can look at some simpler cases to give us some guidance.

1. How many handshakes would there be if there were just two justices?

We can illustrate the situation by drawing two dots, one for each justice, and connecting them with a line, which represents them shaking hands with each other.

You might think that drawing the diagram for two justices was silly, but it gives us an idea of how to figure out the number of handshakes for larger groups.

2. Connect each pair of dots with a line to indicate a handshake that will take place. Then count the number of lines (handshakes) for each group size.

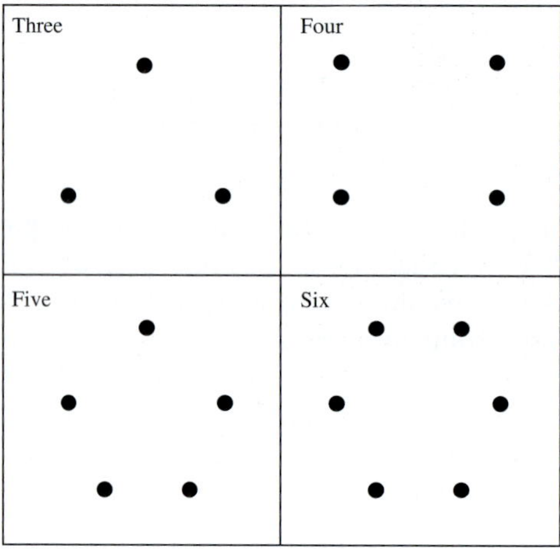

Inductive reasoning is the process of reaching a conclusion based on specific examples. After finding the number of handshakes for two through six justices, can we use those examples to draw a conclusion about how many are needed for all nine?

3. Fill in the table with the results you've already discovered, then see if you can use inductive reasoning to find a pattern and finish the rest of the table.

Number of Justices	Number of Handshakes
2	
3	
4	
5	
6	
7	
8	
9	

Did You Get It

Try this problem to see if you understand the concepts we just studied. The answer can be found at the end of the Portfolio section.

1. A polygon is a closed geometric figure with three or more sides, like a triangle, rectangle, pentagon, etc. Each point where two sides meet is called a vertex. A diagonal of a polygon is a line segment from one vertex to another that isn't one of the sides of the polygon. If you look at your drawings from Question 2, you'll see that a polygon with 3 sides has no diagonals, one with 4 sides has 2 diagonals, one with 5 sides has 5 diagonals, and one with 6 sides has 9 diagonals. Use inductive reasoning to make a prediction for the number of diagonals in polygons with 7, 8, 9, and 10 sides.

In Question 3, you made an educated guess about the number of handshakes necessary for nine justices using inductive reasoning. **Conjecture** is another name for an educated guess based on available evidence. Proving that a conjecture is true almost always involves more than inductive reasoning. For example, after watching a couple of races on TV, you might conjecture that all NASCAR drivers are men. In order to prove that conjecture, you would need to confirm the sex of every single driver in NASCAR; just watching a bunch of races won't prove anything conclusively.

Proving that a conjecture is false is much easier — all you need is a single example that violates the conjecture. If Danica Patrick drives in the next race you watch, she provides a **counterexample** that proves your conjecture was false.

In Questions 4–6, find a counterexample to prove that each conjecture is false.

4. Every time it rains, my softball game gets cancelled. Our game was cancelled last night.

Conjecture: It must have rained last night.

Counterexample:

5. Look at the equation $y = (x + 3)(x - 7)(2x - 1)$.

a. Substitute -3 in for x and find the corresponding value for y.

b. Repeat part a for $x = 7$.

Conjecture: Any number substituted in for x will make $y = 0$.

Counterexample:

6. These three right triangles are all isosceles.

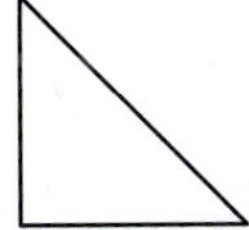  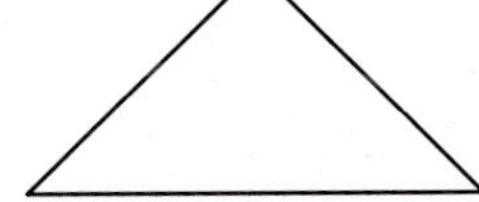

Conjecture: Every right triangle is an isosceles triangle.

Counterexample:

Math Note

A triangle is **isosceles** when exactly two sides have the same length.

Did You Get It

2. Prove that each conjecture is false by finding a counterexample.
   a. Every month of the year has at least four letters in its name.
   b. No U.S. city is farther north than Toronto, Canada.

7. The numbers illustrated here are called **triangular numbers**. Use inductive reasoning to find the fifth and sixth triangular numbers. You can either draw diagrams or notice a pattern.

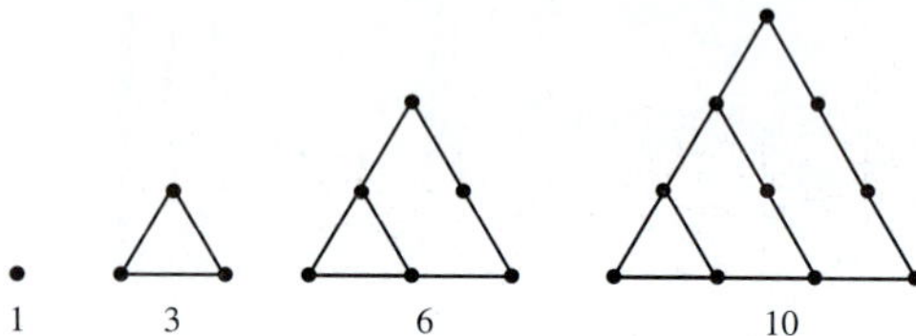

8. This pattern should look familiar. . . . If it doesn't, go back to page 458. Looking at the diagrams, can you think of a reason that the triangular numbers match the number of handshakes needed, starting with two people? Or do you think it's a coincidence?

9. The next triangle is called **Pascal's triangle** in honor of the 17th-century French mathematician Blaise Pascal. Even though it's relatively simple, it turns out to have a wide variety of applications in math. Use inductive reasoning to complete the 6th and 7th rows of the triangle.

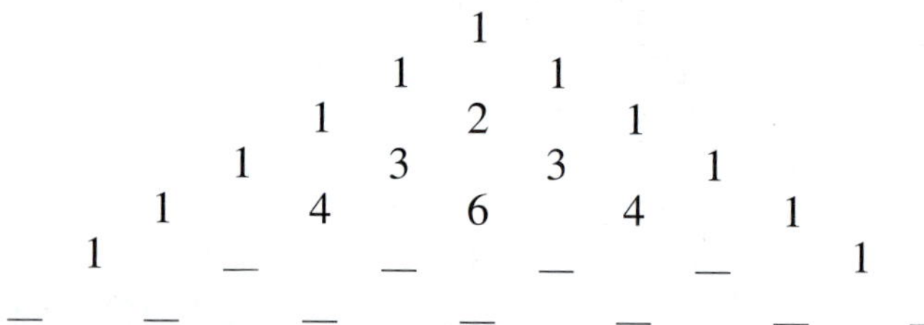

10. Here's a clever number trick. Pick any number you like, and multiply it by 8. Add 40, then divide by 4. Subtract 10 from the result, then subtract your original number. What do you get? Try some specific examples, and use inductive reasoning to make a conjecture as to what the result will always be.

4-1 Class

Now it's time to take a closer look at deductive reasoning. **Deductive reasoning** is the process of reasoning that arrives at a conclusion based on known rules or principles. Inductive reasoning is great for seeing patterns and making educated guesses about general results. For example, after completing Question 10 in the Group section, you're probably pretty darn sure that the result of the number trick is always the original number. But no matter how many different numbers you choose, you can't be 100% certain that it ALWAYS works that way unless you try every number in the world. Of course, that's impossible.

This is where deductive reasoning comes into play. Instead of looking at specific examples, we'll start with an **arbitrary number** and represent it with the letter n. By arbitrary, we mean a symbol that is able to represent ALL possible numbers. That's what makes using a variable so great: Since we want n to represent ANY number, we need it to be a number that can change. In other words, a variable!

1. Starting with a number n, multiply that number by 8.

2. Now add 40 to the result of Question 1.

3. Next, divide the result of Question 2 by 4. Be careful! Make sure you divide BOTH TERMS by 4.

4. Subtract 10 from the result of Question 3.

5. Now subtract the original number n. What is the result?

6. Describe why this is different from the specific calculations we did in the group activity, and why it PROVES that the result will ALWAYS be the orginal number.

Did You Get It

3. Try the procedure with a handful of numbers then make a conjecture as to what the result will always be. Then use deductive reasoning to prove your conjecture.

   Take a number.
   Double it.
   Add 12.
   Subtract 4.
   Divide by 2.
   Subtract your original number.

This is an example of deductive reasoning because we used a general calculation to draw a conclusion, not a handful of specific calculations. Let's see if we can use a similar approach to find a general formula that solves the handshake problem once and for all.

7. Pretend that you're one of the nine justices: How many people's hands do you need to shake?

8. Choose one other justice. How many hands does he or she need to shake?

9. If we number the justices from 1 through 9, this table shows how many hands each needs to shake.

Justice	1	2	3	4	5	6	7	8	9
Number of handshakes	8	8	8	8	8	8	8	8	8

This results in how many handshakes total? (Write as a multiplication.)

10. This result doesn't match the answer we got using inductive reasoning. Explain why this answer is twice as big as it should be. (Think about handshakes, not math!)

11. Conclusion: The number of handshakes needed for nine justices is $\frac{9 \times 8}{2}$. Write a general formula with variable n that would describe the number of handshakes needed for a group of n people. Then verify that your formula gives the values you found for the table on page 458.

Did You Get It

4. Look back at the Did You Get It 1, where we studied the number of diagonals in a polygon. Write a formula with variable n (the number of sides) that describes the number of diagonals. (Hint: The formula we developed for the number of handshakes is a good starting point.) How does the number of diagonals differ from the total number of lines drawn in Question 2?

To wrap up our study of inductive and deductive reasoning, decide if the type of reasoning used in each example is inductive or deductive.

12. After failing the first two tests in math class, John decided that he better start doing the homework regularly.

13. Marlene was having a hard time keeping up with minimum payments on her credit card, so she started to make a budget to make sure she wouldn't spend more money than she was making.

14. In spite of his friends' pleas, Mark tried to drive home after drinking too much. None of them were surprised when he got busted for DUI.

15. My favorite professional team started off the new season 12–0, so I'm sure they're going to make the playoffs.

4-1 Portfolio

Name ______________________________

Check each box when you've completed the task. Remember that your instructor will want you to turn in the portfolio pages you create.

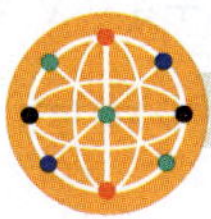

Technology

1. ☐ Complete the following spreadsheet down to the row for 30 people. In the Group section, you noticed a pattern using inductive reasoning: Use that pattern to write formulas in column B. In the Class section, we developed a formula for the number of handshakes required. Use that formula in column E. A template to help you get started can be found in the online resources for this lesson.

	A	B	C	D	E
1	Inductive Reasoning			Deductive Reasoning	
2	Number of people (n)	Number of handshakes (H)		Number of people (n)	Number of handshakes (H)
3	2	1		2	
4	3	3		3	
5	4	6		4	
6	5			5	

Online Practice

1. ☐ Include any written work from the online assignment along with any notes or questions about this lesson's content.

Applications

1. ☐ Complete the applications problems.

Reflections

Type a short answer to each question.

1. ☐ Describe the difference between inductive and deductive reasoning.
2. ☐ Describe a time in your life that you used inductive reasoning, and one when you used deductive reasoning.
3. ☐ What's the point of a counterexample?
4. ☐ Name one thing you learned or discovered in this lesson that you found particularly interesting.
5. ☐ Do you have any questions about this lesson?

Looking Ahead

1. ☐ Complete the Prep Skills for Lesson 4-2.
2. ☐ Read the opening paragraph in Lesson 4-2 carefully and answer Question 0 in preparation for that lesson.

Answers to "Did You Get It?"

1. 7 sides has 14 diagonals, 8 sides has 20, 9 sides has 27, 10 sides has 35.
2. **a.** May has only 3 letters.

   b. The entire state of Alaska is well north of Toronto, as are many cities in the northern portion of the mainland U.S.
3. The answer is always 4. $\frac{2x + 12 - 4}{2} - x = \frac{2x + 8}{2} - x = \frac{2x}{2} + \frac{8}{2} - x = x + 4 - x = 4$
4. Subtract the number of sides from the total number of handshakes and you're left with the number of diagonals, so $\frac{n(n-1)}{2} - n$ will work.

Answers to "Prep Skills"

1. **a.** $8x - 6$ **b.** $-6m^2 + 4m + 5$ **c.** $7y^3 + 4y + 9$
2. **a.** $5x + 4$ **b.** $-2n - 16$

4-1 Applications

Name ______________________________

In the Group section, we looked at triangular numbers. We can also define square numbers and pentagonal numbers.

1. Use inductive reasoning to find the fifth square number.

Square Numbers

1 4 9 16

2. Use inductive reasoning to find the fifth pentagonal number.

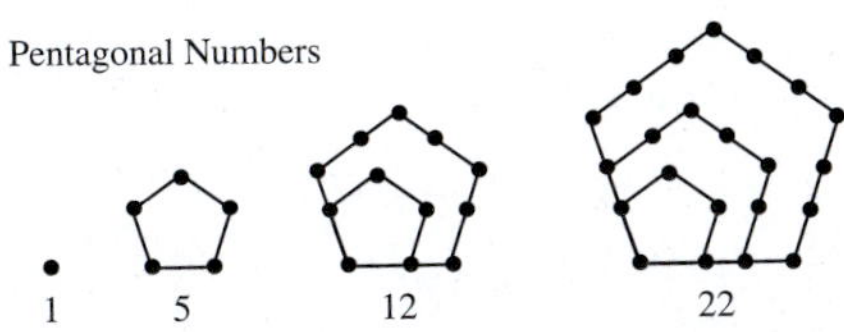

3. A formula for the nth triangular number is $\frac{n}{2}(1n - (-1))$; this simplifies to $\frac{n(n+1)}{2}$.

A formula for the nth square number is $\frac{n}{2}(2n - (0))$; this simplifies to n^2.

A formula for the nth pentagonal number is $\frac{n}{2}(3n - (1))$; this doesn't simplify. Sigh.

Using inductive reasoning, a formula for the nth hexagonal number is $\frac{n}{2}(\underline{4}\,n - (\underline{2}))$.

(Hint: Look at the pattern in the missing numbers for the previous three formulas.)

4. Use the formulas from Question 3 to complete this table. Feel free to use a calculator, or (better still) a spreadsheet.

	Triangular Numbers	Square Numbers	Pentagonal Numbers	Hexagonal numbers
1st				
2nd				
3rd				
4th				
5th				
6th				
7th				

4-1 Applications

Name ______________________________

In Questions 5 and 6, find a counterexample that shows the conjecture is false.

5. Conjecture: Every eight-legged creature is a spider.

6. Conjecture: No number with last digit zero can be divided evenly by 3.

In Questions 7–10, decide if inductive or deductive reasoning was used.

7. Charles Manson has been denied parole 12 times. Applying again would be a big waste of time.

8. In this state, to be a nurse you need at least a two-year degree. So when I was in the emergency room after an accident, I asked the super-cute nurse where she went to college.

9. We got four inches of spring snow last night, so instead of the softball game I had scheduled, I'll make other plans.

10. Last night while playing blackjack, the dealer's first card was an ace four times in a row. So I really couldn't believe it when it happened again on the next hand.

Lesson 4-2 Prep Skills

SKILL 1: COMPUTE OR APPROXIMATE SQUARE ROOTS

A square root of a positive number a is a number you have to square to get a back. For example, since 7^2 is 49, 7 is a square root of 49. Here's another way to think of it: If you're my square, I'm your square root. Notice that $(-7)^2$ is also 49, so -7 is a square root of 49, too. In fact, every positive number has two square roots, one positive and one negative. The symbol $\sqrt{\ }$ is used to represent the *positive* square root of the number enclosed by the symbol. So we would write $\sqrt{49} = 7$.

- $\sqrt{25} = 5$ because $5^2 = 25$
- $\sqrt{144} = 12$ because $12^2 = 144$

Each of 49, 25, and 144 are called **perfect squares** because each is the square of a whole number. When numbers get larger, it can be really hard to tell if a number is a perfect square, and really hard to compute square roots mentally. In that case, a calculator becomes an essential tool. Scientific calculators (nongraphing) have a $\sqrt{\ }$ key: You enter the number you want the square root of, then hit the square root key. On many graphing calculators, you'll press 2nd then x^2 to get the root symbol, then key in the number and press ENTER.

- $\sqrt{841} = 29$

If the number we're finding the square root of isn't a perfect square, the decimal equivalent of the square root will always have infinitely many digits. Since you can't write all the digits, you'll have to round at some point, turning the decimal equivalent into a decimal approximation. This is another situation where a calculator (or spreadsheet, using the "= SQRT( )" command) is pretty much essential.

- $\sqrt{8} \approx 2.83$
- $\sqrt{1{,}243} \approx 35.3$

SKILL 2: FIND DISTANCE ON A GRAPH

When two points on a graph have the same y coordinate, it means they live at the same height. In this case you can find the distance between them by just finding the difference of their x coordinates.

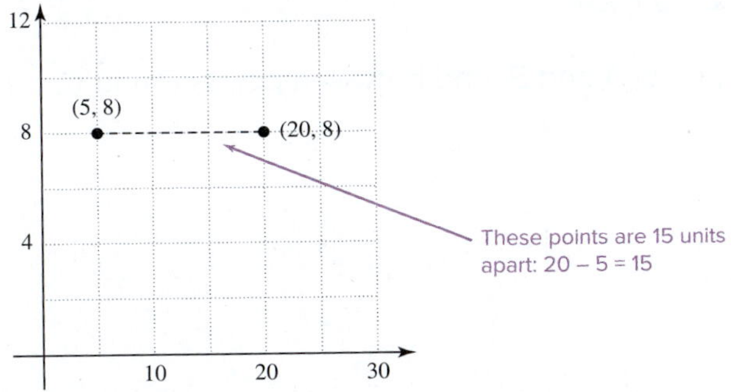

When two points on a graph have the same x coordinate, it means they live on the same vertical line. In this case you can find the distance between them by just finding the difference of their y coordinates.

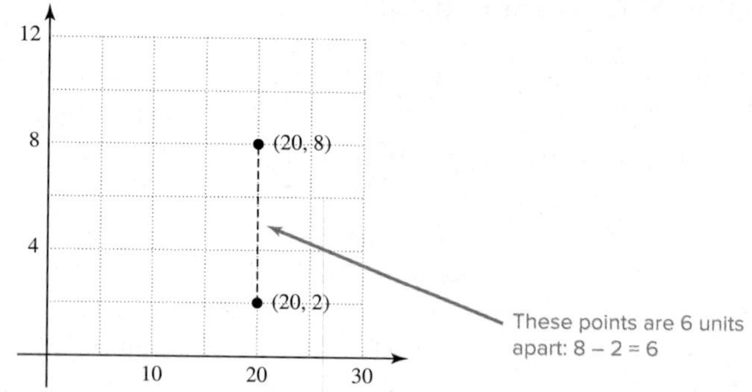

SKILL 3: SET UP AND SOLVE A PROPORTION

In the applications for this lesson you'll set up a proportion to find the size of something that we have a scale model for. We did this in Lesson 3-6 (the famous Dave Sobecki bobblehead problem). The procedure is based on writing a ratio that compares the size of the model to the size of the actual object on both sides of an equation.

- On a map, 1 inch corresponds to 1,000 feet. How far apart are two buildings that are $4\frac{1}{4}$ in. apart on the map?

$$\frac{\text{Map distance}}{\text{Real distance}}: \frac{1 \text{ in.}}{1{,}000 \text{ ft}} = \frac{4\frac{1}{4}\text{in.}}{x \text{ ft}} \Rightarrow x = 1{,}000 \times 4\frac{1}{4} = 4{,}250 \text{ ft}$$

PREP SKILLS QUESTIONS

1. Compute the square root exactly without using a calculator or spreadsheet.

   a. 16 b. 36 c. 100

2. Compute the square root exactly using a calculator or spreadsheet.

   a. $\sqrt{289}$ b. $\sqrt{961}$ c. $\sqrt{12{,}769}$

3. Approximate the square root to two decimal places using a calculator or spreadsheet.

   a. $\sqrt{14}$ b. $\sqrt{140}$ c. $\sqrt{1{,}107}$

4. Find the distance between points A and B, and between points B and C.

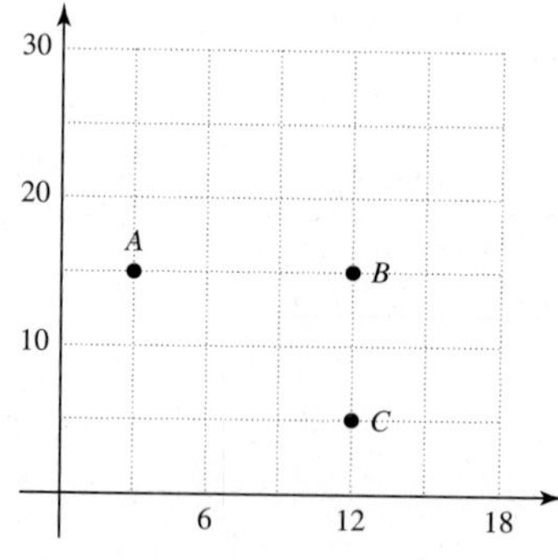

5. While on vacation last year, my wife and I bought a scale model of the space needle in Seattle. The model is 5 in. tall, and the actual building is 605 ft tall. How tall would a model of the Statue of Liberty (305 ft) be at this scale? Round to the nearest hundredth.

Lesson 4-2 A Road Map to Success

LEARNING OBJECTIVES

- ☐ 1. Solve application problems using the Pythagorean Theorem.
- ☐ 2. Solve application problems involving the distance formula.

Map out your future, but do it in pencil. The road ahead is as long as you make it.
—Jon Bon Jovi

©Ilene MacDonald/Alamy RF

While a road map to success is a useful metaphor, in this lesson we'll study a particular type of map that will help us to introduce some key ideas in geometry. When you're looking at a road map, it's very much like looking straight down from a plane, and the roads all look two-dimensional. But of course, we live in a three-dimensional world, and what you're missing is the change in height. Because of this, the actual distance you'd drive on a road isn't always the distance you'd measure on a map. We'll use the change in height for roads to learn about two important results used to study our physical world: the Pythagorean theorem and the distance formula.

0. What do you think the sign in the picture means?

4-2 Class

The road sign above describes a mountain road with a grade of 8%. This is a warning that the road is steep enough to require some extra caution, but what exactly does it mean? The slope of a road is very much like the slope of a line: It's a comparison between how far the road goes up or down for a certain horizontal distance. A grade of 8% means that for any horizontal distance, the change in height is 8% of that distance.

1. What's the change in elevation over a 1-mile horizontal stretch for a road with a 8% grade? Fill in that change in height on the diagram. Why did we write 1 mile as 5,280 feet on the diagram?

5,280 ft

Did You Get It?

Try this problem to see if you understand the concepts we just studied. The answer can be found at the end of the Portfolio section.

1. The steepest road in America is Canton Avenue in Pittsburgh, with a 37% grade. To be fair, it's only that steep for about 25 feet, but whatever. If the road was that steep for a full mile, what would be the elevation change in feet?

The diagram now illustrates the situation we described in the lesson opener: The horizontal distance (5,280 feet) would appear on a map of the road, but the slanted line at the top of the triangle represents the road. So the actual distance you drive is the length of that slanted side, which we don't know based on the horizontal distance and the change in height.

Fortunately, there's a very old and very famous result from geometry that will allow us to find the actual driving distance: the Pythagorean theorem. This provides a mathematical relationship between the lengths of sides in a triangle when two of the sides are perpendicular. In the following diagram, we'll call the lengths of the three sides *a, b,* and *c*. The little square at one of the angles indicates that those two sides are perpendicular: We call that angle a **right angle**, and the triangle a **right triangle**.

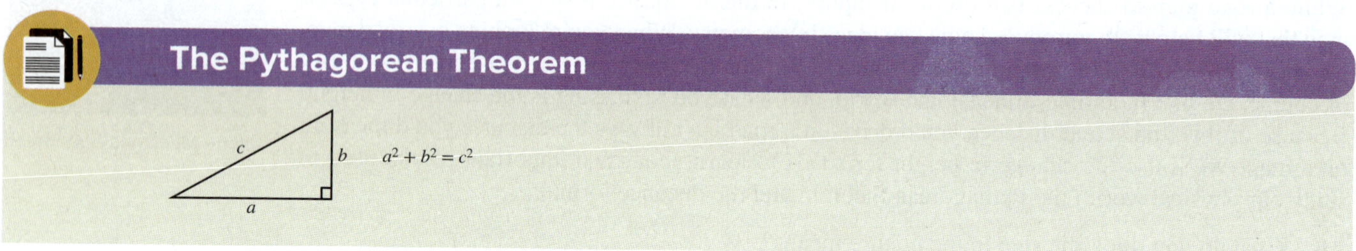

In words, the Pythagorean theorem says that if you square the lengths of the two shorter sides in a right triangle and add the results, you'll get the square of the length of the longest side. The longest side is called the **hypotenuse**, and it's always across from the right angle.

In using the Pythagorean theorem, we're going to work with square roots quite a lot. It turns out that approximation plays a really important role when dealing with roots. For example, we know that $\sqrt{49} = 7$ because $7^2 = 49$. But what is $\sqrt{47}$? Since 47 is a little bit less than 49, it seems reasonable to estimate that $\sqrt{47}$ is a bit less than 7. But how much?

To get more accurate (but still approximate) values for square roots, we can use a calculator or a spreadsheet.

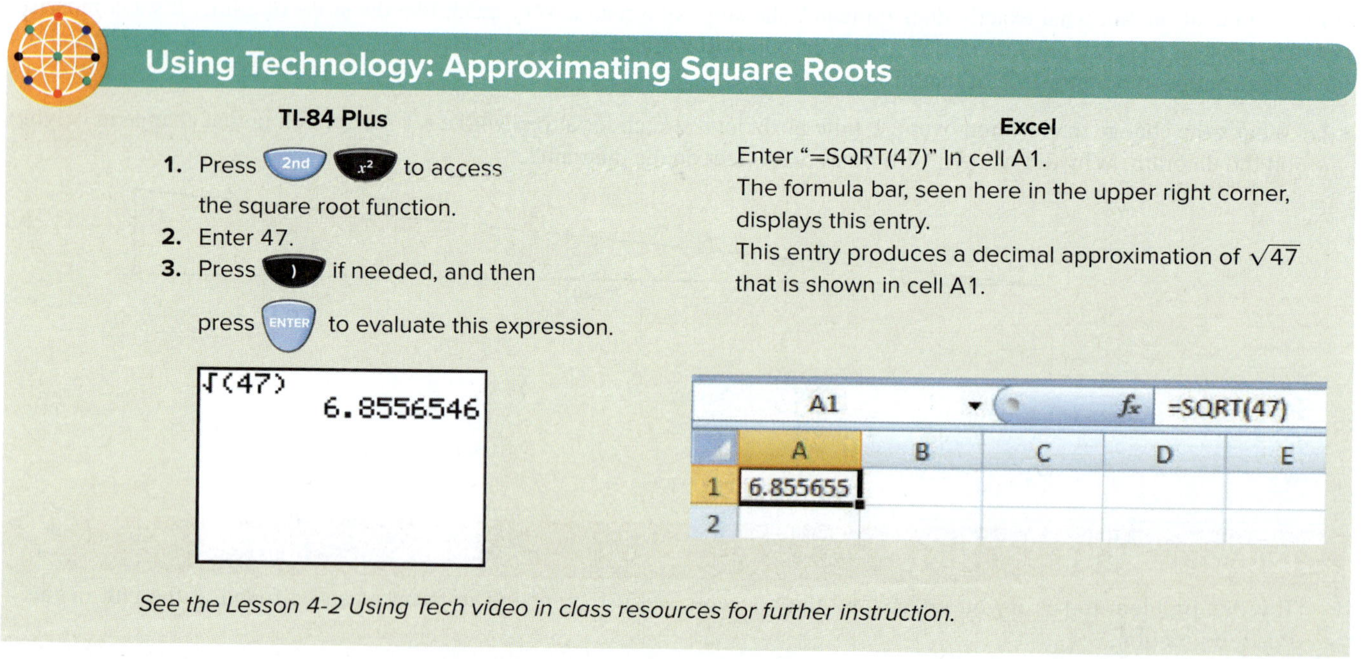

2. Complete the following table of common square roots.

Fill in the correct value to make each equation true. No calculators!		Estimate each square root to the nearest whole number and fill in the relationship between the square root and your estimate with either $<$ or $>$.	Use a calculator or a spreadsheet to approximate the following square roots to the nearest hundredth.
$\sqrt{1} = 1$	$\sqrt{64} = 8$	$\sqrt{11}$	$\sqrt{11} \approx$
$\sqrt{4} = 2$	$\sqrt{81} =$		
$\sqrt{\ } = 3$	$\sqrt{100} = 10$	$\sqrt{78}$	$\sqrt{78} \approx$
$\sqrt{16} = 4$	$\sqrt{121} =$		
$\sqrt{25} = 5$	$\sqrt{\ } = 12$	$\sqrt{99}$	$\sqrt{99} \approx$
$\sqrt{\ } = 6$	$\sqrt{169} =$		
$\sqrt{49} =$	$\sqrt{196} =$	$\sqrt{124}$	$\sqrt{124} \approx$
	$\sqrt{\ } = 15$		

Let's look at how we can use the Pythagorean theorem to study grade.

3. Notice that the diagram on page 471 is a right triangle. Use the Pythagorean theorem to write an equation involving the length of the missing side of the triangle. Then simplify the side with only numbers on it.

4. In order to solve your equation to find the missing length, you'll need to "undo" the square: This is exactly what square roots are designed to do. So apply a square root to both sides to find the missing length.

> **Math Note**
>
> There are actually two numbers that make the equation $c^2 = 28{,}056{,}821.76$ true: 5,296.9 and −5,296.9. But in this setting, c is a physical distance, so we ignore the negative solution in Question 4.

5. How much farther would you drive on that road than the 1 mile that would appear on a map?

Did You Get It?

2. Waipio Road in Hawaii is actually steeper than the road described in Did You Get It 1 (at least in some stretches), but access is restricted to four-wheel-drive vehicles and people on foot, so most don't count it as a legitimate road. According to Fixr.com, this road goes up 800 feet in a horizontal span of 3,170 feet. How far does a hiker cover in walking that span?

4-2 Group

1. Suppose that we know that the distance between points A and B is exactly five miles. If the contour of the road followed the solid path in each of the three figures below, would the distance traveled by the car be more than, less than, or equal to five miles?

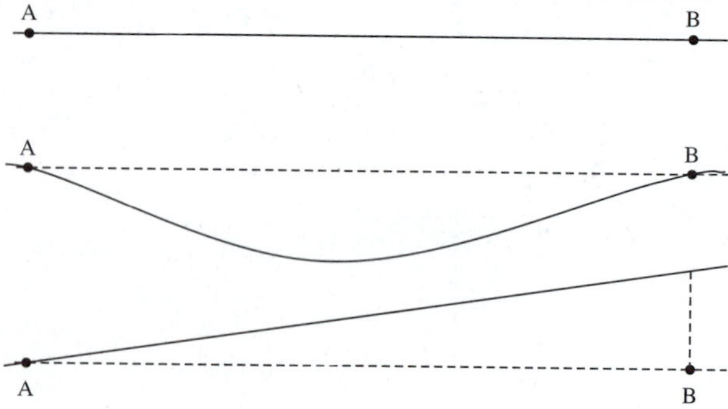

Contour maps might remind you of the isotherm maps we learned about back in Lesson 2-5. But instead of the lines representing locations that have the same temperature, the lines represent locations that have the same altitude. This is a way to overcome the two-dimensional nature of maps, allowing us to see both straight-line distance from above, and changes in height as well. The top part of the next diagram is the contour map; under the map is an illustration of a side view that shows what the elevations look like based on the map.

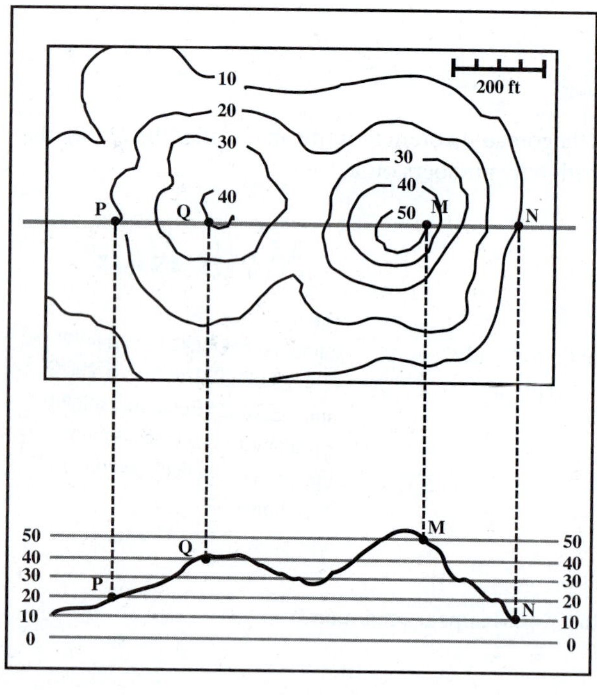

2. How does the distance between points P and Q compare to the distance between points M and N on the map? Use the scale on the map.

3. The lower illustration also shows points P, Q, M, and N, but this time the elevations are shown as well. How does the actual hiking distance from P to Q compare to the hiking distance from M to N? What can you conclude about apparent distance on a contour map?

4. Based on the scale provided on the two-dimensional contour map, the distances from P to Q and from M to N are both about 200 feet. Use the lower illustration to estimate the actual trail length between points P and Q, making sure to describe how you got that estimate. Then repeat for the trail between points M and N.

5. Illustrate the grade of each trail by drawing a right triangle.

6. Use the Pythagorean theorem to estimate the true length of each trail.

7. The road sign at the beginning of this lesson is from a road with an 8% grade for a horizontal span of 8 miles. By how much does the elevation change over that 8-mile span?

8. How far would you actually drive in covering that 8-mile span?

Before selling his house, Brian needs to remove a dead mouse that got trapped under the bathtub. (Clearly, "includes a deceased animal" is not a feature buyers are looking for.) Using the corner of the room as the origin, and using a patented, high-tech rodent corpse location device, he was able to determine that the mouse was located at coordinates (4, −2). An access panel under the sink is located at coordinates (1, −5). Assuming that the units are feet, the goal is to find the distance from the access panel to the mouse.

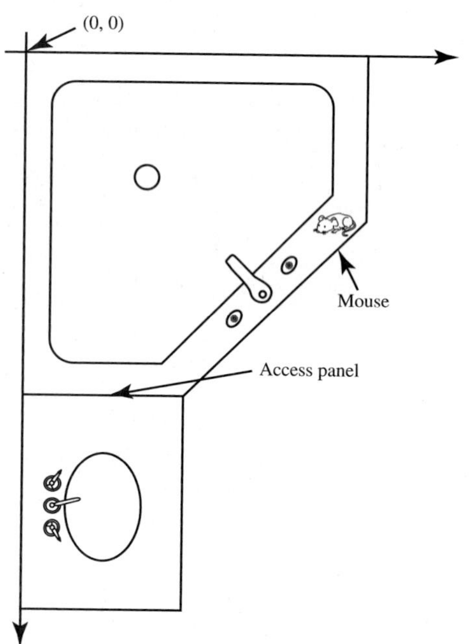

9. Estimate the distance by plotting the two given points on the graph below. If you're creative, you can use a group member's page with the same grid as a ruler to make a good estimate.

10. Using your plot from Question 9, use the Pythagorean theorem to find the exact distance, first in exact form (which will have a square root in it), then in decimal form rounded to the nearest tenth of a foot.

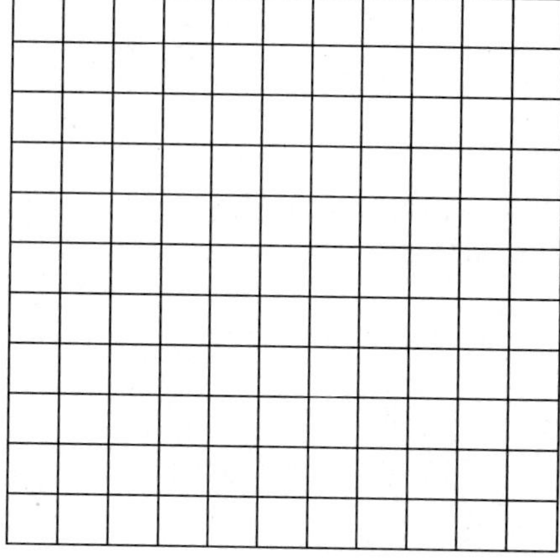

4-2 Class (Again)

The process used to find the distance from the access panel to the mouse in Group Questions 9–10 can be mimicked to develop a generic formula for finding the distance between any two points on a graph. Let's start with two arbitrary points, labeled (x_1, y_1) and (x_2, y_2). Since we're not labeling specific number coordinates, but using symbols, these points can represent ANY pair of points.

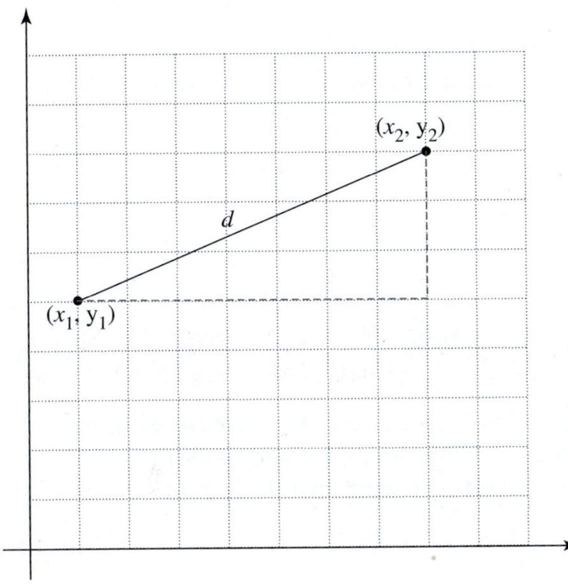

1. What are the coordinates of the point where the two dotted lines meet to form a right angle? (Hint: That point is at the same height as (x_1, y_1) and lives directly below (x_2, y_2).)

2. Use your answer to Question 1 to find the length of the two dotted sides of the right triangle, then label those lengths on the diagram. Taking another look at the mouse problem calculation might help.

3. Use the Pythagorean theorem to set up an equation containing the distance d that we're looking for. (Hint: Putting parentheses around the expression you wrote for the lengths in Question 2 is a good idea.)

4. Solve the equation to find a formula for the distance between any two points.

Math Note

Even though the diagram we're using is in the first quadrant, the fact that we used arbitrary points means that our formula will work for any two points. It also doesn't matter which point you call (x_1, y_1) and which you call (x_2, y_2).

The Distance Formula

The distance between any two points (x_1, y_1) and (x_2, y_2) can be found using the formula:

$$d = \sqrt{(x_2 - x_1)^2 + (y_2 - y_1)^2}$$

5. A map of a national park has a grid on it to help reference locations. The camping office is located at the point (−5, 10) on the grid, and the lodge is at (2,4). Each square on the grid represents a distance of two miles. Use your formula from Question 4 to find the distance from the camping office to the lodge.

Did You Get It?

3. Grids are often used at crime scenes to help in locating evidence, and to document exactly where any evidence was located. At one crime scene, two shell casings were found: one at coordinates (3, 12) and another at coordinates (−5, 6). Each square on the grid represents 3 feet. How far apart were the two shell casings?

4-2 Portfolio

Name ______________________________

Check each box when you've completed the task. Remember that your instructor will want you to turn in the portfolio pages you create.

Online Practice

1. ☐ Include any written work from the online assignment along with any notes or questions about this lesson's content.

Applications

1. ☐ Complete the applications problems.

Reflections

Type a short answer to each question.

1. ☐ What does the Pythagorean theorem say? When can you use it?
2. ☐ Why can't you just find the distance between two points by plotting them on a grid, then counting the number of boxes between them?
3. ☐ Name one thing you learned or discovered in this lesson that you found particularly interesting.
4. ☐ What questions do you have about this lesson?

Looking Ahead

1. ☐ Complete the Prep Skills for Lesson 4-3.
2. ☐ Read the opening paragraph in Lesson 4-3 carefully and answer Question 0 in preparation for that lesson.

Answers to "Did You Get It?"

1. 1,953.6 ft **2.** 3,269.4 ft **3.** 30 ft

Answers to "Prep Skills"

1. **a.** 4 **b.** 6 **c.** 10

2. **a.** 17 **b.** 31 **c.** 113

3. **a.** 3.74 **b.** 11.83 **c.** 33.27

4. Distance between A and B: 9

Distance between B and C: 10

5. About 2.52 inches tall

4-2 Applications

Name ______________________________

Questions 1–3 are based on the contour map provided, which is of Mount St. Helens in Washington. All elevations on the contour lines are given in meters.

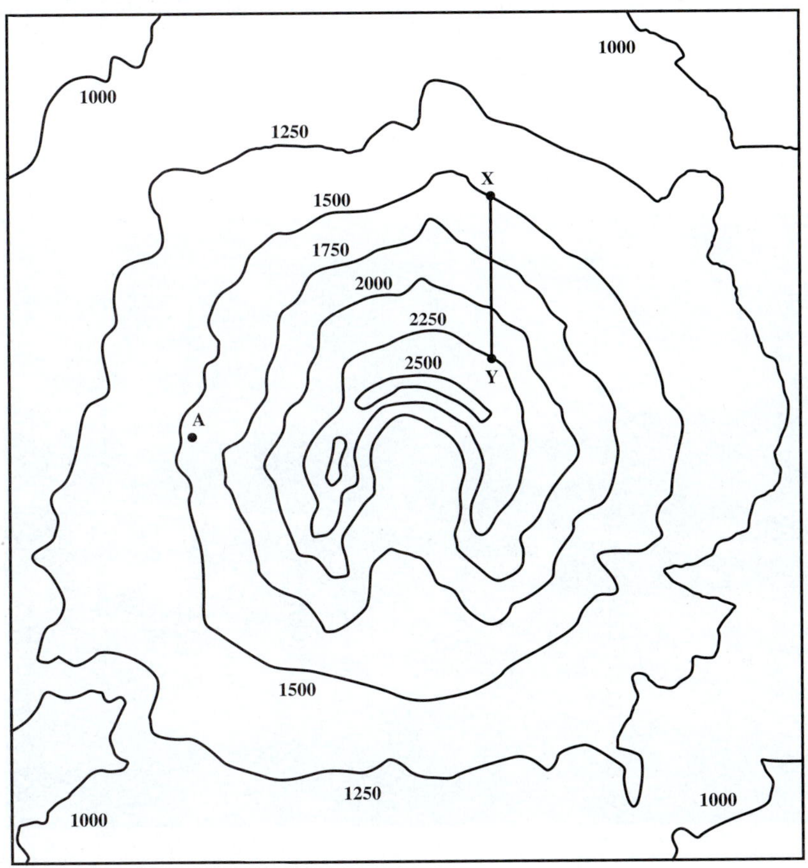

Source: John Pallister/USGS

1. Estimate the elevation of point A.

2. If points X and Y are 5,100 m apart looking straight down at the map, find the grade of a trail that connects those two points. (Hint: Find the change in elevation first.)

3. How far would you walk along the trail from point X to point Y? Assume that the elevation changes in a straight line. Round to the nearest tenth of a meter, then convert the distance to feet and miles.

4-2 Applications

Name ______________________________

4. An archaeological dig is marked with a rectangular grid where each square is 5 feet on a side. An important artifact is discovered at the point corresponding to $(-50, 25)$ on the grid. How far is this from the control tent, which is at the point $(20, 30)$?

Courtesy of Dave Sobecki

I took this photo from the deck of a cruise ship in the harbor just west of downtown Seattle. The road jumped out at me because it looks just like the diagrams we often draw to illustrate slope and grade, and I figured I'd be able to use it for a problem at some point. Well, here's that point.

5. Using proprietary software hacked from the NSA, I was able to tremendously enlarge a second photo that shows more of the bridge without losing any clarity. This allowed me to identify the make and model of every vehicle on the bridge to identify their lengths. (Okay, that's not true — I used Adobe Fireworks to measure the cars in pixels, and looked up the length of an average passenger car. It just sounded cooler the other way.)

   In any case, the average car is 14 feet long, and the average length of the cars I could measure on the bridge was 34.4 pixels. The length of the bridge is 2,709 pixels. How long is that span in feet? (Hint: Set up a proportion that compares photo length in pixels to actual length in feet for cars on one side of the equation, and the bridge on the other.)

4-2 Applications

Name ______________________________

6. I was also able to use the software (and some math, of course) to determine that the grade of the bridge is about 8.15%. How far does the bridge go up vertically over the span you found in Question 5?

7. What distance does a vehicle cover horizontally in driving up the bridge? (Hint: Your answer to Question 5 is the *hypotenuse* of a right triangle, so you'll need to set up the Pythagorean theorem and then do some solving.)

Lesson 4-3 Prep Skills

SKILL 1: FIND THE ANGLES FOR A SECTOR IN A PIE CHART

We studied pie charts a million years ago, way back in Lesson 1-1. The information needed to build a pie chart is percentages: the percentage of the full circle corresponding to each sector. A full circle measures 360°, so to find the angle we need we multiply the given percentage (in decimal form) by 360°.

- In a July 29 Gallup poll, 38% approved of the president's handling of the office, 57% disapproved, and 5% had no opinion. To build a pie chart for this survey you would need:

  $0.38 \times 360° = 136.8°$

  $0.57 \times 360° = 205.2°$

  $0.05 \times 360° = 18°$

SKILL 2: APPROXIMATE QUANTITIES INVOLVING SQUARE ROOTS

When a number is a perfect square, like 25, its square root is a whole number (5 in this case). But if a number isn't a perfect square, it turns out that it's impossible to write an exact decimal form because there will be infinitely many digits. In that case, the best you can do is an approximation, and for that we'll almost always use a calculator or spreadsheet. If you're using a graphing calculator, you'll push 2nd, then x^2 to get the square root feature, then key in the number you're finding the square root of and hit Enter. When using a spreadsheet, pick any cell and enter **=SQRT**(number you're computing the square root of)

- $\sqrt{39} \approx 6.24$
- $\sqrt{1{,}104} \approx 33.23$
- $\sqrt{-18}$ is not a real number because there's no real number whose square is negative.

SKILL 3: SQUARE A RADICAL EXPRESSION

In some sense, the whole point of square roots is that they're the opposite of squares. So if you take any non-negative number, apply a square root, and then square the result, you get right back where you started.

- $(\sqrt{3})^2 = 3$
- $(\sqrt{n+5})^2 = n + 5$ as long as we know that $n + 5$ isn't negative

In short, it doesn't make any difference what the stuff inside the square root is (as long as it isn't negative): If you compute the square of a square root expression, that simply eliminates the square root and you're left with the expression that was inside the radical.

SKILL 4: SOLVE AN EQUATION WITH THE VARIABLE IN THE DENOMINATOR

Students have a tendency to kind of freak out and do some strange things when they need to solve equations that contain fractions. But the truth is that you never actually have to solve equations with fractions, because with one simple step you can always eliminate the fractions and turn an equation with fractions into an easier one without fractions.

And what is this brilliant step? Multiplying both sides by a common denominator for all fractions involved. I promise that this will get rid of all fractions.

- $\frac{12}{x} = 8$ Common denominator is x: Multiply both sides by x

$x \cdot \frac{12}{x} = 8 \cdot x$ Simplify left side: $x/x = 1$

$12 = 8x$ Divide both sides by 8

$x = \frac{12}{8} = \frac{3}{2}$

PREP SKILLS QUESTIONS

1. In a pie chart, one sector represents a data value that's 47% of a whole. Find the angle that sector covers in the pie chart.
2. Use a calculator to find an approximate value for $\sqrt{18.4}$. Round to two decimal places.
3. Use a calculator to find an approximate value for $\sqrt{\frac{0.4(1-0.4)}{800}}$. Round to two decimal places.
4. Find the square of $\sqrt{\frac{12}{n}}$. You can assume that n represents a positive number.
5. Solve the equation: $\frac{36}{n} = 4$
6. Solve the equation: $\frac{0.027}{n} = 3.5$

Lesson 4-3 The Error of Your Ways

LEARNING OBJECTIVES

- ☐ 1. Determine the margin of error in a given poll.
- ☐ 2. Explain the meaning of the margin of error in a given poll.
- ☐ 3. Calculate the number of poll respondents needed for a given margin of error.

Better to trust the man who is frequently in error than the one who is never in doubt.

—Eric Sevareid

Courtesy of Dave Sobecki

Do you know what percentage of Americans think that Bigfoot exists? Me neither. But I bet you can find that information online somewhere. In our information society, it seems like there's a poll for everything, from approval of the president to favorite type of underwear. And since so much information is provided to citizens and consumers in the form of polls, an understanding of how polls operate has become a pretty important survival skill. In this lesson, we'll study a key aspect of interpreting the results of a poll, the margin of error. And we'll find out that understanding error in polls can prevent errors in judgment.

0. When a poll says that a certain percentage of college students feels a certain way on a subject, how do you think they determine that?

4-3 Group

Polling has become big business, and two of the biggest names in that business are Gallup and Pollster. One of the more widely quoted polls they run is used to determine the president's current approval rating among citizens. But of course, they can't ask everyone in the country every week whether they approve or disapprove of the president's performance, so instead they ask a sample of citizens and use sampling and statistical techniques to try to decide if the results are truly indicative of the general population.

1. Go to the website www.pollster.com, then find the link for the presidential approval survey and click on it. What is the president's most current approval rating? Write a couple of sentences explaining what that means.

2. How do you think the approval rating was calculated? Be specific. Make sure your answer contains both of these key words: "population" and "sample."

3. Draw a pie chart describing the most current approval rating. Your chart should have three sectors.

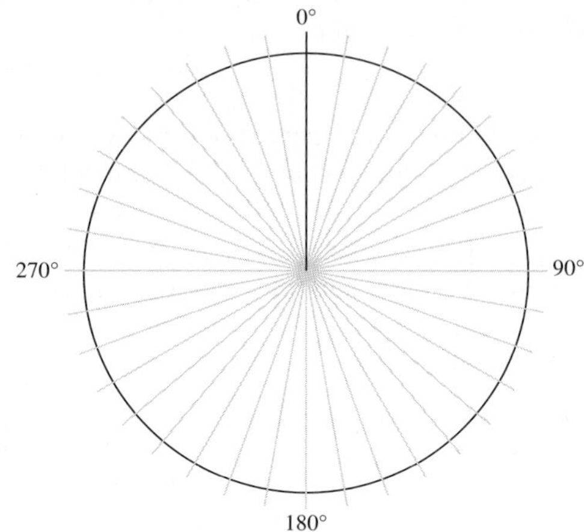

4. How would you describe the trend in approval rating over the last six months?

5. Scroll down a bit and find the first link to the Gallup survey and click on it. You'll see an approval rating and a brief description of how the poll was conducted, including a line that says "Margin of error," followed by something like "±3 percentage points." What do you think the term "margin of error" means in this context?

6. Can you think of another situation in life, school, or work where you've heard either "margin of error" or "margin for error"? Describe.

7. What do you think is better to have, a small margin of error or a big one? Why? Does it depend on the situation?

Did You Get It

Try this problem to see if you understand the concepts we just studied. The answer can be found at the end of the Portfolio section.

1. Back at the Pollster website, find and click on a link for "Congress Job Approval."
   a. What is the approval rating for Congress?
   b. What's the margin of error?
   c. Compare the approval rating to the one for the president and discuss what you think that says.

4-3 Class

Because it's not possible to poll every individual all the time to ask their opinions, sampling plays a really important role in polling. As you know, sampling is the process of choosing a portion of a population in such a way that the individuals chosen are in some way representative of an entire population. The downside of sampling is that it introduces uncertainty in the results of a poll, since to some extent the pollster is making an educated guess.

The **margin of error** for a poll is a statistical measure that provides a range of values that the true outcome of the poll is likely to be inside. A margin of error of 5% means that the true value of the statistic reported by the poll is likely to be within 5% of the reported value.

Let's look at an example. If someone reports that 84% of Americans like pizza (which sounds awfully low to me, by the way), and the margin of error is ±3%, this means they're confident that the actual percentage of Americans that like pizza is somewhere between 81% and 87% (since those are the percentages that are within 3% of 84.)

Here's the catch: How confident is confident? For margins of error, confidence is measured as a percentage. One way of computing margin of error is with a 95% confidence level. In our example above, a 95% confidence level would mean that the polling institution is 95% sure that the accurate percentage of Americans that like pizza is between 81% and 87%.

1. Take another look at the Gallup presidential approval rating, and find the margin of error. Then describe exactly what the poll result means, referring to both the approval rating and the margin of error.

2. An Associated Press poll indicated that 63% of teachers below college level feel like the amount of homework students get is not excessive. The poll had a margin of error of ±3.5% with a 95% confidence level. Write a sentence or two explaining what that result means.

A formula for computing margin of error is provided in the colored box. Sadly, it means absolutely nothing until you know what all of the symbols stand for, which is why you should read everything in the colored box.

Computing Margin of Error

When n people are polled, and $\hat{p}$ give a particular response, the margin of error with a 95% confidence level is given by the formula

$$\text{Margin of Error} = 2\sqrt{\frac{\hat{p}(1-\hat{p})}{n}}$$

Note that both the margin of error and $\hat{p}$ (which is often called the **sample proportion**) are percentages written in decimal form. The symbol $\hat{p}$ is usually read "p hat."

The margin of error means that if this poll were conducted many times, we would expect that at least 95% of the time the actual percentage of the entire population that would give a particular response is somewhere between $\hat{p}$ − the margin of error and $\hat{p}$ + the margin of error.

©Marka/Alamy

3. On December 18, 2015, an organization called Public Policy Polling released the results of a poll with 532 responses. Of those respondents, 181 supported one particular candidate for the presidency in 2016. Among those folks, 41% were in favor of bombing Agrabah, which as it turns out is the fictional country from the Disney movie *Aladdin*. Use the formula in the colored box to find the margin of error for this reported percentage.

4. Write a sentence describing exactly what your answer to Question 3 tells us about all American voters who supported that candidate. More detail is always better.

5. Among all respondents to the survey, 30% supported bombing a cartoon country, and Public Policy Polling reported a margin of error of 4.3%. Does that number match what you get from the formula above?

Did You Get It

2. A *Wall Street Journal* poll conducted in December 2014 asked 1,000 adult Americans if they or a family member had ever been notified by a credit card company of a possible data breach involving their personal information. Forty-five percent reported that they had been notified. Use the formula in the colored box to find the margin of error for this result, then describe what the poll tells us.

6. For this question, don't look at the margin of error formula. Seriously, no peeking or I'll know. Based on what margin of error measures, do you think the margin of error for a poll should go up or down as the number of people surveyed gets larger? Justify your answer.

7. Now look at the margin of error formula. Based strictly on that formula, what should happen to the size of the margin of error if the number of people surveyed gets larger? Does this match your answer to Question 6? Explain.

8. What if the folks from Public Policy Polling who conducted the Agrabah poll got a little bit lazy, and decided to only survey 100 people, then go bowling. Recalculate the margin of error for the overall poll. Did it get bigger or smaller?

While it's absolutely fine and dandy to calculate the margin of error after conducting a poll, in many cases it's more useful to think about what the margin of error might be like BEFORE you're done sampling. This would allow you to decide how many people need to be surveyed in order to make your result reliable.

Suppose that you begin surveying students on your campus about their attitudes on race relations in the United States. After 70 surveys are returned, you find that 32% of respondents think that race relations are a serious problem facing our society.

9. What is the margin of error at that point in the survey?

10. If you decide that you'll only consider your survey a success if the margin of error is at most 4%, it would be perfectly reasonable to want to know how many more people you'd have to survey. Assuming that the result holds at 32%, set up the margin of error formula with a 4% margin of error, a 32% sample proportion, and variable n representing the number of people surveyed. (Remember, percentages need to be written in decimal form.)

In order to find the sample size needed to get that 4% margin of error, we'd need to be able to solve the equation you wrote in Question 10 for n. There are two issues to address: First, the quantity we want to solve for is inside a square root, and second, it's in the denominator of a fraction.

Let's begin by eliminating the radical. To accomplish this, we can square both sides of the equation, because we know that squaring a square root eliminates the radical. But first, it's helpful to rearrange so that the square root is the only thing on one side of the equation.

11. Starting with the equation $0.04 = 2\sqrt{\frac{(0.32)(0.68)}{n}}$ (Aw man, I just gave away the answer to the last question . . .), divide both sides by 2 to isolate the radical.

12. Now square both sides of the equation.

13. Keeping in mind that the goal is to solve for n, what can we do to BOTH SIDES of the equation in order to get n out of the denominator? Explain what we can do, then, you know, do it.

14. Now you should be able to solve for n, so write how many MORE PEOPLE you'd need to survey to get that 4% margin of error we were shooting for.

Math Note

Remember, n represents the **MINIMUM** number of people that would be needed to get a particular margin of error. So if n doesn't work out to be a whole number, you'll need to round UP to the next whole number.

15. Do you find that number surprising? Discuss.

Did You Get It

3. Preliminary results of a survey show that 11% of Americans lack health insurance. How many people need to be surveyed in order to have a margin of error of 2%?

16. Now let's go back to the presidential poll we started this lesson with. Use the approval rating you discovered, along with the margin of error, to find the number of people that were polled. Now look again at the description of the poll. How does your result compare to the number of people that were actually surveyed? What do you think accounts for any discrepancy?

17. Suppose that the result of a public opinion poll shows that just 9% of people would be willing to give up their cell phone for one week in exchange for $100. How many people would need to be surveyed in order to have a margin of error of ±2%?

©Hero/Corbis/Glow Images RF

18. Repeat Question 17, but this time for a poll where 50% of respondents say they'd be willing to give up their phone for a week. How does this affect the number of responses needed to get the 2% margin of error?

Math Note

Remember, the formula we're using provides 95% confidence in our margin of error. For other confidence levels, the coefficient in front of the root changes.

19. Bonus Question: Look carefully at the margin of error formula. What about it, mathematically, makes your conclusion from Question 18 make perfect sense? Detailed explanation if you want bonus points!

20. Now design your own problem. Find a survey online that interests you, and using the result and the reported margin of error, calculate the number of respondents.

4-3 Portfolio

Name ______________________________

Check each box when you've completed the task. Remember that your instructor will want you to turn in the portfolio pages you create.

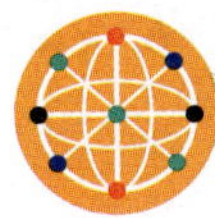

Technology

1. ☐ Find an article on either Gallup.com or Pollster.com that has the term "margin of error" in it, and briefly explain what that means in the context of that particular poll.
2. ☐ As we saw in this lesson, being able to find the number of people that are needed to get an acceptable margin of error is really useful. Build a spreadsheet like the one here that uses a formula to compute the value of n needed to achieve a certain margin of error given a value of $\hat{p}$. (Hint: You'll need to solve the equation that computes margin of error for n. Questions 10–14 in the Class portion will be a big help.)

	A	B	C
1	p	Margin of error	n
2	0.32	0.01	8704
3	0.32	0.02	2176
4	0.32	0.03	967
5	0.32	0.04	544

Online Practice

1. ☐ Include any written work from the online assignment along with any notes or questions about this lesson's content.

Applications

1. ☐ Complete the Applications problems.

Reflections

Type a short answer to each question.

1. ☐ What is the margin of error for a poll? When is it used, and why?
2. ☐ Describe how being familiar with margin of error might help you in the future.
3. ☐ Name one thing you learned or discovered in this lesson that you found particularly interesting.
4. ☐ What questions do you have about this lesson?

Looking Ahead

1. ☐ Complete the Prep Skills for Lesson 4-4.
2. ☐ Read the opening paragraph in Lesson 4-4 carefully and answer Question 0 in preparation for that lesson.

Answers to "Did You Get It?"

1. Answers vary depending on recent poll numbers. **2.** 3.1%; Descriptions vary.
3. 979 people

Answers to "Prep Skills"

1. 169.2° **2.** 4.29 **3.** 0.02 **4.** $\frac{12}{n}$ **5.** $n = 9$ **6.** $n \approx 0.008$

4-3 Applications

Name ______________________________

A stats class was interested in determining the percentage of students on their campus who are opposed to a new policy eliminating all paper books from the campus bookstore in favor of eBooks. After surveying 90 students, the class finds that 33% are opposed to the policy.

1. Calculate the margin of error for this survey. Round the percentage to one decimal place.

©Fancy/Veer/Corbis/Glow Images RF

2. Write a sentence or two describing exactly what your answer to Question 1 tells you about the percentage of the overall student body that is opposed to the policy.

3. The class conducting the poll was unhappy with the margin of error (as well they should be), so they decide to survey another 90 students. With the additional data, the percentage of students opposed to the policy goes up to 35%. What's the margin of error now?

4. If surveying more students keeps the percentage at 35%, how many more would need to be surveyed in order to be 95% confident that the correct percentage for the entire student body is between 32% and 38%? Don't forget to consider the number of students that had already been surveyed. (Hint: What would the margin of error be in this case?)

4-3 Applications

Name ______________________________

You may have noticed that the margin of error formula only factors in the *sample* size, not the *population* size. If you think about it, this sounds kind of fishy: If you survey 30 members of a club out of a population of 40, you'd expect your results to be much more reliable than if you survey 30 people out of the entire U.S. population of 325 million. But the margin of error formula would treat both polls the same.

It turns out that the margin of error formula we've been using works consistently well as long as the sample surveyed is less than 5% of the overall population. In cases where the sample is larger than 5% of the population, we would want to adjust the margin of error to reflect the fact that we're actually polling a large sample relative to the population. For that purpose, statisticians developed a statistic called **finite population correction** (FPC) to account for the smaller margin of error that should result.

5. Suppose that the campus in Questions 1–4 is a small satellite campus, with only 400 students. After surveying 180 students (Question 3), almost half of everyone on campus would have been surveyed. The formula for finite population correction is below. Find the value, and in the next question we'll learn what to do with that number.

$\text{FPC} = \sqrt{\dfrac{N-n}{N-1}}$ where N is the population size, and n is the sample size.

6. The FPC acts as a multiplier: To get the adjusted margin of error, multiply the value of the FPC by the margin of error that came from the usual formula. Find the new margin of error for the situation where 180 students have been polled out of a campus population of 400. (Use the margin of error you found in Question 3.)

7. Without referring to any formulas, describe what should happen to the margin of error as the size of the sample gets really close to the size of the population. Then use the FPC formula to argue that your conclusion matches what would happen mathematically.

Lesson 4-4 Prep Skills

SKILL 1: USE DIRECT VARIATION EQUATIONS

We learned all about direct variation in Lesson 3-6. Here are some of the important ideas:

The most notable characteristic of quantities that vary directly is that if one doubles, the other does as well, and if one is cut in half, the other is too. There's more to it than that—the quantities go up or down proportionally—but this is a nice basic way to think about what direct variation means.

More specifically, if two quantities x and y vary directly, the relationship between them can be described by an equation of the form $y = kx$, where k is some constant. This number is known as the constant of variation.

- When you fill your gas tank, the cost C varies directly with the number of gallons g. So we know that these quantities can be modeled by the equation $C = kg$.

In order to find the constant k in a variation situation, you need values for each quantity. This results in an equation where k is the only unknown, so you can solve for it.

- Continuing the gas example, if it costs \$30.69 to pump 14.2 gallons:

$$30.69 = k \cdot 14.2$$
$$k = \frac{30.69}{14.2} \approx 2.16$$

Now we know that the equation relating cost and number of gallons is $C = 2.16g$.

SKILL 2: MAKING A TABLE OF VALUES FOR AN EQUATION

This is another skill that we've practiced multiple times in this course. It was originally introduced in Lesson 2-2. More recently, we practiced this a lot in Lesson 3-7. Focus on questions where you're given an equation with two variables, and you build a table of inputs and the associated outputs. If you're not sure how to do the one in the Prep Skills Questions, look back at Lessons 2-2 and 3-7.

SKILL 3: MAKING A SCATTER PLOT

This important skill was introduced in Lesson 1-5, and of course we've used it several times since then. There are additional practice questions in the Prep Skills for Lessons 3-1, 3-2, and 3-9. Are you starting to get the idea that we think being able to draw scatter plots is a pretty big deal?

PREP SKILLS QUESTIONS

1. The force required to move an object varies directly as the resulting acceleration of that object. Write a direct variation equation that describes the force in terms of acceleration.
2. More fun with physics: The direct variation equation $F = kx$ describes a relationship between the distance in meters that a spring stretches or compresses (x) when a force of F newtons is applied. If one particular spring stretches 0.3 meters when a force of 60 newtons is applied, find the specific direct variation equation for this spring.

3. For the equation $y = \frac{200}{x}$, fill in the table of values.

x	1	5	10	20	40	150
y						

4. Make a scatter plot using your table from Question 3.

Lesson 4-4 Where's My Jetpack?

LEARNING OBJECTIVES

- ☐ 1. Identify situations where inverse variation occurs.
- ☐ 2. Solve problems involving direct and inverse variation.

Nothing is too small to know, and nothing is too big to attempt.
— William Cornelius Van Home

Courtesy Everett Collection

If you talk about modes of travel with most Americans who grew up before the 1990s, you'll likely find that they feel kind of cheated. Based on movies, TV shows, and books from the 1950s through the 1980s, we all pretty much figured we'd have flying cars and jetpacks by the time 2017 rolled around, just like James Bond. What a letdown! For the most part, we're still stuck with modes of transportation that have been around for over 100 years. Sigh. Jetpacks exist of course, but they're not exactly widely available. In this lesson, we'll use modes of transportation, including the long-promised jetpack, to study the relationship between speed and time for a given trip. Clearly those quantities are related, but it turns out that exactly how they're related will lead us to study a new type of variation.

0. Think about the direct variation that we studied in Lesson 3-6. Do you think speed and time for a trip will vary directly? Why or why not?

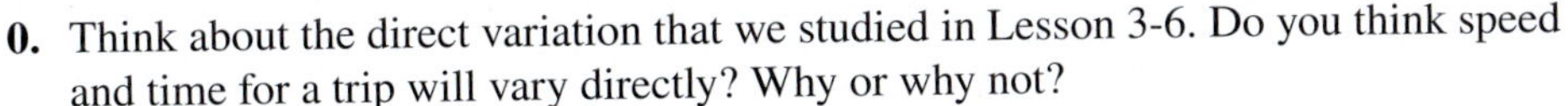

4-4 Group

In studying direct variation in Lesson 3-6, we found that when two quantities vary directly, if one goes up the other does as well by the same factor. But think about speed and time when traveling: The faster you go, the sooner you get there. So as speed goes up, time goes down. The goal of this activity is to study that relationship in depth. Let's say you need to travel 40 miles, and you have several different choices of how to get there.

1. For each mode of travel, find how long it will take to make the 40-mile trip.

Mode	Speed (mi/hr)	Time (hr)
Mosey	1	
Walk	3	
Jog	5	
Bike	10	
Bus	20	
Car	40	
Jetpack	80	

2. The speed doubles when going from jogging (5 mi/hr) to biking (10 mi/hr). What happens to the time?

3. The speed doubles when going from bus (20 mi/hr) to car (40 mi/hr). What happens to the time?

4. The speed is cut in half when going from bus (20 mi/hr) to bike (10 mi/hr). What happens to the time?

5. The speed is divided by four when going from jetpack (a zippy 80 mi/hr) to bus (20 mi/hr). What happens to the time?

6. Does time depend on speed in this problem, or does speed depend on time? Explain your answer.

Did You Get It

Try this problem to see if you understand the concepts we just studied. The answer can be found at the end of the Portfolio section.

1. The amount of time it takes to boil a pan of water varies inversely with the wattage of the stove it's sitting on.
   a. If a stove is turned up so that the wattage is doubled, what happens to the time required to boil water?
   b. If it took 4 minutes for a pan of water to boil, then the wattage was reduced to one-third of the previous wattage, how long would it take another pan with the same amount of water to boil?

7. Graph the values in your table and connect them with a curve. Put the independent variable (as decided in your answer to Question 6) on the x axis and the dependent variable on the y axis. Include labels on each axis.

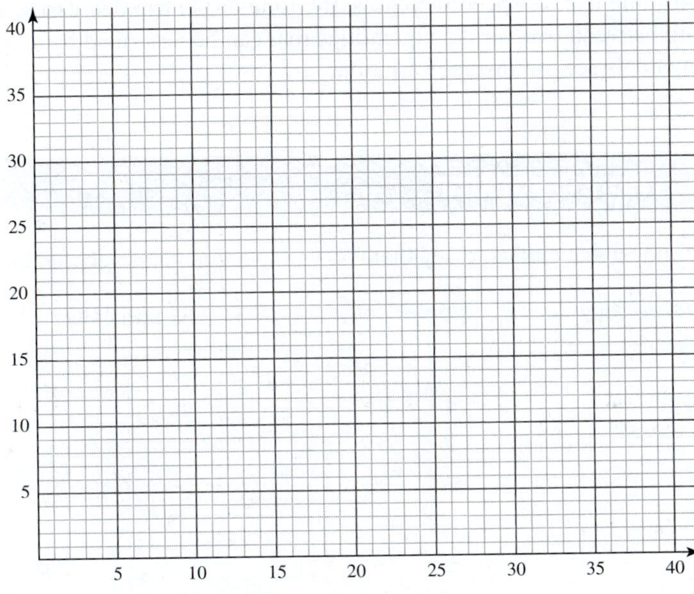

8. Use your graph to estimate the time it would take to travel 40 miles at 7 mi/hr.

9. Use your graph to estimate the speed needed to make the 40-mile trip in two and a half hours.

Did You Get It

2. According to the graph, about how long would it take to travel 40 miles at 32 miles per hour? What speed would be needed to make the trip in 6 hours?

10. (This one will impress your instructor.) Find an equation that relates speed to time for a 40-mile trip.

11. Use your equation to find how long it would take to make the trip at 65 miles per hour.

4-4 Class

In the Group portion of this lesson, we saw that for a 40-mile trip, as speed increases, the time of the trip decreases, and vice versa. This is typical of **inverse variation,** a relationship between two variable quantities that can be described by an equation of the form $y = \frac{k}{x}$, where k is some constant. Let's compare direct and inverse variation.

Comparison of Direct and Inverse Variation

The following are various ways of illustrating what it means to say that two variable quantities vary directly.

Verbally

The quantity y varies directly as the quantity x, and the constant of variation is k.

Algebraically

$y = kx$

Example:
$y = 3x$

Numerically

x	$y = 3x$
1	3
2	6
3	9
4	12
5	15

Graphically

And these are various ways of illustrating what it means to say that two variable quantities vary inversely.

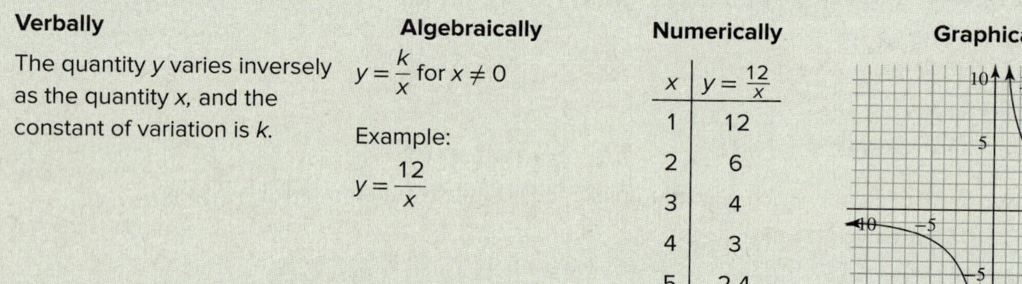

Verbally

The quantity y varies inversely as the quantity x, and the constant of variation is k.

Algebraically

$y = \frac{k}{x}$ for $x \neq 0$

Example:

$y = \frac{12}{x}$

Numerically

x	$y = \frac{12}{x}$
1	12
2	6
3	4
4	3
5	2.4

Graphically

1. When you exchange currency, the value of currency you trade in varies directly with the amount you get back. So as the amount you trade in increases, the amount you get back ________________.

2. When you drive a certain distance, the time it takes varies inversely with your speed. So if your speed increases, the amount of time for the trip ________________.

In Questions 3–5, write an equation of variation for the situation, using k as the constant of variation.

3. The price P at which a manufacturer is willing to sell a certain product varies inversely as the number of items that consumers are willing to buy u.

4. When a store has discounted gallons of milk by a certain amount, the total savings T varies directly as the number of gallons bought g.

5. Isaac Newton discovered that the gravitational force between two bodies F varies inversely as the square of the distance between them d.

Did You Get It

3. Write a variation equation using constant k for the situation described in Did You Get It 1.

6. The gross pay P for working a certain hourly job varies directly as the number of hours worked h. Use that fact to fill in the table.

h	P
5	\$56
10	
15	
20	
25	

7. The number of days d it takes for a construction crew to repave a certain road varies inversely as the number of workers w that work on the project. Use this to fill in the table.

w	d
6	30
12	
18	
24	
30	

The weight of an object on Mars varies directly as its weight on Earth. Suppose that a 170-lb astronaut becomes the first human to set foot on Mars. (Note: Matt Damon didn't actually go there. That was a movie.) He or she would weigh 64.6 lb on Mars.

8. Find the constant of variation.

9. Write an equation relating the weight of an object on Mars to its weight on Earth.

10. Use your equation from Question 9 to fill in the table of values.

Weight on Earth (lbs)	Weight on Mars (lbs)
120	
135	
150	
165	
180	
195	

11. Suppose that a second excursion to Mars happens, and the landing craft holds an extra astronaut and more equipment, which makes it twice as heavy on launch as the first manned trip. How much heavier would this craft be when it lands on Mars?

12. On the next trip advances in lightweight materials lead to the landing craft weighing 1/3 as much as the previous trip. How do the weights compare on Mars?

When a wheel rolls along, the number of times it completes a full revolution to cover a certain distance varies inversely as the circumference of the wheel.

13. Explain why this makes perfect sense.

14. The wheels on my Explorer have a circumference of just about 240 cm. If they make 150 revolutions in traveling a certain distance:

a. Find the constant of variation.

b. Write an equation relating the number of revolutions made by a wheel to its circumference in centimeters for this particular distance.

Did You Get It

4. Refer back to Did You Get It 3. With a stove set on high, it provides 1,400 watts of power, and boils a quart of water in 320 seconds. Write a variation equation that models the relationship between wattage and time to boil a quart of water.

15. Use your equation from Question 14 to complete the table of values.

Circumference (cm)	Number of revolutions
100	
150	
200	
250	
300	

16. What happens to the number of revolutions if the circumference is tripled?

17. What happens to the number of revolutions if the circumference is cut in half?

Did You Get It

5. Use the equation you wrote in Did You Get It 4 to find how many watts would be needed to boil a quart of water in 2 minutes.

4-4 Portfolio

Name ______________________________

Check each box when you've completed the task. Remember that your instructor will want you to turn in the portfolio pages you create.

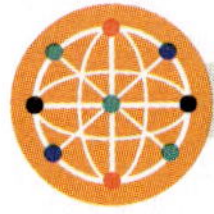

Technology

1. ☐ Use Excel to create a table and graph illustrating the rate-time problem at the beginning of this lesson. You should use a formula to calculate the times corresponding to a variety of speeds. Begin by using a total distance of 40 miles. Then create a second worksheet using a parameter that allows you to change the distance to anything you like in a single cell and have that automatically change the table and graph. There's a template to get you started in the online resources for this lesson.

Online Practice

1. ☐ Include any written work from the online assignment along with any notes or questions about this lesson's content.

Applications

1. ☐ Complete the Applications problems.

Reflections

Type a short answer to each question.

1. ☐ Describe the differences between direct and inverse variation from as many different aspects as you can think of.
2. ☐ Explain why the topic of inverse variation is in this unit. (Hint: Look at the unit title.)
3. ☐ Body mass index (BMI) is essentially a comparison between a person's weight and height. A BMI between 18.5 and 24.9 is considered healthy; over 25 is considered overweight, and over 30 is considered obese. How do you think a person's height varies with BMI? What about weight with BMI? Explain.
4. ☐ Name one thing you learned or discovered in this lesson that you found particularly interesting.
5. ☐ What questions do you have about this lesson?

Looking Ahead

1. ☐ Complete the Prep Skills for Lesson 4-5.
2. ☐ Read the opening paragraph in Lesson 4-5 carefully and answer Question 0 in preparation for that lesson.

Answers to "Did You Get It?"

1. **a.** The time is cut in half. **b.** It would take 12 minutes.
2. A little over an hour; a little over 7 mph
3. $t = \frac{k}{w}$, where t is time to boil and w is wattage.
4. $t = \frac{448{,}000}{w}$
5. About 3,733 watts

Answers to "Prep Skills"

1. $F = ka$, where F is the force and a is the acceleration
2. $F = 200x$
3.

x	1	5	10	20	40	150
y	200	40	20	10	5	1.33

4.

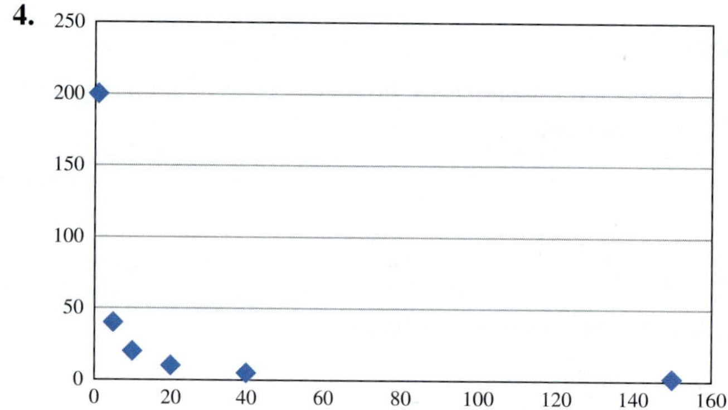

4-4 Applications

Name ______________________________

For each problem, the game plan is to use the information to write an equation of variation. The information provided will allow you to find the constant of variation. Then use your equation to solve the problem.

1. The UV (ultraviolet light) index is a measure of how intense the sun's rays are at a given location and time. The UV index would be low on a cloudy day and much higher on a sunny day. Fill in the blank with either *directly* or *inversely:*

   The time it takes to get a sunburn varies ______________ as the UV index.

2. Write a general variation equation relating the time it takes to get a sunburn and the UV index. Make sure you define what any variables stand for.

3. According to *CBS News* online Consumer Tips, at a UV rating of 6, an average person can get a sunburn in as little as 15 minutes. Use this fact to find the constant of variation and rewrite your variation equation using that value of k.

4. How long would it take to get a sunburn when the UV rating is 8.5?

5. Fill in the blank with either *directly* or *inversely:*

   The amount of newsprint used by a newspaper varies ______________ as the number of people in the area served by that paper.

4-4 Applications

Name ____

6. According to the North American Newsprint Producers Association, the newsprint used to supply the annual needs of 1,000 people is 34,800 kg. How many kilograms would be needed to supply Tampa, Florida, which has a population of about 353,000? Write a variation equation and use it to solve this problem. How much is that in pounds?

7. Fill in the blank with either *directly* or *inversely:* The length of skid marks left when a driver slams on the brakes varies _______________ as the square of the speed of the car v.

8. A child was struck by a car in a crosswalk. The driver of the car had slammed on his brakes and left visible skid marks that were 31 feet long on the pavement. He told the police he had been driving at 25 miles per hour. The police know that, under the conditions at that time, skid marks would be 62 feet long for a car traveling at 40 miles per hour. If the man is telling the truth, how long should the skid marks be? Write a variation equation and use it to solve this problem.

9. The illumination provided by a car's headlight (L) varies inversely as the square of the distance from the headlight (d). A headlight produces 19 foot-candles (fc) at a distance of 26 ft. What will the illumination be at 60 ft?

Lesson 4-5 Prep Skills

SKILL 1: CONVERTING BETWEEN PERCENTS AND DECIMALS

To write a decimal in percent form, move the decimal point two places to the right and include the percent symbol.

- $0.435 = 43.5\%$

To write a percent in decimal form, move the decimal point two places to the left and remove the percent symbol.

- $8.2\% = 0.082$

SKILL 2: RECOGNIZING LINEAR OR EXPONENTIAL GROWTH OR DECAY

When a quantity grows according to linear growth, the same number is repeatedly ADDED to the total.

- 8, 11, 14, 17, 20, 23, . . . is a sequence of numbers growing linearly because every new value is obtained from adding 3 to the previous value.
- 100, 95, 90, 85, 80, 75, . . . is a sequence of numbers decaying linearly because every new value is obtained from subtracting 5 from the previous. We can think of this as adding −5.

When a quantity grows according to exponential growth, the total is repeatedly MULTIPLIED by the same number. We call this multiplier the **growth factor**, or **multiplication factor**.

- 1, 3, 9, 27, 81, 243, . . . is growing exponentially because every new value is obtained from multiplying the previous value by 3.
- 10, 8, 6.4, 5.12, 4.096, . . . is decaying exponentially because every new value is obtained from multiplying the previous value by 0.8.

SKILL 3: FINDING RELATIVE CHANGE

This skill was the main topic of Lesson 2-6. Recall that when a quantity changes, the actual change is simply the new value minus the original value. The relative change is the actual change divided by the original value. Written as a formula (kind of):

$$\text{Relative change} = \frac{\text{New value} - \text{Original value}}{\text{Original value}}$$

SKILL 4: INTERPRETING INFORMATION FROM A GRAPH

The key to interpreting information from ANY graph is recognizing what quantity is represented by the values on each axis. Finding the coordinates of a point does you no good unless you know what each coordinate represents. Once you've done that, it's simply a matter of reading the scale on each axis, and locating points corresponding to values along each axis.

- The following graph represents the value of an investment account in dollars (vertical axis) in terms of years (horizontal axis) since the account was opened.

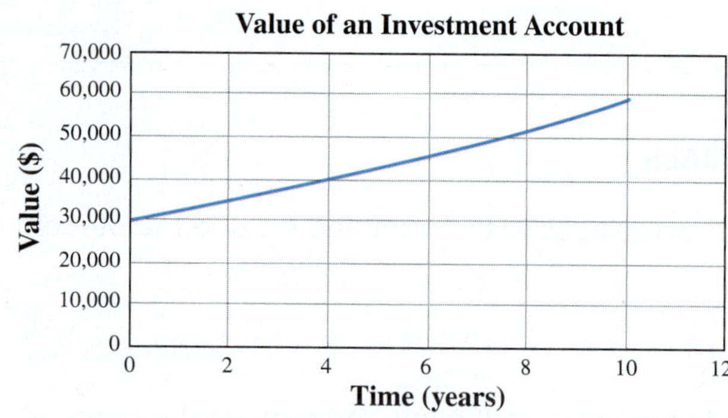

- The value of the account was $30,000 at the time it was opened because the point (0, 30,000) is on the graph. We can see from the labels that the first coordinate represents years after the account was opened, and the second represents the value in dollars.
- It took the account just about 4 years to pass $40,000 in value because the point where the graph goes above height 40,000 is just a bit to the right of 4 on the horizontal axis.

PREP SKILLS QUESTIONS

1. Convert each percent to decimal form.

   a. 4% b. 81% c. 11.4%

2. Convert each decimal to a percent.

   a. 0.37 b. 0.914 c. 0.019

3. Does this sequence represent linear growth, exponential growth, or neither?

   7, 16, 25, 34, 43, 52

4. Does this sequence represent linear growth, exponential growth, or neither?

   2, 3, 4.5, 6.75, 10.125, 15.1875

5. Write the growth factor for any sequence in Questions 3 and 4 that is exponential.

6. In one year, Chad's salary was raised from $31,500 to $35,250. Find the relative change.

7. The graph here shows the concentration of a certain drug in a patient's bloodstream in milligrams per liter in terms of the number of hours since the drug was administered. Use the graph to answer the questions.

   a. What was the original concentration at the time the drug was administered?

   b. How long did it take for the concentration to drop below 4 mg/L?

   c. This particular patient needs a concentration of at least 2 mg/L for the drug to be effective. About how long was this dose effective?

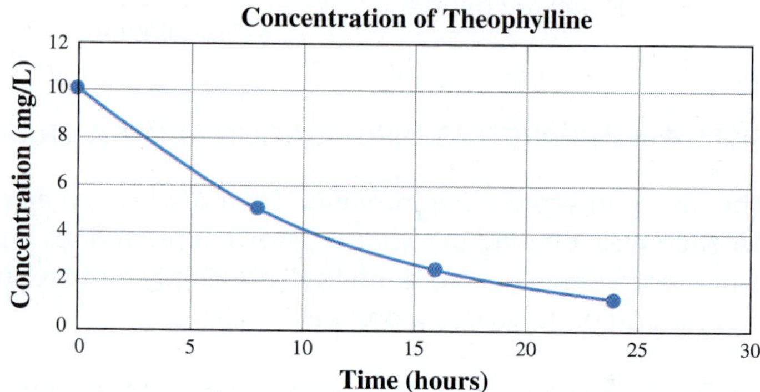

Lesson 4-5 Sit Back and Watch Your Money Grow

©Picturenet/Getty Images RF

LEARNING OBJECTIVES

- ☐ 1. Define function and use function notation.
- ☐ 2. Identify the significance of a and b in an equation of the form $y = ab^x$.
- ☐ 3. Find exponential models.
- ☐ 4. Compare exponential models using graphs, tables, and formulas.

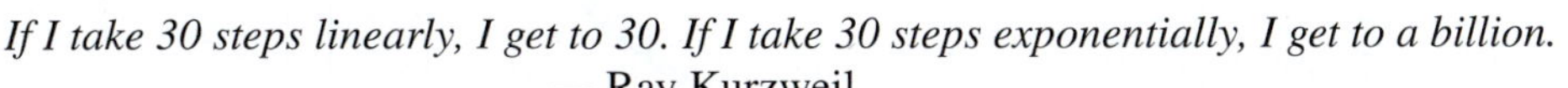

If I take 30 steps linearly, I get to 30. If I take 30 steps exponentially, I get to a billion.
— Ray Kurzweil

Raise your hand if you hope to be rich someday. There's a pretty good chance that your hand is up right now: Financial independence is a goal for a lot of people. We've already studied the differences between quantities that grow linearly and those that grow exponentially. Now that our skills at interpreting graphs have grown, well, exponentially, it's a good time to revisit exponential growth to learn a little bit more about its magic. We already know that the long-term growth of money is an ideal way to study exponential growth, so we'll start there, and on the way start you on the path to riches!

0. If a person invested $15,000 today in a reasonably successful fund, how much do you think it could be worth in 40 years?

4-5 Class

1. Describe the differences between linear and exponential growth based on what we know from earlier in the course.

If most people won $15,000 on a game show, they'd go on a spending spree. But not you! You're too smart for that. Instead you'd decide to invest that money to make it grow. I'm not sure how patient you are, but what if you were super-patient? Like patient enough to let that money grow for 40 years in a fixed-rate investment? Let's take a look. Make sure that you answer every question with a complete sentence.

Time (years)	Value ($)
0	$15,000.00
1	$16,050.00
2	$17,173.50
3	$18,375.65
4	$19,661.94
5	$21,038.28
6	$22,510.96
7	$24,086.72

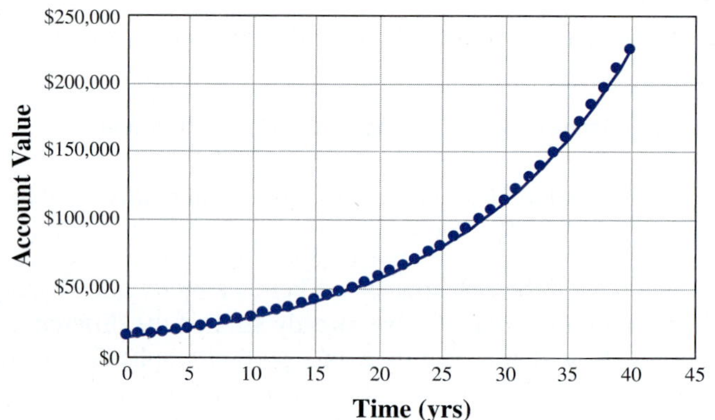

2. For each value in the table after the initial investment, divide the amount by the value of the account the previous year. What do you notice?

3. What does your result in Question 2 say about the type of growth this account exhibits?

4. Find the relative change in value for each year in the table compared to the previous year. What can you conclude?

5. Find the value of the account after 10 years. Don't multiply by 1.07 ten times … that's what exponents are for!

6. Use your answer to Question 3 to write an equation in the form "$y =$ expression," where the expression describes the amount of money in the account after x years. (Hint: Starting with $15,000, how many times have you multiplied by your answer to Question 3 after x years?)

7. Verify that the equation you wrote generates the account values in the table. Ideally, you would use either the table feature on a graphing calculator or spreadsheet, but in any case, make sure you describe how you got your values.

Did You Get It

Try this problem to see if you understand the concepts we just studied. The answer can be found at the end of the Portfolio section.

1. Elaine keeps an investment whose value x years after opening the account can be described by the equation $y = 12{,}500(1.05)^x$. What was her initial investment? By what percentage does it grow each year?

We'll continue studying your genius plan to grow your winnings in a bit. For now, we'd like to point out that we've been working with a really important idea in math for quite a while now without giving it a name. Look at the spreadsheet, which shows tournament scores for five members of the Findlay College golf team at the 2016 NCAA championships. If you wanted to find the total team score in cell B7, what formula would you use?

	A	B
1	Player	Score
2	Kasey Petty	69
3	Shelby Warner	73
4	Makenzie Torres	75
5	Samantha Hatter	80
6	Kelsey Koesters	81
7		

If you said "=SUM(B2:B6)," you win. The key elements of that command are SUM, which is the name of the formula that you're asking Excel to use, and B2:B6, which tells Excel what numbers to apply that formula to. This may remind you of the terminology we used back in Lesson 2-2: inputs and outputs. The input here is the range of cells, and the output is the sum.

This process of matching up inputs and outputs in math is given a special name: function.

A **function** is a relationship between two quantities where each input produces a unique output.

Our Excel sum qualifies as a function, because when you specify the input (cells B2 through B6) there is a unique output (the sum). The equation we wrote describing the value of the $15,000 investment is also a function: The input is the number of years, and for each number of years there is a unique output (the value of the account after that many years). It's not like the bank would let you choose between two different amounts: There's a single value, and that's that.

In order to distinguish that a formula defines a function, we use **function notation.**

Function Notation

The output of a function named f is described by the symbol $f(x)$, which is read "f OF x." Warning: NEVER read $f(x)$ as "f TIMES x." It's NOT a multiplication.

So we would write the investment equation as $f(x) = 15{,}000(1.07)^x$ and read this as "f of x equals 15,000 times 1.07 to the x power." It's not a coincidence that this looks JUST like the syntax for Excel formulas. In fact, if you look up Excel documentation, it refers to things like =SUM, =AVERAGE, and =MEDIAN as functions. Everything ties together and life is good.

8. For the spreadsheet displayed earlier, find the output for the function =SUM(B3:B4).

9. For the function describing the value of our $15,000 investment, find $f(12)$, and describe what it means.

Did You Get It

2. a. For the spreadsheet on the previous page showing golf scores, find the output for the function =SUM(B2:B5).
   b. For the function describing the value of our $15,000 investment, find $f(23)$ and describe what it means.

10. Look back at the graph that illustrates the value of the account. Trace along the graph slowly from left to right, and describe what the changing steepness of the graph tells you about how the value is growing.

11. How much interest will you earn from zero to five years into the investment? This will be the difference in the value after five years and the initial investment. You could use the formula for the function, but the table of values printed earlier is a better choice.

12. How much interest will you earn between years 35 and 40? This time the formula is your only option.

13. Describe what you can learn from comparing the answers to Questions 11 and 12. Also describe how that fits in with your answer to Question 10.

Did You Get It

3. Elaine from Did You Get It 1 never touches the money in that account, and it gets passed down through three generations. After 94 years in the bank, how much would the account be worth?

4-5 Group

We've now studied exponential growth a couple of different times in this course. But there's a flip side to exponential growth, known as **exponential decay**. The thing that distinguishes exponential growth is that the growth starts off relatively slowly, then speeds up dramatically as time passes. Let's see how exponential decay compares.

©Ingram Publishing/Alamy RF

Coffee is a popular morning drink because the caffeine it contains acts as a temporary stimulant. Once the caffeine gets into your system, the amount dissipates exponentially (assuming that you don't add more, of course). Answer every question (except for 2 and 6) with a full sentence.

Time (hours)	Caffeine (mg)
0	180.0
1	144.0
2	115.2
3	92.2
4	73.7
5	59.0
6	47.2
7	37.7

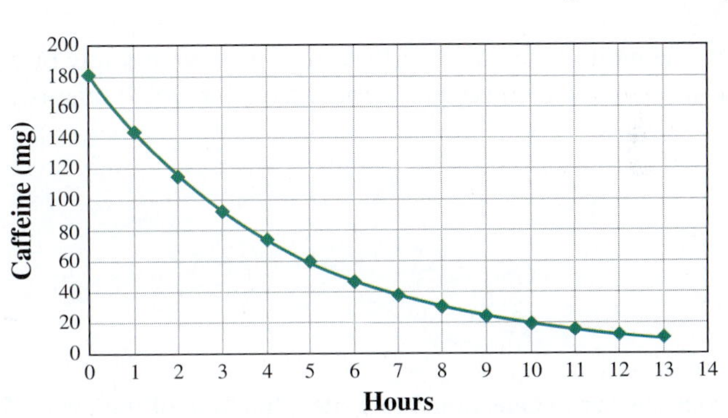

1. If you're the coffee drinker represented by the table and graph, how much caffeine was initially in your system?

2. Find the relative change in caffeine for each one-hour period listed in the following table. Round to the nearest full percent.

Hours	Relative change
0–1	
1–2	
2–3	
3–4	
4–5	
5–6	
6–7	

3. Complete this important statement about exponential change: A quantity either grows or decays exponentially when the _______________ _______________ in that quantity is _______________.

4. Looking at the graph of the caffeine remaining in your system, describe what exponential decay looks like compared to exponential growth.

5. What's the multiplication factor that you'd need to multiply by a previous hour's amount to get the next hour's amount? (Hint: If 10% of something goes away each hour, 90% remains.)

6. Use your answer to Question 5 to write a function of the form "$f(x)$ = expression" where the expression describes the amount of caffeine left in your system after x hours. Your answer to Question 6 in the Class portion is a good place to look if you need help.

Did You Get It

4. The radioactive elements used in medical imaging are good examples of substances that decay exponentially. Iodine-123 is a radioactive substance used in diagnosing thyroid conditions. This particular isotope decays in such a way that 95% of the initial sample remains after each one-hour period. If an initial sample is 200 mg, write a function that describes the amount remaining after x hours.

7. Use a graphing calculator to verify that the equation you wrote in Question 6 gives you the values in the table, and the given graph. Describe your results.

8. Use the function you wrote to find the amount of caffeine left in your system after 150 minutes.

9. Use the graph to estimate the amount of time needed for the amount of caffeine to drop to 40 mg. Describe how you found your answer.

10. Use either TABLE or TRACE commands on your calculator to try and come up with a more accurate estimate of the time needed for the caffeine level to reach 40 mg.

Did You Get It

5. Use the equation you wrote in Did You Get It 4, along with a graphing calculator, to answer each question.
   a. How much of the initial 200-mg sample of Iodine-123 remains after 8 hours? One week?
   b. About how long does it take until half of the original sample remains?

11. Make up a situation that could be modeled by each of the following exponential graphs. As always, extra points for creativity!

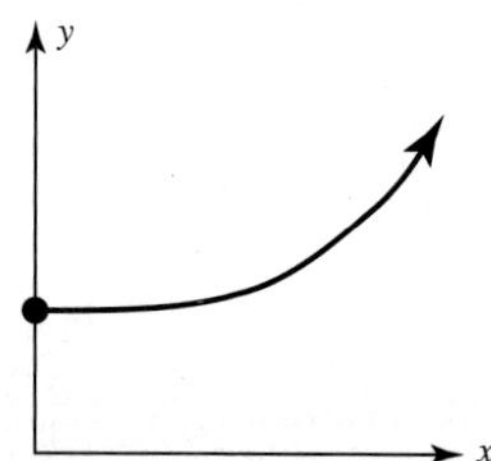

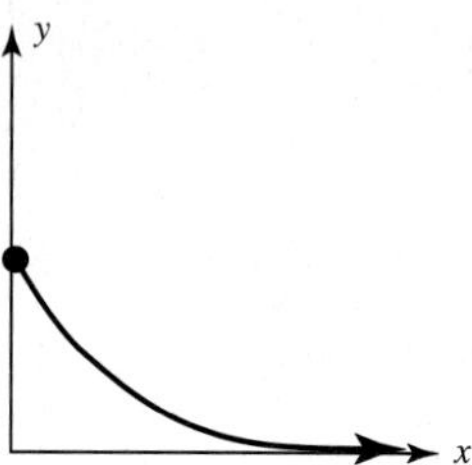

4-5 Portfolio

Name ______________________________

Check each box when you've completed the task. Remember that your instructor will want you to turn in the portfolio pages you create.

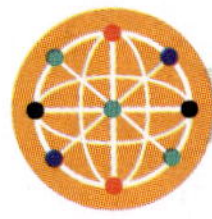

Technology

1. ☐ Use a spreadsheet to extend both the table and the graph for the caffeine problem (p. 521) out to the point where there's less than 1 mg left in your system. How long does this take? A template to help you get started can be found in the online resources for this lesson.

Online Practice

1. ☐ Include any written work from the online assignment along with any notes or questions about this lesson's content.

Applications

1. ☐ Complete the Applications problems.

Reflections

Type a short answer to each question.

1. ☐ Compare exponential growth and exponential decay. What do they have in common? How are they different? A really good answer will say something about relative change.
2. ☐ What did you learn in this lesson about the value of leaving money invested for long periods of time? How can you apply this to retirement?
3. ☐ Name one thing you learned or discovered in this lesson that you found particularly interesting.
4. ☐ What questions do you have about this lesson?

Looking Ahead

1. ☐ Complete the Prep Skills for Lesson 4-6.
2. ☐ Read the opening paragraph in Lesson 4-6 carefully and answer Question 0 in preparation for that lesson.

Answers to "Did You Get It?"

1. \$12,500, 5%

2. a. 297 **b.** \$71,107.95: This is the value of the account after 23 years.

3. \$1,226,603.29

4. $y = 200(0.95)^x$

5. a. 8 hours: 132.7 mg, 1 week: 0.036 mg **b.** 13.5 hours

Answers to "Prep Skills"

1. a. 0.04 **b.** 0.81 **c.** 0.114

2. a. 37% **b.** 91.4% **c.** 1.9%

3. linear growth

4. exponential growth

5. Growth factor for Question 4 is 1.5

6. 11.9%

7. a. 10 mg/L **b.** about 11 hours **c.** about 18 hours

4-5 Applications

Name ____

We'll begin this assignment by looking at the savings plan for Joe, a 30-year-old with a steady job making $30,000 a year, with an annual raise of 4%. Looking ahead to the future, Joe commits to saving 12% of his earnings every year. He chooses a fixed fund investment that returns 9% annually. The following graph shows Joe's earnings, the amount of money he spends, and the amount of money that is generated in interest on his investment each year. The purple curve isn't the amount he's saved; it's the amount of interest earned each year. The values on the *x* axis represent Joe's age in years.

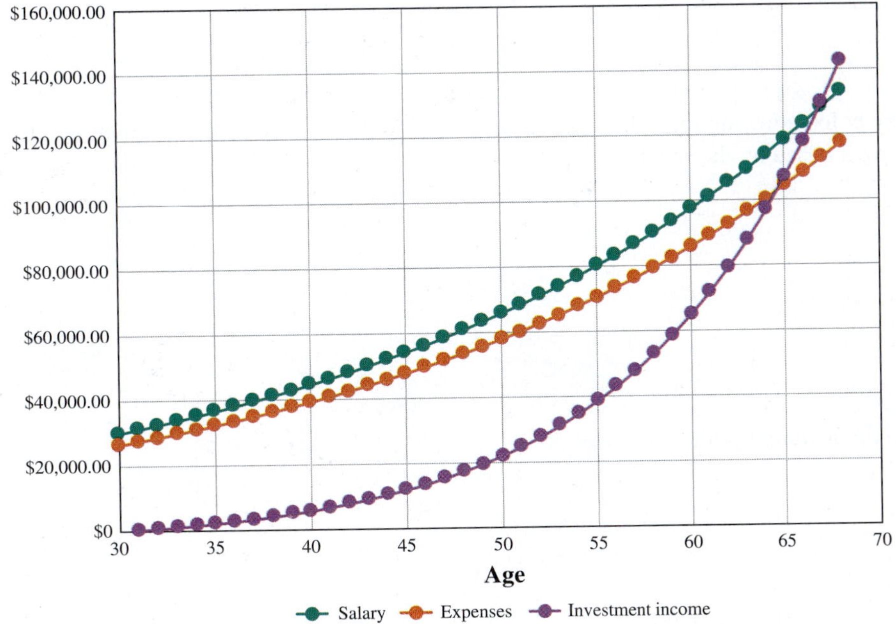

The investment income graph displays the true magic of exponential growth: The income is limited when Joe's a young guy, but as he gets closer to thinking about retirement, the income really takes off. Notice that at age 65, something awesome happens: The income that Joe's making from his investments is more than his living expenses. At that point, he can retire and live as well as ever without, you know, working for a living. Sweet! This age is known in financial circles as the **crossover point**.

Use the information provided to answer Questions 1–5.

1. How much money is Joe earning when he's 30?

2. How much money is he allowing himself to spend when he's 30? Show a calculation, then describe whether or not your answer seems correct based on the graph.

4-5 Applications

Name ______________________________

3. How much is Joe earning when he's 35? Show a calculation based on the annual relative growth rate of 4%. Then describe whether or not your answer seems correct based on the graph.

4. How much money is Joe allowing himself to spend when he's 35? Show a calculation, then describe whether or not your answer seems correct based on the graph.

5. Carefully explain the meaning of the crossover point.

Now turn your attention to the graph to answer Questions 6–8.

6. How much is Joe earning in salary when he's 60?

7. How much is Joe allowing himself to spend when he's 60?

4-5 Applications

Name ______________________________

8. How much is Joe earning from his investments at 60?

9. Write a function of the form "$f(x)$ = expression" where the expression describes the amount Joe earns x years after age 30. If you need guidance, look at the answers to Applications Question 3 and Class Question 6.

10. Use your answer to Question 9 to find how much Joe earns at age 60. How does it compare to the estimate you got from the graph?

11. Write a function of the form "$f(x)$ = expression" where the expression describes the amount Joe spends x years after age 30.

12. Use your answer to Question 11 to find how much Joe spends at age 60. How does it compare to the estimate you got from the graph?

4-5 Applications

Name ______________________________

Now let's look at a coworker of Joe's named Fran. She has the same job as Joe, makes the same amount of money, and in an amazing coincidence happens to be the same age. Sounds like she and Joe should get together and go bowling or something. Anyhow, Fran is much more frugal than Joe: She commits to spending only 70% of her salary and saving the rest in the same retirement plan that Joe is invested in. Let's see what Fran's graph looks like. Again, the horizontal axis is her age.

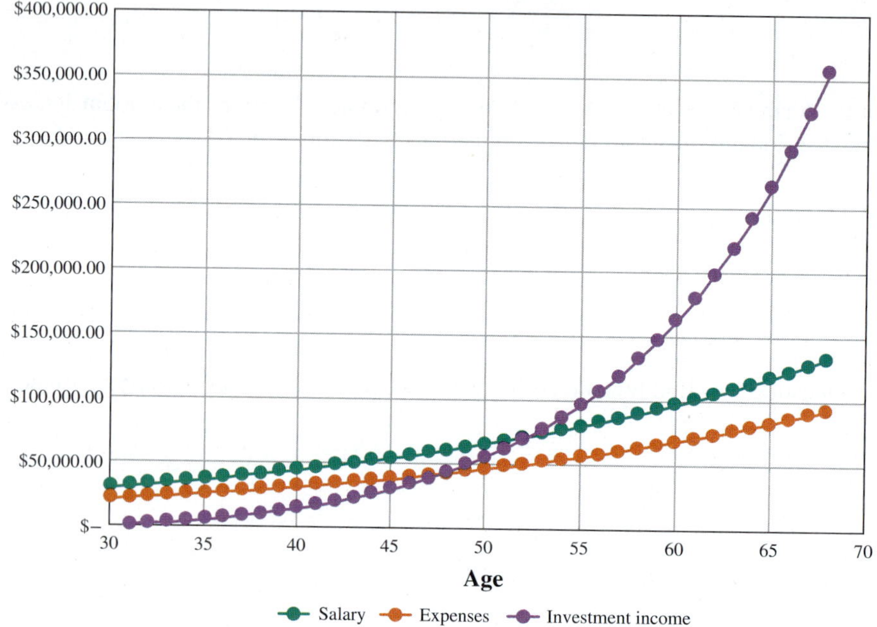

13. How much less money does Fran spend at age 30 than Joe?

14. How much earlier does Fran reach her crossover point?

15. Describe what impact this article might have on your saving vs. spending plan in the future.

Lesson 4-6 Prep Skills

SKILL 1: RECOGNIZE LINEAR OR EXPONENTIAL GROWTH/DECAY

This skill was covered earlier in the course, and was reviewed in the Prep Skills for Lesson 4-5.

SKILL 2: FIND A LINE OF BEST FIT

This skill was the topic of Lesson 3-9, where we learned how to use both calculators and spreadsheets to find regression lines. Refer to the Tech boxes in that lesson along with the accompanying videos in online resources if you need a refresher. In Lesson 4-6, we'll be using very similar procedures to find different types of equations.

SKILL 3: INTERPRET PARAMETERS IN AN EXPONENTIAL EQUATION

We learned how to do this in Lesson 4-5. Here's a quick refresher. For an exponential equation of the form $y = a(b)^x$ that describes the size of some quantity in terms of time, the coefficient a tells us the initial size of the quantity being represented by the equation. This is pretty easy to see: If you input time zero, b^x is just 1, so the output is a. The base b of the exponent describes the percentage by which the quantity is changing each time period. For exponential growth, the base is of the form $1 + r$, where r is the growth rate as a percentage in decimal form. For exponential decay, the base is of the form $1 - r$, where r is the decay rate as a percentage in decimal form.

- The population of a town is described by the formula $P = 12{,}400(1.04)^t$, where t is the number of years after 2000. The 12,400 tells us that the population was 12,400 in 2000. The 1.04, which can be written as $1 + 0.04$, tells us that the growth rate is 0.04 in decimal form, making it 4%.

- The amount of a drug in milligrams in a patient's system is described by the formula $A = 110(0.72)^x$, where x is the number of hours after the drug has entered her system. The 110 tells us that there were initially 110 mg of the drug in her bloodstream. If we write 0.72 as $1 - 0.28$, we can see that the amount of the drug was decreasing at the rate of 28% per hour.

SKILL 4: FIND RELATIVE CHANGE

This skill was covered in Lesson 2-6, and reviewed in the Prep Skills for Lesson 4-5.

SKILL 5: USE FUNCTION NOTATION

Function notation was just introduced in the last lesson, but in case you didn't catch it, here are the essentials. We've used the terms *input* and *output* fairly often to describe quantities in modeling. The input is often represented by x and the output by y. Not always, mind you—we sometimes use letters that are indicative of the quantity they represent, like C for cost, or t for time. But if we're using x for the input, we can also use the symbol $f(x)$ for the output. This is read as "f of x" and indicates that we're defining a function named f whose input is the variable x. Basically, instead of writing an equation like $y = 2.4 + 0.8x$, we'd write it as $f(x) = 2.4 + 0.8x$. To indicate the output that goes along with input 5, for example, we'd use $f(5)$, and find that $f(5) = 2.4 + 0.8(5) = 6.4$

PREP SKILLS QUESTIONS

1. Does this sequence represent growth that is close to linear, exponential, or neither?

   12, 33, 53, 75, 98

2. Does this sequence represent decay that is close to linear, exponential, or neither?

   400, 320, 254, 200, 162, 130

3. Use a calculator or spreadsheet to find the line of best fit for the data in the table.

x	12	24	36	48	60	72
y	$312	$600	$1,040	$1,279	$1,703	$2,041

4. A quantity is described by the equation $y = 26(1.2)^x$. What was the initial size of the quantity? By what percentage is it increasing each time period?

5. The first time a new course is offered at a college, 73 people signed up for it. The following semester, only 60 signed up. Find the relative change.

6. For the function $f(x) = 3 \cdot 5^x$, find $f(0)$, $f(2)$, and $f(-1.2)$.

Lesson 4-6 Follow the Bouncing Golf Ball

LEARNING OBJECTIVES

- ☐ 1. Gather and organize data from an experiment.
- ☐ 2. Find an exponential curve of best fit for an experimental data set.
- ☐ 3. Study the decay rate for exponential decay.

It's just a glorious day. The only way to ruin a day like this would be to play golf on it.

—David Feherty

Courtesy of David Sobecki

Avid golfers have a classic love-hate relationship with the game. They enjoy it enough to spend a lot of time and money pursuing it, but they also know how downright maddening it can be. Golf is a whole lot harder than it looks, and sometimes it seems like the ball just moves and bounces with a mind of its own—an evil mind intent on ruining the golfer's day.

Even so, like everything else in our world, golf balls are bound by the laws of physics and their motion actually is predictable—at least in a lab setting, if not on the golf course. In this lesson, we'll follow a bouncing golf ball to help us further our study of exponential growth and decay, relying on technology to find exponential equations that best fit a data set.

0. Suppose that you drop a golf ball from a height of 10 feet onto a hard surface, and the ball bounces back up to a height of 8 feet. Could you make a prediction on how high it would bounce if dropped from 5 feet? Why or why not?

4-6 Group

Here's the list of supplies needed for our golf ball experiment:

1. A golf ball; without this, it's not a golf ball experiment, is it?

2. A meter stick or tape measure with metric units

3. A chair or short stepladder

4. Masking tape and a marking pen

5. A hard, smooth floor area next to a wall

6. A graphing calculator or spreadsheet program with a regression feature

Part 1: The Bounce Lab

Location is everything here: You'll need a relatively high ceiling, and a smooth concrete or other hard floor. Tile with grout lines won't work: The grout lines will affect the bounces of the ball. The key is to get consistent bounces straight up with no deflection. Have the tallest member of your group stand on a chair or short stepladder, and run a strip of masking tape vertically along the wall from the floor to the highest point they can reach. (Ideally, the starting point will be at least 250 cm above the floor.) Using your meter stick/tape measure and the marking pen, label heights on the masking tape every 10 cm starting with 0 cm at floor level.

Now you're ready to bounce. Hold the golf ball a few inches from the wall with the bottom of the ball at the highest height marked on the tape. Drop the ball and have someone in the group note the approximate height the ball reaches at the top of its first bounce. This will give you a rough idea of where to look.

Now repeat the drop, and have the person who watched the height of the first bounce put their finger on the tape at the height the ball reaches on the first bounce (the bottom of the ball, that is). Then mark that height with the pen. Repeat twice more FROM THE SAME HEIGHT, then calculate and record the *average* of the three bounce heights. (Use the table to record your results. Once you add the bounce heights in the Total bounce height column, you just need to divide by 3 to get the average bounce height.)

Next, we'd like to calculate the height of the second bounce. But rather than having the ball bounce twice, use the height of the first bounce that you calculated as the new release point. Clever, right? Repeat three times, recording the bounce heights in the table and calculating the average. Then use that average height as release point to calculate the height of the third bounce, and so on. Continue until you have heights for the first 8–10 bounces.

1. Record the data from your experiment in the table.

x Bounce #	Trial 1 bounce height (cm)	Trial 2 bounce height (cm)	Trial 3 bounce height (cm)	Total bounce height (cm)	y Average bounce height (cm)
0					
1					
2					
3					
4					
5					
6					
7					
8					
9					
10					

2. Use these data to create a scatter plot. Use the average bounce height as your y coordinates.

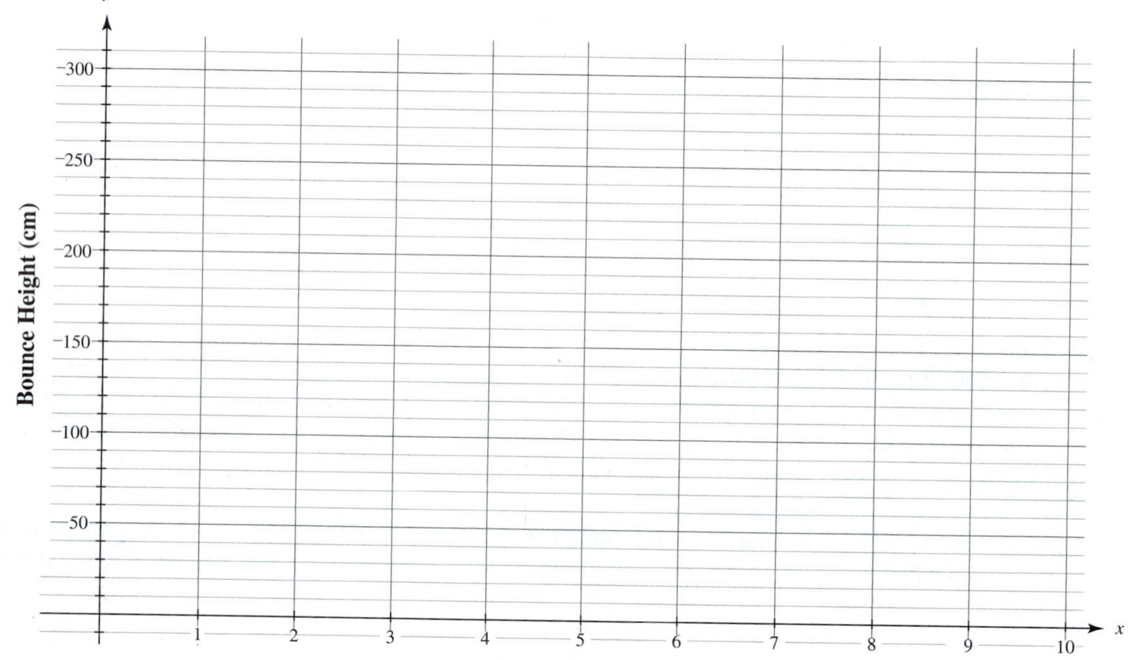

Part 2: Modeling Your Data

Hopefully, you can see that a linear model would be a bad choice for our bounce height data. The scatter plot should remind you of the exponential decay models we saw in Lesson 4-5. In order to find a model, we'll use exponential regression commands on either a graphing calculator or spreadsheet.

Finding an Exponential Equation of Best Fit

Whether using a graphing calculator or spreadsheet, the setup is the same as when finding the line of best fit: Enter the data as before. Then:

Graphing Calculator

Press STAT followed by the right arrow key to access the **STAT CALC** menu, and choose **ExpReg,** which is choice 0 on that menu. Then press ENTER to calculate the exponential equation of best fit.

Spreadsheet

Create a scatter plot, then point the cursor at one of the data points. Right click (Windows) or option click (Mac) and choose "Add Trendline" from the contextual menu, and choose Exponential. To display the equation, you'll need to locate the Format Trendline menu. For some spreadsheets, you can find it by right or option clicking on one of the data points again. On others, the menu will automatically appear to the right after you choose Add Trendline. In any case, you'll find check boxes for Display Equation on Chart and Display R-squared value on chart.

Here's where things get funky: Spreadsheets provide exponential equations in the form $y = ae^{kx}$, rather than the $y = ab^x$ provided by a graphing calculator. The number e is known as the *natural base:* It's a bit more than 2.7. To convert the spreadsheet formula into the form we want, compute 2.7^k, where k is the coefficient of x in the equation provided by the spreadsheet. The result is the base you need (b) in the formula $y = ab^x$.

See the Lesson 4-6 Using Tech video in online resources for further information.

3. Use either a calculator or spreadsheet to find the exponential functions of best fit for your bounce data. Write your answer in the form $f(x) = ab^x$.

4. Use your function to approximate the height of the 7th bounce. How does the result compare to the value in your table? Does this surprise you?

5. Find $f(0)$. What does your answer mean?

Did You Get It

Try this problem to see if you understand the concepts we just studied. The answer can be found at the end of the Portfolio section.

1. When trying this experiment at my campus, I got the model $f(x) = 263(0.84)^x$.
   a. What was the approximate height I dropped the ball from originally?
   b. About how high was the seventh bounce?

Now let's analyze the significance of every symbol in our equation.

6. What does the variable x represent?

7. In the function $f(x) = ab^x$ that models your data, what is the value of a? What does it tell us about the experiment? Your answers to Questions 5 and 6 should help. Does it match the results in your table? Why do you think that is?

8. In the function $f(x) = ab^x$ that models your data, what is the value of b? What does it tell us about the experiment?

Did You Get It

2. Look back at the bounce height function I got in Did You Get It 1. What is the significance of the 0.84 in that function? What does it tell you about the difference between my golf ball experiment and yours?

Part 3: The Connection Between Exponential Decay and the Rate of Decay

We've put a considerable amount of energy into studying exponential growth or decay. The independent variables we've worked with have typically been whole numbers: years after an account was opened, hours since caffeine was introduced to the bloodstream, number of bounces for a golf ball, etc. In each case, the distinguishing feature is that each new output comes from multiplying the previous by the same fixed number. In the bounce lab, we see that same pattern: Each bounce is a certain percentage of the height the ball was dropped from. The result is a function where the input is an exponent, and that's why we call this type of function an **exponential function**.

Now let's study the rate of decay.

9. Using data from your original table, fill in this new table. The drop height corresponds to the height of the previous bounce; start with the original height the ball was dropped from. Use the average bounce heights from the original table. Note: The difference between this table and the previous one is that the input is no longer the bounce number: It's the *height of the previous bounce.*

x Drop height (cm)	y Bounce height (cm)	x Drop height (cm)	y Bounce height (cm)

10. Use these data to create a second scatter plot.

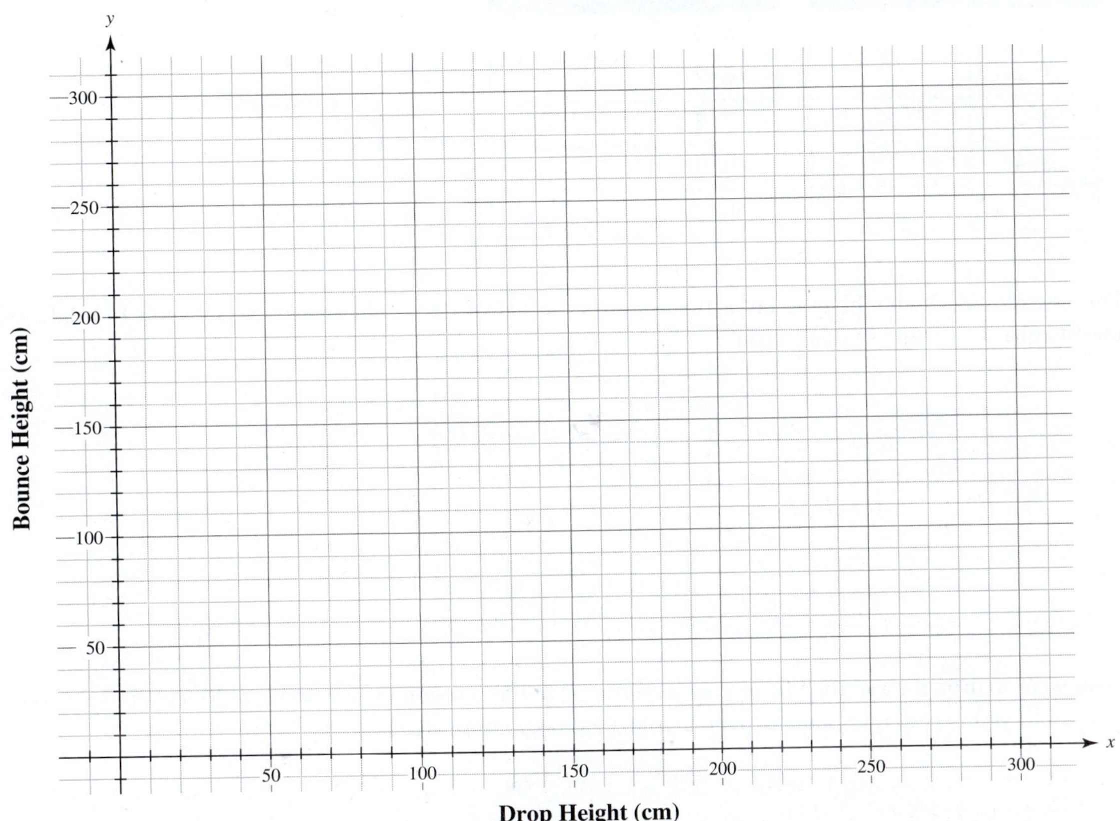

11. Based on the scatter plot, what kind of equation seems to be a good choice for modeling this data?

12. Use your graphing calculator to find the equation of best fit. Round to two decimal places.

$y =$ ____________________

13. Your equation has a variable in it, but also has **parameters**: These are constants in an equation of a certain form that are determined by the data for a particular model. For example, when you found a function of the form $f(x) = ab^x$ earlier, x was the variable, while the values you found for a and b were parameters. Identify the variables and parameters in your answer to Question 12, then describe what each means in this setting.

14. Find the relative change in height from one bounce to the next. If you need a reminder, we studied relative change in Lesson 2-6. Then find the average relative change for all bounces.

Bounce	Rel. Change
0–1	
1–2	
2–3	
3–4	
4–5	

Bounce	Rel. Change
5–6	
6–7	
7–8	
8–9	
9–10	

15. (This one requires some thought, but is the key question.) How does your answer to Question 14 relate to EACH of the best-fit equations found in this lesson?

16. If you were to drop the golf ball from your experiment out of a second story window 20 feet above a concrete patio, how high would you expect it to bounce? Don't make this too complicated!

Did You Get It

3. Recall that the function giving the height of the bounces in terms of bounce number (from Did You Get It 1) was $f(x) = 263(0.84)^x$.
   a. What should the relative change in bounce heights work out to be? In other words, if I'd made a table comparing the drop height to the bounce height, like in Question 9, what should I have gotten for the average relative change (Question 14)?
   b. If my experiment had worked out ideally, with perfectly consistent bounces and completely accurate measurements, what would I have gotten for the equation of the line of best fit for the data comparing drop heights to bounce heights? Your answers from Questions 13 and 15 will save the day here.

17. List as many factors as you can think of that contribute to the data from this experiment not working out to be a perfect exponential equation.

According to the International Federation of Tennis, a regulation tennis ball should bounce about 55 inches when dropped from a height of 100 inches.

18. What's the bounce height as a percentage of the drop height for a regulation tennis ball?

19. Write an expression that calculates the bounce height for a regulation tennis ball dropped from 200 inches.

20. Write an equation that describes the bounce height (y) of a regulation tennis ball when it's dropped from x inches.

21. Using your answer to Question 18, it's possible to write an equation that describes the height of the xth bounce when a regulation tennis ball is dropped from a height of 100 inches. Do it! The things we learned in the golf ball experiment will definitely come in handy.

22. If you were to toss a tennis ball a bit forward from 10 feet above the ground, draw a graph of what you think the path of the ball would look like. The horizontal axis should measure lateral distance from where it was tossed, and the vertical axis should measure height. In essence, your graph should be a side view of watching the ball bounce forward.

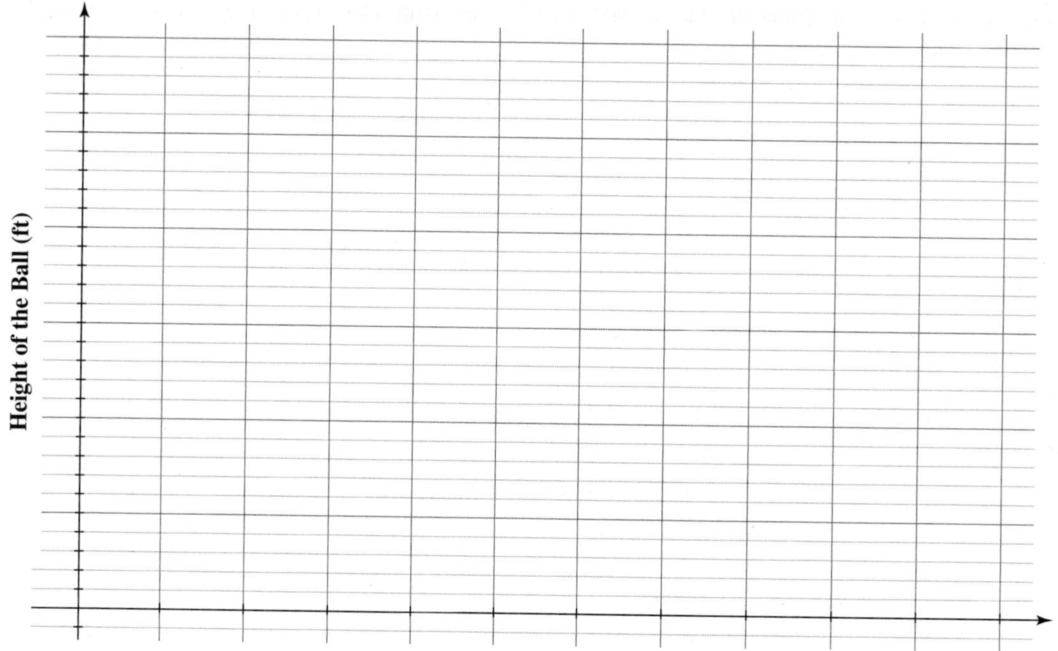

4-6 Portfolio

Name ______________________________

Check each box when you've completed the task. Remember that your instructor will want you to turn in the portfolio pages you create.

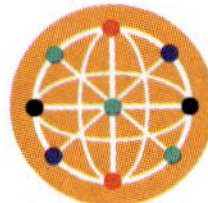

Technology

1. ☐ Questions 1, 2, and 3 in the Group section are great examples of the three ways we've modeled data: numerically, graphically, and algebraically. Build an Excel spreadsheet that illustrates each of those models, using a table, connected scatter plot, and a formula. A template to help you get started is available in the online resources for this lesson. Remember that Excel will give you a formula in the form $y = ae^{kx}$. You can leave the equation in that form, or convert it to the form $y = ab^x$, as described in the Using Technology box on page 535.

Online Practice

1. ☐ Include any written work from the online assignment along with any notes or questions about this lesson's content.

Applications

1. ☐ Complete the Applications problems.

Reflections

Type a short answer to each question.

1. ☐ We've studied linear and exponential equations of best fit in this course. If you're looking at a set of data, how can you decide which of those would be most appropriate for the data?

2. ☐ How could you tell if neither of those types of equations is a good fit for a data set?

3. ☐ Name one thing you learned or discovered in this lesson that you found particularly interesting.

4. ☐ What questions do you have about this lesson?

Looking Ahead

1. ☐ Complete the Prep Skills for Lesson 4-7.

2. ☐ Read the opening paragraph in Lesson 4-7 carefully and answer Question 0 in preparation for that lesson.

Answers to "Did You Get It?"

1. a. 263 cm **b.** 77.6 cm **2.** It's an average of what percent of the previous height each bounce reaches. By comparing to your result, you can tell if my golf ball bounces more or less than yours. **3. a.** −0.16 **b.** $y = 0.84x$

Answers to "Prep Skills"

1. linear growth **2.** exponential decay **3.** $y = 29.03x - 56.8$ **4.** 26; 20% **5.** −17.8% **6.** $f(0) = 3; f(2) = 75; f(-1.2) \approx 0.43$

4-6 Applications

Name ____

1. The function $f(x) = 81.15(1.013)^x$ can be used to model the population of the United States in millions of people based on the number of years since 1900. Identify the variables and the parameters in this equation, and describe what the significance of each is in terms of population.

2. Use the function in Question 1 to complete the table, then use the table to draw a graph of the function. Don't forget to label an appropriate scale on each axis.

Years after 1900	Population (Millions)
0	
20	
40	
60	
80	
100	

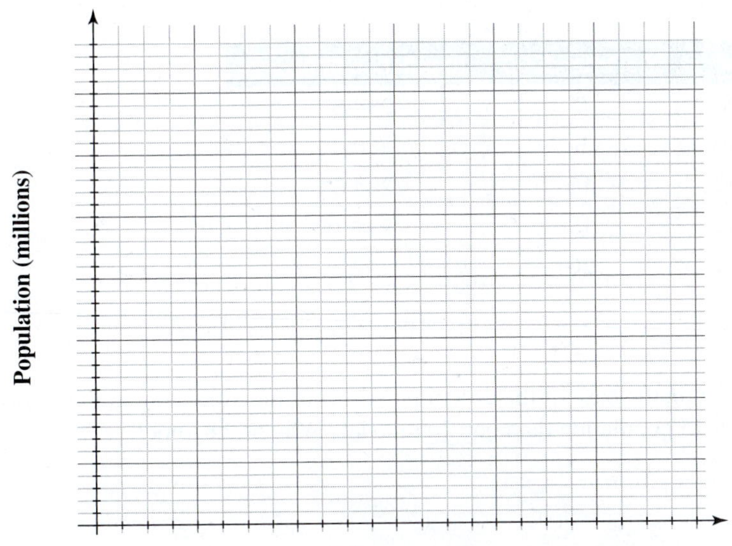

3. The function $f(x) = 81.15(1.013)^x$ illustrates ______________ growth. The population was growing by ______% each year.

4. The model used in Questions 1–3 was obtained using official data for every census from 1900 to 2010. (The census is conducted every 10 years.) Use your model to predict the population in 2015. Then find the population of the United States in 2015 online and compare to your result. How accurate is it?

4-6 Applications

Name ______________________________

5. What do you think the result of your answers to Question 4 say about population growth in the 21st century?

6. What if we use only more recent data? Since the census is conducted only every 10 years, we'll have to use a different source. I used estimates from a site called Worldometers, which are based on multiple data sources that are likely to make them more accurate than the census. In fact, the census is historically not the best measure of the true population for a wide variety of reasons. The table has data for years from 1990 to 2010. Use the data to find the exponential function of best fit. Include FIVE digits in all parameters. *Note that x now represents years after 1990.*

Year	Yrs. after 1990	Population (Millions)
1990	0	254.51
1995	5	268.04
2000	10	284.59
2005	15	298.16
2010	20	312.25

7. Based just on the base of the exponential in your function, do you think this equation is likely to give a better estimate for the 2015 population than our first model? Why or why not?

8. Use your function from Question 6 to estimate the population in 2015 and 2050. Compare the 2015 estimate to the population you already found on the Internet. Then search for a site that does population projections and see how your prediction compares. What does the comparison say about that site's population growth forecast?

Lesson 4-7 Prep Skills

SKILL 1: USE PERIMETER AND AREA FORMULAS FOR A RECTANGLE

A rectangle has two dimensions: length and width. If we use l to represent length and w to represent width, then the perimeter of the rectangle is $P = 2l + 2w$. This is really just adding up the lengths of each individual side. The area of the rectangle is the product of the length and the width: $A = lw$.

- This rectangle has perimeter $P = 2(8) + 2(12) = 40$ cm. It has area $A = 8(12) = 96$ cm^2.

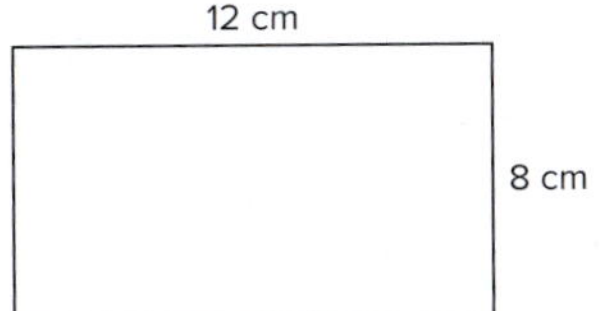

- If a rectangle has perimeter 240 inches and the length is 50 inches, we can find the width by subtraction. Since the length is 50 in., there are 2 sides of that length, which accounts for 100 in. That leaves 140 in. left over for the other two sides, so each is 70 in.

SKILL 2: USE THE DISTRIBUTIVE PROPERTY

This skill was reviewed in the Prep Skills for Lesson 1-4. In this lesson, we'll be distributing a variable into an expression, so you also need to be familiar with rules for exponents, as in this example. We'll first apply the distributive property, multiplying the y by both terms in the parentheses.

- $y(40 - 5y) = y \cdot 40 - y \cdot 5y$ When multiplying y by itself, add the original exponents (both 1) to get y^2

  $= 40y - 5y^2$

SKILL 3: SUBSTITUTE AN EXPRESSION INTO AN EXPRESSION

This skill was covered in the Prep Skills for Lesson 3–10. We're all very comfortable with substituting NUMBERS into an expression by now. Just do the same thing when substituting in an expression! Replace every occurrence of the variable with that expression, making sure to keep the expression in parentheses, then simplify if possible.

SKILL 4: FIND AND INTERPRET INTERCEPTS

Intercepts were introduced in Lesson 3-1. Here's a quickie review: The intercepts of a graph are the points where the graph crosses one of the axes. The y intercept is the point where the y axis is crossed. This happens exactly when the x (or first) coordinate is zero. So this intercept tells us the output of the equation when the input value is zero. The x intercept is the point where the x axis is crossed. This happens exactly when the y (or second) coordinate is zero. So this intercept tells us what values of the input will result in an output of zero.

In order to effectively interpret the story told by intercepts, you have to be completely aware of what quantity each of the input and output represents. Then just think of each intercept as what happens when one of those quantities is zero.

- This graph represents the total amount owed on a student loan, where x is the number of years since the borrower graduated from college.

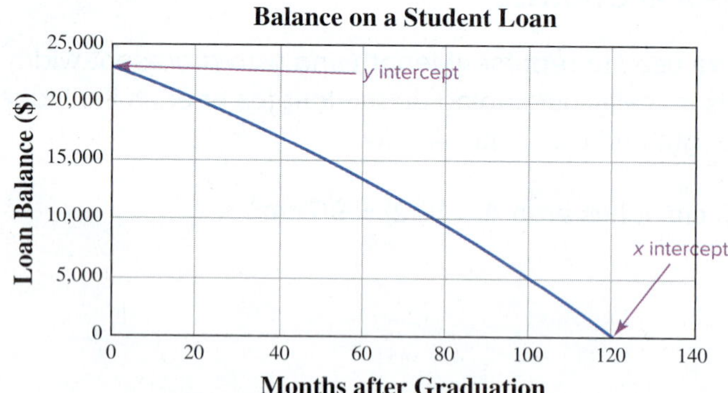

The graph crosses the y axis at about 23,000, so the y intercept is (0, 23,000), and the initial amount owed at the time of graduation was $23,000. The graph crosses the x axis at 120, so the x intercept is (120, 0), and the balance of the loan falls to $0 after 120 months.

PREP SKILLS QUESTIONS

1. Find the perimeter and area of a rectangle that is 13 ft wide and 4.5 ft long.

2. A rectangle with perimeter 148 m is 42 m long. How wide is it?

3. Multiply out the parentheses:

   a. $3w(w + 5)$

   b. $10x(8 - x + 2x^2)$

4. Substitute $90 - 2w$ for l in the expression $P = 2l + 2w$, then simplify the result.

5. This graph describes the height of a drone in terms of number of seconds since it took off. Find all intercepts, and describe what each tells us.

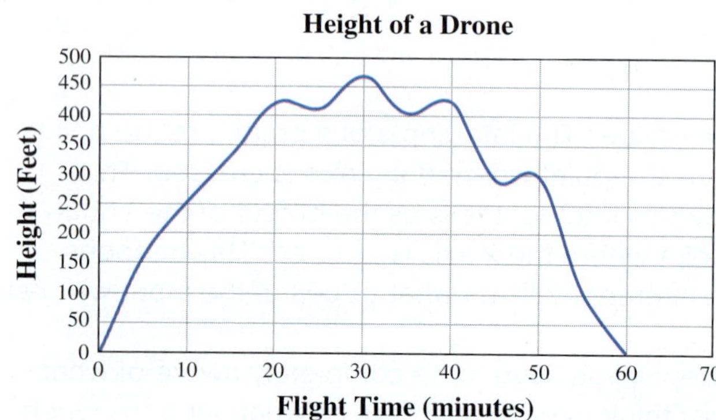

Lesson 4-7 Irate Ducks

LEARNING OBJECTIVES

- ☐ 1. Identify a parabolic graph.
- ☐ 2. Solve problems using the graph of a quadratic equation.

I can't change the direction of the wind, but I can adjust my sails to always reach my destination.

— Jimmy Dean

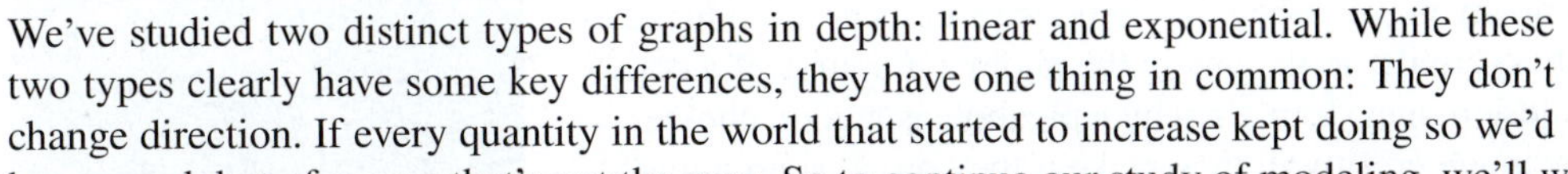

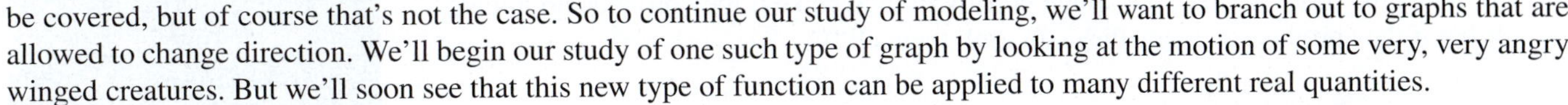

We've studied two distinct types of graphs in depth: linear and exponential. While these two types clearly have some key differences, they have one thing in common: They don't change direction. If every quantity in the world that started to increase kept doing so we'd be covered, but of course that's not the case. So to continue our study of modeling, we'll want to branch out to graphs that are allowed to change direction. We'll begin our study of one such type of graph by looking at the motion of some very, very angry winged creatures. But we'll soon see that this new type of function can be applied to many different real quantities.

0. Write down at least three real quantities you can think of that are likely to change direction when graphed.

4-7 Class

1. Ducks seem awful mild-mannered, always floating placidly on a pond, or waddling in a little line across the street. But like anyone else, push them too far and it turns out that they can be downright irate! (Especially when you couldn't get legal permission to use a different group of angry birds.) A group of beavers has started to build a dam that will flood our ducks' nests, and it's time to strike back! Draw four flight paths: one from the duck cannon on the left to each of the beavers on the right. Keep in mind that gravity exists! It's not possible to fly in a straight-line path. And birds can't fly through wooden posts. Use your imagination to change the launch angle of the cannon as necessary.

This lesson was inspired by Jake Mercer, who commandeered his dad's phone to play a certain video game involving birds. Afterwards, Dad (Brian) noticed that the smudges left on the screen resembled a well-known shape to math folks. (Young boys are not exactly known for clean hands.)

2. Describe the shape of each path in your own words. How are they similar? How are they different?

The shape of the path that an object moving through the air under the influence of gravity follows is called a **parabola**. This is illustrated in the time-lapse photo of a bouncing ball. Parabolas have several distinguishing features, but the two most obvious are the change of direction when reaching a high (or in some cases low) point, and the fact that the two halves on either side of the direction change are mirror images. This is known as **symmetry**.

3. The amount of profit that a company makes when they produce x units of a product can often be represented by a parabola similar to the one in the photo. Of course, if a company makes no product they'll lose money. Explain why a parabola is a good choice to model the profit. (Hint: How much profit will they make if they produce no items? What if they produce so many they can't sell them all?)

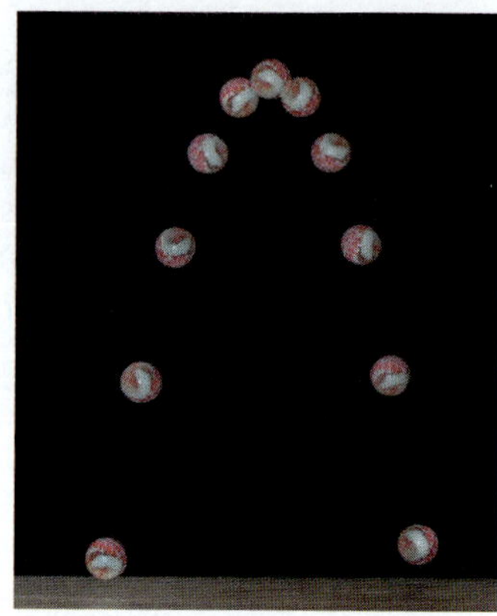

©Custom Life Science Images/ Alamy

Did You Get It

Try this problem to see if you understand the concepts we just studied. The answer can be found at the end of the Portfolio section.

1. Suppose you were modeling the temperature on a certain location from midnight until noon the following day. Do you think a parabola might make a reasonable model? Why or why not?

4-7 Group

A homeowner is planning to fence in a play area for his dog Moose, and he decides that he'd like to spend no more than the cost of 160 feet of fence. But since he wants Moose to have plenty of room for chasing squirrels, he decides to use the full 160 feet. He also wants the play area to be rectangular, as shown in the diagram. **Answer each question (except 3 and 8) with a full sentence.**

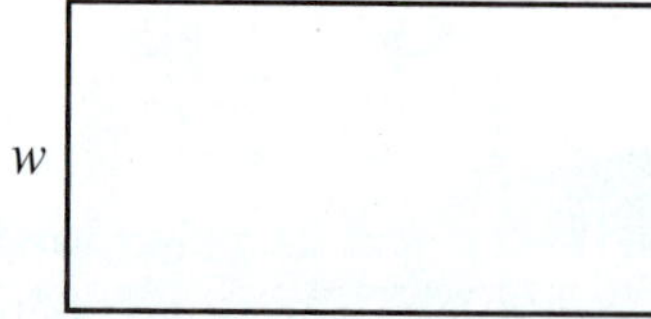

Courtesy of Dave Sobecki

1. If he decides to make the play area 5 feet wide, how long would it be? Don't forget to consider all four sides.

2. What would the length be if the play area is 50 feet wide?

3. Complete the table, then plot the points; your ordered pairs should look like (width, area), or (w, A). Recall that the area of a rectangle is the length times the width.

Width	Length	Area	Ordered Pair (w, A)
5			
10			
20			
30			
40			
50			
60			
70			
80			

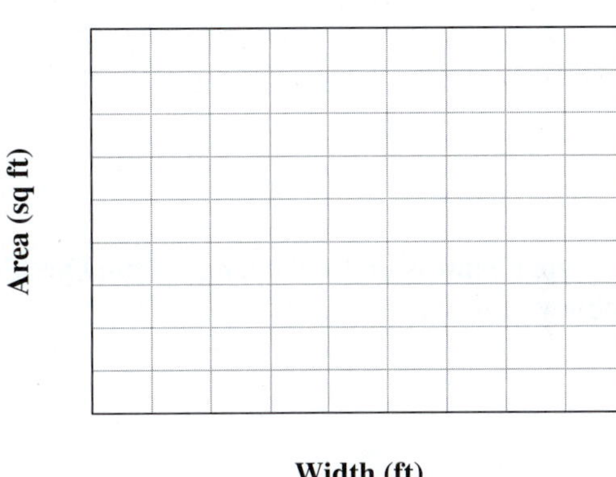

4. Based on the graph, what width makes the area as large as possible?

5. What's the area of the largest possible pen he can make out of 160 feet of fence?

6. Describe the shape that the play area should be to make the area as big as possible.

7. Write an explanation of the process used to calculate the length needed for each width.

8. Write an expression with w representing the width to describe the length when the width is w feet. Your answer to Question 7 should be a big help.

9. Write the formula for the area of a rectangle with length l and width w.

10. Substitute your expression for the length from Question 8 into the area formula for l, then simplify the result. What's the largest power of w that appears?

Did You Get It

2. Suppose that the homeowner in Questions 1–9 doesn't love Moose nearly enough, and decides to only use 100 feet of fence for the play area.
   a. What would the area be if the width is 20 feet? What about 35 feet?
   b. Write an expression for the length of the play area when the width is w feet.
   c. Use part b to write an expression for the area of the play area when the width is w feet.

The profit made by a company that manufactures wind turbines is illustrated by the following graph. The first coordinates represent the monthly number of kilowatt hours of electricity generated by their turbines in millions. We'll use the letter x to represent this variable quantity. The second coordinates of points are the company's profit in dollars. Answer every question with a complete sentence.

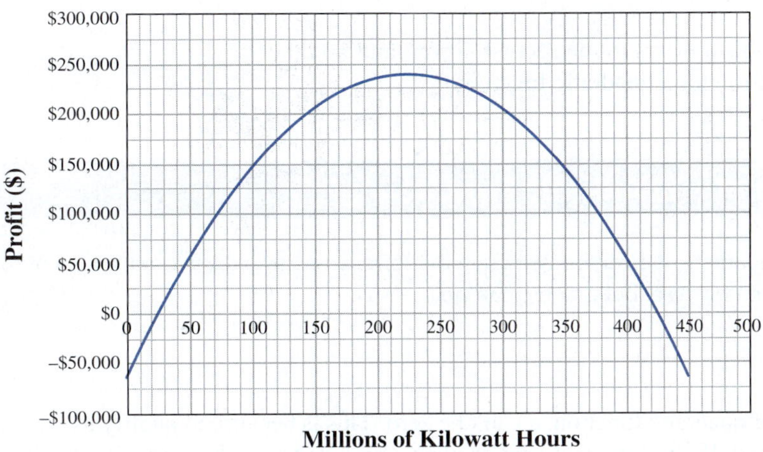

11. What's the y intercept of the graph? What does it tell us?

12. What are the x intercepts of the graph? What do they tell us?

13. How much electricity would need to be generated in order to make a profit of $100,000?

14. What's the maximum profit the business can earn? How much electricity needs to be generated to reach that profit?

15. The function that describes the profit made by the company is $f(x) = -6x^2 + 2{,}700x - 63{,}750$, where x represents the amount of electricity generated in millions of kilowatt hours, and $f(x)$ is the associated profit. Use a graphing calculator or computer program to recreate the graph from this problem, and use a trace and/or a table feature to verify your answers to Question 14.

A function in which the variable appears to the second power, and no other power except perhaps first, is called a **quadratic function**. Here's the general form for a quadratic function:

Definition of a Quadratic Function

A function is quadratic if it can be written in the form $f(x) = ax^2 + bx + c$, where all of a, b, and c are numbers. Note that a can't be zero, but b and c can be.

Notice that in our definition of quadratic function, a can't be zero. This is because to qualify as quadratic, a function has to have a term with the variable squared. If a is zero, that term is gone. But b and c can be zero, which tells us that the other two terms may or may not be there. Notice also that the definition says "can be written in the form." Sometimes you may have to do a bit of simplifying to recognize a quadratic function.

16. Which of the following functions is quadratic?

a. $f(x) = 4x^2 + 16x - 3$

b. $f(x) = 8x - x^2$

c. $f(x) = x^2 + 2x^3$

d. $f(x) = 2x(x - 4)$

e. $f(x) = 2x^2 + 3\sqrt{x}$

Math Note

Other than the algebraic form of the function, every quadratic function has one main thing in common: The graph is a parabola.

Did You Get It

3. a. Adapt your answer from Did You Get It 2c into a function describing the area when the width is w feet. Did you get a quadratic function?

   b. Use a calculator or computer to graph your function and use the graph to find the width that will provide the greatest play area if the cheapskate homeowner buys 100 feet of fence.

4-7 Portfolio

Name ______________________________

Check each box when you've completed the task. Remember that your instructor will want you to turn in the portfolio pages you create.

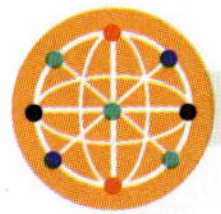

Technology

1. ☐ Use Excel to create a graph describing the width and area of the rectangular play area in the Group portion of this lesson. Use the table on page 549 as a model for your spreadsheet. A smooth marked scatter plot should get the job done nicely. Print your graph and add labels on the key points that were discussed in Questions 1, 2, 4, and 5.

Online Practice

1. ☐ Include any written work from the online assignment along with any notes or questions about this lesson's content.

Applications

1. ☐ Complete the Applications problems.

Reflections

Type a short answer to each question.

1. ☐ If you're looking at a graph, what are some key things that will tell you that it might be a parabola?
2. ☐ How can you tell if a function is quadratic?
3. ☐ Think of some quantities that could be modeled with quadratic graphs. Did you do better than you did in Question 0 at the beginning of this lesson?
4. ☐ Name one thing you learned or discovered in this lesson that you found particularly interesting.
5. ☐ What questions do you have about this lesson?

Looking Ahead

1. ☐ Complete the Prep Skills for Lesson 4-8.
2. ☐ Read the opening paragraph in Lesson 4-8 carefully and answer Question 0 in preparation for that lesson.

Answers to "Did You Get It?"

1. It depends on the particular day, but on average this would probably be reasonable. You'd expect the temperature to decrease overnight, reaching a low early in the morning, then increase after the sun rises.

2. a. $w = 20, A = 600$ sq ft; $w = 35, A = 525$ sq ft **b.** $50 - w$ **c.** $w(50 - w)$

3. a. $f(w) = 50w - w^2$; this is a quadratic function **b.** 25 ft

Answers to "Prep Skills"

1. perimeter: 35 ft; area: 58.5 sq ft **2.** 32 m **3. a.** $3w^2 + 15w$ **b.** $80x - 10x^2 + 20x^3$

4. $P = 180 - 2w$

5. (0, 0); this tells us that the drone was on the ground before it took off (flight time 0 minutes).

(60, 0); this tells us that the drone landed back on the ground after being in flight for 60 minutes.

4-7 Applications

Name ____________________

After a football is kicked by a punter, it's subjected to gravity, and follows a parabolic path. The height of one particular punt is shown in the graph. The first coordinates represent the horizontal distance from the punter, and the second coordinates represent the height of the ball above the ground. Both measurements are in feet. **Answer each question with a full sentence.**

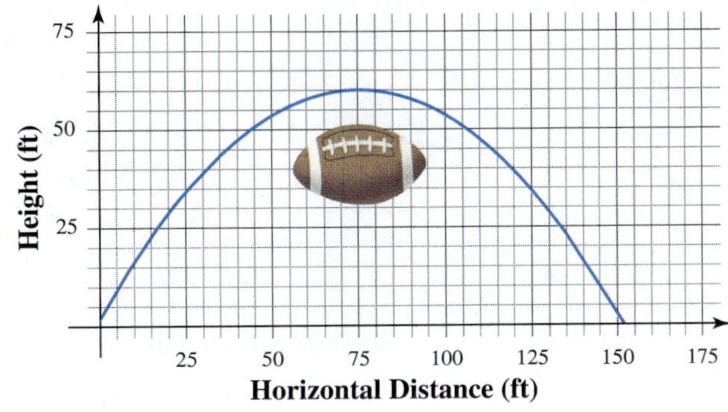

1. About how high above the ground was the ball when it was kicked?

2. How far away from where it was kicked did the ball land?

3. How high did the punt go?

4. Use the graph and some ingenuity to estimate the actual distance the punt traveled. (This is different from how far away it landed! Think of the parabola as a road, and estimate how far you'd travel on that road.) There's no right answer here: A big part of the question is describing how you got your estimate.

4-7 Applications

Name ______________________________

The next graph also shows the height of the ball on the y axis, but this time the x coordinates are the number of seconds since the ball was kicked.

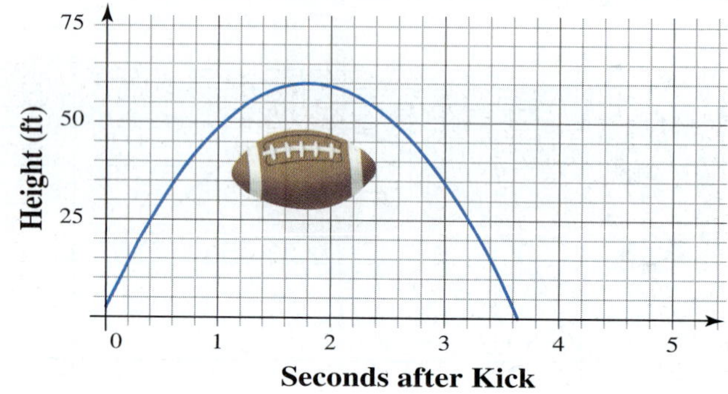

5. The hang time of a punt is how long it's in the air before returning to the ground. What was the hang time for this punt?

6. After how many seconds was the ball 45 feet above the ground?

7. The equation of the parabola representing the ball's height is $f(x) = -18(x - 1.8)^2 + 60$, where $f(x)$ is the height in feet and x is the number of seconds after it was kicked. Find the exact height after the number of seconds you found in Question 6. How accurate were your answers to Question 6?

4-7 Applications

Name ______________________________

8. If you find the value $f(4)$ in that function, why does the result not apply to the punt?

9. Explain why the shape of the graph on the previous page does NOT tell you what the flight path of the ball looked like from a side view.

Lesson 4-8 Prep Skills

SKILL 1: FIND THE EQUATION OF A LINE GIVEN TWO POINTS

We've previously studied two ways to find the equation of a line. If you know the slope of the line and the *y* intercept, you can use the formula $y = mx + b$, where *m* is the slope and *b* is the second coordinate of the *y* intercept. This allows you to just write the equation directly. That formula is called slope-intercept form.

- The equation of the line with slope $\frac{2}{3}$ and *y* intercept $(0, -8)$ is $y = \frac{2}{3}x - 8$.

If you know the slope and one point on the line, you can use the formula $y - y_0 = m(x - x_0)$, where *m* is the slope, and x_0 and y_0 are the *x* and *y* coordinates of the point that you know. This is called point-slope form.

- The equation of the line with slope -2 that goes through the point $(1, 5)$ is $y - 5 = -2(x - 1)$. You can then simplify to write in the standard form $y = mx + b$.

If you know two points on the line, you have two choices. First, you definitely need to find the slope using the slope formula $m = \frac{\text{difference of } y \text{ coordinates}}{\text{difference of } x \text{ coordinates}}$. Then you can either choose one of the two points you know and use the point-slope form, or you can use the procedure outlined below to find the equation using the slope-intercept form.

- Find the equation of line through the points $(2, 5)$ and $(-3, 15)$.

  The slope is $m = \frac{15 - 5}{-3 - 2} = \frac{10}{-5} = -2$. If we want to use the point-slope formula, we can choose either point. I'll go with $(2, 5)$.

  $y - 5 = -2(x - 2) \quad \Rightarrow \quad y - 5 = -2x + 4 \quad \Rightarrow \quad y = -2x + 9$

  If we want to use the slope-intercept formula, we know the equation looks like $y = -2x + b$, but we don't know what *b* is. Pick one of the points and substitute in for *x* and *y*. This allows us to solve for *b*. If we again choose the point $(2, 5)$, we substitute 5 for *y* and 2 for *x* into our equation. $5 = -2(2) + b \Rightarrow$ $5 = -4 + b \quad \Rightarrow \quad b = 9$

  Again, we get $y = -2x + 9$.

SKILL 2: COMBINE LIKE TERMS

Way back early in Unit 1, we talked about how you can only add or subtract quantities when they're like. For example, everyone knows that 12 grams of protein + 8 more grams of protein is 20 grams of protein, and that if you went toys-in-the-attic crazy and bought 42 pineapples, if someone not quite as loco as you stole 30 of them, you'd have 12 left. But 12 grams of protein + 8 grams of fat is just 12 grams of protein + 8 grams of fat: The quantities can't be added.

This idea also applies to expressions involving variable quantities. For example, $12x + 8x$ is $20x$ for the same reason that 12 grams of protein + 8 grams of protein is 20 grams of protein. We call $12x$ and $8x$ **like terms** because they each represent a certain number of the same thing: the variable quantity *x*. We could also subtract $42y - 30y$ to get $12y$. But $12x$ and $8y$ aren't like terms, so they can't be added or subtracted.

- $5x + 12 - 2x - 17 = 3x - 5$ — $5x$ and $2x$ are like terms and can be subtracted; so are 12 and 17.
- $13x - 4y + 8x - 7y + 10 = 21x - 11y + 10$

SKILL 3: MULTIPLY FACTORS

When multiplying two factors of the form ax^n, where a and n are numbers and x is a variable, you multiply the numeric coefficients together, then multiply the variable factors by adding the exponents.

- $(9y)(3y) = 27y^2$
- $(-10x^2)(7x^4) = -70x^6$

SKILL 4: USE THE DISTRIBUTIVE PROPERTY

This skill was reviewed in the Prep Skills for Lesson 1-4. In this lesson, we'll be distributing a variable into an expression, so you also need to be familiar with rules for exponents, as in this example. We'll first apply the distributive property, multiplying the y by both terms in the parentheses.

- $y(40 - 5y) = 40 \cdot y - y \cdot 5y$
  $= 40y - 5y^2$

PREP SKILLS QUESTIONS

1. Find the equation of the line that has slope −5 and goes through the point (10, 2). Write your answer in the form $y = mx + b$.

2. Find the equation of the line that goes through the points (7, −10) and (4, 32).

3. Simplify each expression by combining like terms.

   a. $9x + 12 + 3x - 9$ b. $2a - 3b - 4a - 5 + 6b$

4. Perform each multiplication.

   a. $(12t)(5t)$ b. $\left(-\frac{5}{2}x^2\right)(4x^3)$

5. Perform each multiplication.

   a. $x(x + 5)$ b. $-4y(12 - y)$

Lesson 4-8 Minding Your Business

LEARNING OBJECTIVES

- ☐ 1. Combine algebraic expressions using addition or subtraction.
- ☐ 2. Demonstrate the relationships between revenue, cost, and profit functions.
- ☐ 3. Combine algebraic expressions using multiplication.
- ☐ 4. Write a revenue function from a demand function by multiplying algebraic expressions.

©Erik Isakson/Blend Images RF

An economist is an expert who will know tomorrow why the things he predicted yesterday didn't happen today.

— Laurence J. Peter

At its core, business is simple: come up with a product or service that people want, and sell it to them. It gets complicated by competition, though. If you're the only person selling a product that people want or need, you're golden. But how often does that happen? Pretty much never. One of the most important aspects of business is studying financial figures and using them to predict future conditions. And since those figures are numbers, it should come as no surprise that we can use algebraic equations to model them. In this lesson, we'll use some common economic concepts to study some ways that we can combine expressions involving variables in order to model more complicated things in our world.

0. How do you think studying financial figures might impact how much a company decides to charge for a certain product or service?

4-8 Group

After studying two years' worth of sales records, the owner of a popular boutique has found that she can sell 400 jars per month of a brand of bath beads that was featured on *Shark Tank* when she prices them at \$12. When she temporarily lowered the price to \$10, she sold 480 per month.

1. If the relationship between price and sales is linear, write an equation that relates the price p to the number of jars of bath beads sold x. Your equation should be in the form $p = mx + b$. Hint: Start by writing ordered pairs of the form (jars sold, price), then find the slope.

In economics, the word **revenue** refers to the amount of money that a business takes in as a result of selling a product or service. Revenue is pretty easy to calculate, especially if you know the amount of sales and the price: Revenue is simply the selling price times the number of units sold. Using the symbols from Question 1, revenue R is number of units sold x times price p.

2. Write an equation that represents the revenue made on bath beads for our boutique owner by using the equation $R = xp$. Then simplify your expression by multiplying out the parentheses using the distributive property. Your answer to Question 1 will be helpful.

The **costs** of doing business for a company come in two flavors: fixed and variable. Fixed costs are things like rent, utilities, insurance, and wages, which stay the same whether the company makes and sells anything or not. Variable costs refer to what it costs to produce the items being sold, which varies depending on the number of items produced.

3. The portion of the boutique's fixed monthly costs attributed to bath beads is \$325; the cost of each jar of bath beads from the supplier is \$3.50. Write an equation that represents the total cost C to the boutique of acquiring x jars of bath beads. Your equation should be in the form $C = mx + b$.

4. Since revenue is the amount of money a business takes in, and cost is the amount the business pays out, the difference between revenue and cost is the **profit** made by the business. That is, profit P is revenue R minus costs C. Write an equation that represents the profit made on bath beads by the boutique. (Hint: Substitute your answers to Questions 2 and 3 into the formula $P = R - C$.) Don't simplify your expression yet.

5. The costs consist of two distinct terms, and when we subtract we need to make sure to subtract ALL of the costs, not just some of them. To make sure this happens, we put parentheses around the entire cost expression; if you didn't already do this, add parentheses to your answer from Question 4. Then perform the subtraction, keeping in mind that the parentheses are there to remind you to subtract EVERY term in the expression for cost. The result is an equation describing profit in terms of jars of bath beads sold that's relatively easy to work with.

Math Note

In algebra, **terms** are individual parts of an expression that are added or subtracted. For example, in $250x + 30$, the terms are $250x$ and 30.

Did You Get It

Try this problem to see if you understand the concepts we just studied. The answers can be found at the end of the Portfolio section.

1. A company that makes portable Bluetooth speakers can sell x units if they price them at $p = 32 - 0.04x$ dollars. Find an equation that describes their revenue R in terms of the number of units they sell.
2. The cost function for the company in Did You Get It 1 is $C = 11.2x + 785$. Find the profit equation in the form $P =$ expression.

6. Find the profit the boutique makes if it sells 200 jars of bath beads per month.

7. Find the profit made from selling 600 jars per month.

8. How in the world can the boutique make LESS money for selling MORE bath beads? Explain.

9. What's the profit if the boutique sells no bath beads in a month? What does the result mean?

10. Use a calculator or computer to graph the profit equation you wrote and then use your graph to estimate the number of jars that must be sold to make the biggest possible profit.

11. What's the largest monthly profit that can be made on bath beads? (Use your answer to Question 10.)

12. What selling price results in the largest possible profit?

Suppose that our boutique owner decides to donate the proceeds from five sales of bath beads to a local women's shelter each month. In that case, the revenue isn't $x(-0.025x + 22)$ anymore: It's $(x - 5)(-0.025x + 22)$ because the number of items that we need to multiply by the price is five less (the proceeds from five sales donated). Could we still use the distributive property to multiply out the parentheses? We surely can. Let's have a look.

The expression that results is the product of two **binomials**, which is a fancy math term for an expression with two terms. To understand how we would perform this multiplication, let's first pretend that the $(x - 5)$ is just a number, say 10. Then the multiplication would look like

$$10(-0.025x + 22)$$

and the distributive property tells us that we can expand this as

$$10 \cdot (-0.025x) + 10 \cdot 22 = -0.25x + 220$$

No big deal, right? Now let's put the $(x - 5)$ back where it belongs, and do exactly the same thing.

$$(x - 5)(-0.025x + 220) = \underbrace{(x - 5) \cdot (-0.025x)}_{\text{Multiply }(x-5)\text{ by first term}} + \underbrace{(x - 5) \cdot 220}_{\text{Multiply }(x-5)\text{ by second term}}$$

13. Now you can use the distributive property twice because you have two terms that each have something multiplied by a difference. Multiply out both sets of parentheses. Don't combine any like terms yet. (Don't let it bother you that the term being distributed is on the right side of the parentheses! You can move it to the left if it makes you feel better.)

Using multiple applications of the distributive property is a routine you can always use to multiply two expressions that each have two or more terms. And that's fine. But if you look a little deeper, you can see a shortcut pattern. Notice that there are four terms in our answer, and those terms come from multiplying every term in the first expression by every term in the second. That's a nice general rule for this type of multiplication.

14. Combine like terms to simplify your answer to Question 13.

Multiplying Multiple-Term Expressions

To multiply two expressions that each have two or more terms, you have two choices.

1. Treat the first expression as if it were just a number, and distribute it to each term in the second expression. Then you can use the distributive property to multiply out any remaining parentheses, and finally combine any like terms.
2. Multiply every single term in the first set of parentheses by every term in the second set, then combine like terms.

15. In designing the boutique's parking lot with a 4-foot landscaping strip on three sides, an architect found that the total area could be expressed as $(w + 4)(2w + 8)$, where w is the width of the paved portion. Multiply out the parentheses in this expression and simplify.

Math Note

If you're familiar with the FOIL procedure for multiplying two binomials, feel free to use it. But remember it only works if you have two binomials!

Did you get it

3. Multiply out the parentheses: $(x + 3)(x - 5)$

The form $(w + 4)(2w + 8)$ is called the **factored form** of an expression. Notice that it's written as a product: $(w + 4)$ TIMES $(2w + 8)$. Each of $(w + 4)$ and $(2w + 8)$ is called a **factor** of the expression. Factors are things that are multiplied together; terms are things that are added or subtracted. Your answer to Question 15 is called the **expanded form** of an expression, and it's written as a sum or difference. These are two different forms of the same expression: If you input ANY value for the variable quantity x, you should get the same result from either expression. (We say that the expressions are **equivalent**.)

4-8 Class

An algebraic expression consisting of one or more terms that look like either a real number, or a real number times a variable raised to a whole number power is called a **polynomial**. In this lesson, we've studied operations on polynomials: combining polynomials using addition, subtraction, and multiplication. Let's summarize what we've learned.

In Questions 1–4, simplify each expression. Aside from what we covered in the Group portion of this lesson, here are some things you should keep in mind:

- Order of operations: Multiplication always comes before addition or subtraction.
- The distributive property is needed when multiplying a number or variable by a sum or difference.
- When subtracting, don't forget to subtract EVERY term in an expression. That's what the parentheses are there for!

1. $(3x^2 + 3x + 5) + (2x^2 - 3x - 10)$

2. $(5x^2 + 3x + 2) - (x^2 - 2x + 7)$

3. $3(x^2 + 3x + 5) + 2(x^2 - 3x - 10)$

4. $2(3x^2 + 4x - 5) - 3(4x^2 - 5x + 1)$

Did You Get It ?

4. Simplify the expression as much as possible: $3(4y^2 + 7y - 3) - (8 - 4y + 6y^2)$.

You probably already noticed that multiplying expressions is more challenging than adding and subtracting them. For that reason we're going to get some more practice on multiplying and also develop a method to check and see if your result is correct when you multiply expressions. (It also works for adding and subtracting, by the way.)

5. Perform each multiplication, keeping in mind that our procedure involves multiplying every term in the first factor by every term in the second factor.

a. $(y - 5)(y + 7)$ **b.** $(2x + 6)(3x - 8)$

6. Use the distributive property to multiply: $2x(3x + 4)$. Does this match the description in Question 5? Explain.

Now about checking our answers. Earlier we pointed out that the goal when simplifying is to get another expression that provides the exact same output as the original no matter what input you choose.

7. If two expressions satisfy that condition, what should the relationship between their graphs be? Think about what the graph of an equation really is.

8. Use a calculator or computer program to graph both $(2x + 6)(3x - 8)$ and your answer to Question 5b. Do you think your answer is correct? Explain.

9. Our next goal is to multiply $(x - 5)$ by $(x^2 + 3x + 4)$.

a. First, multiply x by each term of $x^2 + 3x + 4$, and list the three products.

b. Now multiply -5 by each term of $x^2 + 3x + 4$, and again list the three products. (Careful with the negative sign!)

c. Add all six products from parts a and b, combining any like terms. This is (we hope!) the product of $(x - 5)$ and $x^2 + 3x + 4$.

10. Use a graph to decide if your answer to Question 9c is right. Explain.

Did You Get It ?

5. Perform the multiplication and check your answer with a graph: $(2x + 1)(x^2 - 3x + 2)$.

4-8 Portfolio

Name ______________________________

Check each box when you've completed the task. Remember that your instructor will want you to turn in the portfolio pages you create.

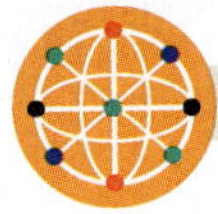

Technology

1. ☐ Create an Excel spreadsheet to explore whether or not your answer to Question 9c in the Class portion of this lesson is correct. Make sure to input a wide variety of x values: positive, negative, fractions, decimals, even irrational numbers if you want to impress your teacher. Can you be sure that the two expressions are equivalent by inputting a variety of values? Explain. (Bonus points for doing some online research about when two polynomials are equivalent. . . .) There's a template to help you get started in the online resources for this lesson.

	A	B	C
1	Input (x)	(x–5)(x^2+3x+4)	Answer to 9c
2			
3			
4			

Online Practice

1. ☐ Include any written work from the online assignment along with any notes or questions about this lesson's content.

Applications

1. ☐ Complete the Applications problems.

Reflections

Type a short answer to each question.

1. ☐ What is a polynomial? What is meant by the phrase "operations on polynomials"?

2. ☐ Describe the significance of each of these words in the context of this lesson: revenue, costs, profit.

3. ☐ Describe the process we developed for multiplying two expressions that each have two or more terms.

4. ☐ Name one thing you learned or discovered in this lesson that you found particularly interesting.

5. ☐ What questions do you have about this lesson?

Looking Ahead

1. ☐ Complete the Prep Skills for Lesson 4-9.
2. ☐ Read the opening paragraph in Lesson 4-9 carefully and answer Question 0 in preparation for that lesson.

Answers to "Did You Get It?"

1. $R = 32x - 0.04x^2$
2. $P = -0.04x^2 + 20.8x - 785$
3. $x^2 - 2x - 15$
4. $6y^2 + 25y - 17$
5. $2x^3 - 5x^2 + x + 2$

Answers to "Prep Skills"

1. $y = -5x + 52$ **2.** $y = -14x + 88$ **3. a.** $12x + 3$ **b.** $-2a + 3b - 5$

4. a. $60t^2$ **b.** $-10x^5$ **5. a.** $x^2 + 5x$ **b.** $-48y + 4y^2$

4-8 Applications

Name ______________________________

Based on the results of an extensive market research study that I totally made up, an economist hired by a company that makes a popular line of indoor grills predicts that consumers in one market will buy x units per week if the price is $p = -0.27x + 61$ dollars.

1. Recall that revenue is price times number of units sold, and write an equation that represents the revenue the company will make from selling x units of the grill in that market. Simplify your answer.

2. It costs the manufacturer $C = -0.12x^2 + 14x + 570$ dollars to make x units of this particular grill. Write an equation describing the profit made from selling x units of the grill in this market. The simpler your answer is, the easier it will be to work with in the next couple of questions.

3. Find the profit made if 10 grills are sold per week. What does this mean?

4. Find the profit made for sales of 100 and 250 grills. How do they compare?

5. Use a calculator or computer program to graph the profit equation you wrote in Question 2 and use it to estimate the number of grills they'd need to sell to make the largest possible profit.

4-8 Applications

Name ______________________________

6. What's the largest profit the company can make from indoor grills in that market, and what should they set the price at to make that happen?

7. Suppose that one month the company runs a promotion where they send a rebate of the full purchase price to four randomly selected customers. To find the revenue in Question 1, you multiplied x (the number of units) by the price formula. What would you need to multiply the price by in this case?

8. Write and multiply out the revenue equation in this case.

9. Find the new profit equation using your revenue equation and the original cost equation.

10. Graph this new profit equation and see what effect this promotion has on the largest profit the company can make.

Lesson 4-9 Prep Skills

SKILL 1: IDENTIFY WHEN A LINEAR EXPRESSION IS ZERO

This is a skill that too many people try to commit to memory, rather than learning a simple procedure so that you don't have to memorize anything. To decide when an expression like $x - 7$ is zero, the most reliable method is to set up an equation with that expression on one side and zero on the other, then solve that equation.

- $x - 7$ is zero for any x that makes the equation $x - 7 = 0$ true.

  $x - 7 = 0$ Add 7 to both sides

  $x = 7$

- $y + 12$ is zero for any y that makes the equation $y + 12 = 0$ true.

  $y + 12 = 0$ Subtract 12 from both sides

  $y = -12$

Of course, if you're able to just look at $x - 7$ and figure out that it will be zero when x is replaced by 7, that's fine. But if you have ANY doubt whatsoever, set up and solve the equation.

SKILL 2: EVALUATE ALGEBRAIC EXPRESSIONS

This is a skill that we've used repeatedly throughout this book. All you really need to keep in mind is that when you're substituting in a number for a variable, you need to replace every occurrence of that symbol with the given number, then use order of operations to complete the calculation.

- Evaluate $12x^2 - 6x + 1$ for $x = 3$

  $12(3)^2 - 6(3) + 1 = 12(9) - 18 + 1 = 108 - 18 + 1 = 91$

Notice the use of parentheses around the 3. It's usually a good idea to put parentheses around the number you're substituting in, but it's especially important to do so when negatives are involved.

- Evaluate $12x^2 - 6x + 1$ for $x = -3$

  $12(-3)^2 - 6(-3) + 1 = 12(9) + 18 + 1 = 108 + 18 + 1 = 127$

Without the parentheses, some common mistakes often result:

$$12 \cdot -3^2 - 6 \cdot -3 + 1$$

First, -3^2 doesn't mean the same thing as $(-3)^2$: Order of operations says that -3^2 means -9, not 9. Second, it's easy to mistake the $-6 \cdot -3$ for subtraction rather than multiplication.

SKILL 3: MULTIPLY BINOMIALS

We saw in Lesson 4-8 that to multiply two algebraic expressions with more than one term, you need to multiply every term in the first expression by every term in the second, then combine like terms.

- $(x-5)(x+2) = x \cdot x + x \cdot 2 - 5 \cdot x - 5 \cdot 2 = x^2 + 2x - 5x - 10 = x^2 - 3x - 10$

x from first expression multiplied by both terms in second expression

-5 from first expression multiplied by both terms in second expression

PREP SKILLS QUESTIONS

1. Identify all x values that make the output of each expression equal to zero.

   a. $x - 4$ b. $y + 3$ c. $2x + 5$

2. Evaluate each expression for the given values of the variable.

   a. $4x - 12$; $x = 0, 4, -3$ b. $2x^2 + 3x - 1$; $x = 0, 2, -3$ c. $-3(y+4)(y-6)$; $y = -4, 0, 1, 6$

3. Perform each multiplication.

   a. $(x+3)(x+6)$ b. $(t-4)(t+4)$ c. $(2y-3)(3y+7)$

Lesson 4-9 The F Word

LEARNING OBJECTIVES

- ☐ 1. Explain why factoring is useful in algebra
- ☐ 2. Explain the connection between zeros and x-intercepts.
- ☐ 3. Factor a trinomial.

©BDS/123RF

Confidence is the most important single factor in this game, and no matter how great your natural talent, there is only one way to obtain and sustain it: work.
— Jack Nicklaus

Relax— the F word we're talking about is perfectly acceptable by FCC standards. It's only in math classes that some people feel it's a dirty word: factoring. Depending on who you ask, factoring can be thought of as anything from an essential part of every college student's curriculum, without which they would surely perish, to an archaic procedure with very limited and artificial applications in the modern world. The truth is that you can type any polynomial into Wolfram Alpha and get its factored form. The trick is in understanding what the point is, and being able to efficiently use the factored form of a polynomial to solve problems. So that will be our first order of business. We'll look at the actual process of factoring later in the lesson, just in case your school is in the "you really, really, really, really need to be able to do this" camp.

0. Without looking ahead at all, provide a description of what you think factoring is.

4-9 Group

1. Perform the multiplication: $5 \times 7 =$ ________

2. Turn your answer from Question 1 around backwards to write 35 as a product: $35 =$ ________ $\times$ ________

Congratulations—you have just factored. Isn't it bizarre that a topic that has such a bad reputation is based on such a simple idea?

> **Factoring** is simply the process of writing a number or expression as a product. In that regard, it's essentially the opposite of performing multiplication.

Here's an example from our study of economics. In Lesson 4-8, we multiplied a variable quantity (number of units sold, represented by x) by the expression $-0.025x + 22$, which represented price, to get an equation describing revenue: $R = -0.025x^2 + 22x$. If we "undo" that multiplication to recover the price:

$$-0.025x^2 + 22x = x(-0.025x + 22)$$

we have factored, because we've written $-0.025x^2 + 22x$ as a product. Simple!

As you probably know, computers have a master volume setting that controls the volume of everything it sends through the speakers. But many applications have volume control as well, like web browsers when playing videos from YouTube. If the master volume is set to 100% and the volume in the YouTube window is set to 60%, the sound in the video will play at 60% of its max. Easy. Now let's look at some other combinations.

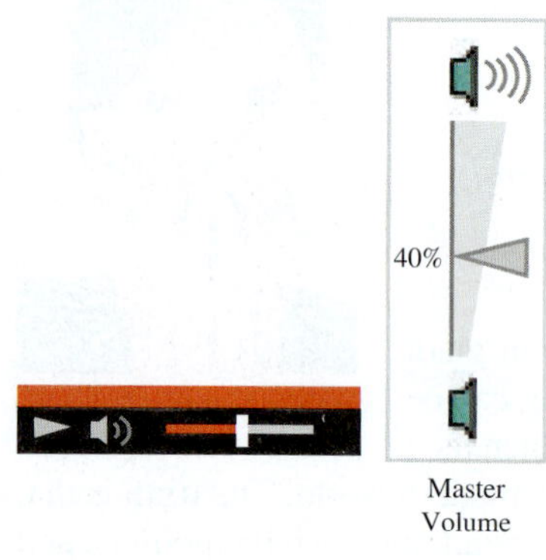

3. If the volume in the YouTube window is set at half and the master volume for the computer is at 40%, what percent of the max volume do you think you'll hear from your speakers? Explain your answer.

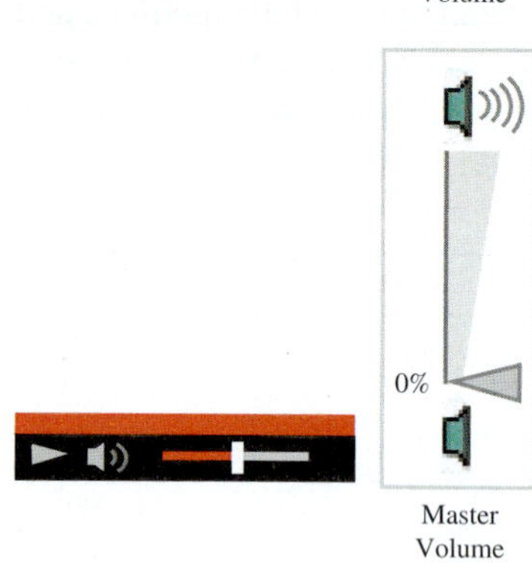

4. If the volume in the YouTube window is set at half and the master volume for the computer is at 0, what percent of the max volume do you think you'll hear from your speakers? Explain your answer.

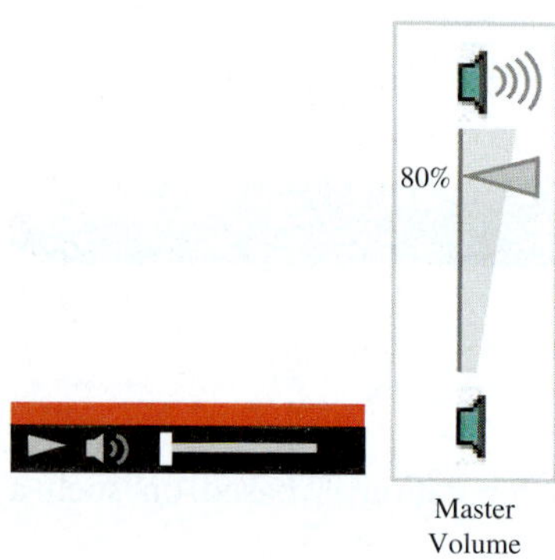

5. If the volume in the YouTube window is set all the way to the left and the master volume for the computer is at 80%, what percent of the max volume do you think you'll hear from your speakers? Explain your answer.

Conclusions:

6. To get the overall volume, we ______________ the percentages in decimal form from the master volume and the YouTube volume.

7. You can only get zero as the overall volume if ____.

So what does this have to do with factoring?

8. For what value of the variable quantity x does the expression $x - 4$ have output zero? What about the expression $x + 5$? (Don't over-think these. They're really easy.)

9. Now look at the product $(x - 4)(x + 5)$, which is a polynomial in factored form. Find the value of the product when . . .

a. $x = 2$ **b.** $x = 4$ **c.** $x = 0$ **d.** $x = -1$ **e.** $x = -5$

Conclusion:

10. The only way a product of algebraic expressions can have value zero is if ________________.

Did You Get It

Try this problem to see if you understand the concepts we just studied. The answer can be found at the end of the Portfolio section.

1. Find all values of y for which the output of the expression $(y - 8)(y + 5)$ is zero.

And THAT'S why factoring is a big deal in algebra. It provides an important technique for solving equations that have zero on one side.

11. Write the number 37 as a product.

12. Write the polynomial expression $3x + 5$ as a product.

In Questions 11 and 12, there's only one way to write the given number or expression as a product (unless you're willing to use fractions, which we are not at the moment): one times the number or expression itself. In the case of 37, we call it a **prime number**. For the polynomial $3x + 5$, we say that it is a **prime polynomial**.

The point: Not every expression can be factored if we restrict our attention to integer coefficients.

4-9 Class

Other than what it means to factor, the main point of the Group portion of this lesson is why factoring is a big deal: It makes it simple to see when an expression has output equal to zero, which has important applications in solving equations.

If $x = c$ is an input value for which the output of a function $f(c)$ is zero, we call the input c a **zero** of the function. There's a useful relationship between the factors of a polynomial, the zeros of the related function, and the x intercepts of the graph.

The Connection Between Factors, Zeros, and Intercepts

For a given polynomial $P(x)$ and a real number c, each of these statements say the same thing:

The point $(c, 0)$ is an x intercept of the graph of $P(x)$. **(Graphical)**	The input c is a zero of $P(x)$, which means that $P(c) = 0$. **(Numerical)**	The expression $(x - c)$ is one of the factors of $P(x)$. **(Algebraic)**

1. For the polynomial $P(x) = (x - 1)(x + 5)$, find each output. Use P in this factored form.

a. $P(-5)$

b. $P(-3)$

c. $P(-1)$

d. $P(1)$

2. Use the table and graph provided for $P(x) = (x - 1)(x + 5)$ to answer each question.

x	P(x)
−5	0
−4	−5
−3	−8
−2	−9
−1	−8
0	−5
1	0

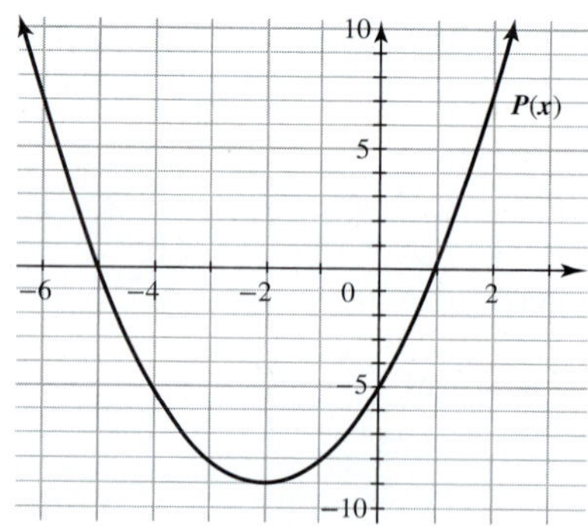

a. List the factors of $P(x)$.

b. List the zeros of $P(x)$.

c. List the x intercepts of the graph.

3. Now let's look at the function $f(x) = x^2 + 2x - 3$.

x	$f(x)$
−4	5
−3	0
−2	−3
−1	−4
0	−3
1	0
2	5

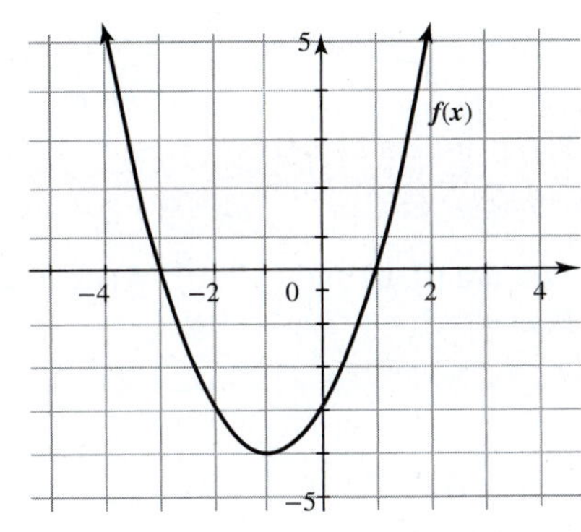

a. List the zeros of $f(x)$.

b. List the x intercepts of the graph.

c. Use parts a and b to write the factored form of $f(x) = x^2 + 2x - 3$. If you're not sure how to do this, look back at the colored box that comes before Question 1.

Use the factored form of each polynomial to list the zeros and x intercepts of the polynomial's graph.

4. $P(x) = 2(x - 3)(x - 4)$

Zeros:

x intercepts:

5. $y(x) = (x + 5)(x + 4)(x - 2)$

Zeros:

x intercepts:

Did You Get It

2. The x intercepts of the parabola $f(x) = x^2 + 2x - 35$ are $(5, 0)$ and $(-7, 0)$. Write a factored form of the function.
3. List the zeros and x intercepts for the function: $P(x) = -3(x + 2)(x - 10)$.

At this point you should have a pretty solid grasp on what factoring means, and some of the information you can get from the factored form of a polynomial. That leaves the \$100,000 question: *How do you get that factored form?* Factoring is a topic that you could spend a couple of weeks studying in depth, but we just want to give you a quick look at one particular type of factoring that pops up a lot in the study of quadratic equations and functions.

There are many different theories and techniques on how to do this type of factoring, and in many ways we're going to leave it up to your instructor to decide how (or even if) to cover factoring. We'll just make some general suggestions, which are based on a key skill we learned in the last lesson: multiplying two binomials.

6. Perform the multiplication: $(x + 2)(x + 12)$. DO NOT combine like terms.

7. When you multiply two binomials, you'll always get four terms at first. If both of those binomials have a first power term and a constant, two of the four terms will be like terms, so you can combine them to get a final answer with three terms. Do that now.

8. What you did in Question 6 was quite literally the *opposite* of factoring. Why?

The final result of the multiplication you did is a quadratic expression with three terms: one with the variable squared, one with the variable to the first power, and one constant term. An expression of this type is called a **trinomial** because it has three terms, and the key to factoring expressions of this form is recognizing that they're the ANSWER to a "multiply binomials" problem.

Now let's look carefully at the multiplication:

$$(x+2)(x+12) = x^2 + 14x + 24$$

The first term of the resulting trinomial (x^2) came from multiplying the first terms of the binomials ($x \cdot x$). The last term of the trinomial (24) came from multiplying the second terms of the binomials ($2 \cdot 12$). And the middle term ($14x$) came from a combination of the outside terms ($12 \cdot x$) and the inside terms ($2 \cdot x$). We can take advantage of these observations to turn this process around backwards.

The expression $x^2 + 5x + 6$ can be factored into the product of two binomials (not obvious, but trust us for a moment). In that case, we can set up an equation as follows:

$$x^2 + 5x + 6 = (\_\_ + \_\_)(\_\_ + \_\_)$$

Our job is to fill in the blanks.

9. What product would give us x^2? Write your answers in the first blank in each factor.

$$x^2 + 5x + 6 = (\_\_ + \_\_)(\_\_ + \_\_)$$

10. One pair of numbers that could give us a constant term of 6 is $1 \cdot 6$. Write those in the second blanks, then multiply out the parentheses and see if you get $x^2 + 5x + 6$.

11. Think of another pair of numbers with product 6, write them on the blank lines and multiply out the parentheses again to see if you get the right product this time.

$$(x + \_\_)(x + \_\_)$$

This is a reliable process for factoring trinomials. I like to think of it as a game of hangman: Set up the parentheses with four blanks to fill in, then figure out what needs to go in each blank to get the right product. When you practice the process a bit, you'll be able to figure out what to put in the blanks much more efficiently. (If your instructor thinks it's important for you to do so, that is.)

Let's try another example: Factor $y^2 - 3y - 10$.

12. How can you get y^2 as a product? Set up parentheses like we did in Question 9, and use your results to fill in the first blank in each set of parentheses.

13. List all pairs of numbers whose product is the constant term, -10. Don't forget to think about the sign!

14. Find a pair of numbers from your list that you can put into the second blanks in each set of parentheses to get a product of $y^2 - 3y - 10$, then write the factored form of that polynomial.

15. Now try one more factoring problem. Don't let the coefficient of x^2 bother you: The same process works just fine, but you'll have to be careful when placing the pairs of numbers with product -20 in the second blanks: Try them in both orders.

$$2x^2 - 3x - 20$$

Did You Get It

4. Factor each trinomial.

a. $x^2 + 6x + 8$ b. $y^2 - y - 30$ c. $2x^2 - 17x - 9$

Dude, what's the point?

I'm glad you asked. We have seen that when you know the factored form of a polynomial, it's super easy to find the x intercepts of its graph. And this can be used to solve many problems, like this one.

If you make your living from growing produce, say apples, the more apples you grow, the more you can sell. Duh. So why not just plant a million apple trees? Space and resources. If the trees are too densely planted, they compete with each other for resources, and each tree produces less. So too many trees could mean less apples. And it's pretty obvious that not enough trees means less apples. Hey—maybe it's useful to find the ideal number of apple trees to plant.

In one orchard, there are 50 trees, each of which produces 800 apples. For every additional tree planted, the average output per tree drops by 10 apples. Bummer. How many additional trees should be planted to get the largest total number of apples?

16. If x more trees are planted, how many trees will there be total?

17. If the current yield per tree is 800, and will go down by 10 for each new tree planted, what will the yield per tree be if 5 more trees are planted?

18. If the current yield per tree is 800, and will go down by 10 for each new tree planted, what will the yield per tree be if x more trees are planted?

19. The total number of apples produced will be the total number of trees times the yield per tree. Write a function $T(x)$ describing the total number of apples, using your answers to Questions 16 and 18.

20. Find the x intercepts of the function you wrote in Question 19.

21. The graph of this function is a parabola that opens down, and parabolas are symmetric. That means the high point we're looking for has to be halfway between the x intercepts. How many new trees should they plant, and how many apples will be produced?

4-9 Portfolio

Name ______________________________

Check each box when you've completed the task. Remember that your instructor will want you to turn in the portfolio pages you create.

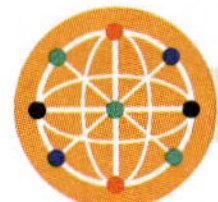

Technology

1. ☐ As we mentioned in the opening paragraph of this lesson, the website Wolfram Alpha (the URL, shockingly, is wolframalpha.com) can factor pretty much anything that is factorable. So go to the site and ask Wolfram Alpha to factor the polynomials from Questions 14 and 15 in the Class portion of this lesson. In exchange for your effort, you'll get back a factored form of the polynomials, and a whole lot more. Put a printout of the result in your portfolio, then type a brief report on some of the other things you can learn about the polynomials this way.

Online Practice

1. ☐ Include any written work from the online assignment along with any notes or questions about this lesson's content.

Applications

1. ☐ Complete the Applications problems.

Reflections

Type a short answer to each question.

1. ☐ What is the value of having the factored form of a polynomial? Think of as many different aspects as you can.
2. ☐ Describe the relationship between factoring and multiplication.
3. ☐ Name one thing you learned or discovered in this lesson that you found particularly interesting.
4. ☐ What questions do you have about this lesson?

Looking Ahead

1. ☐ Complete the Prep Skills for Lesson 4-10.
2. ☐ Read the opening paragraph in Lesson 4-10 carefully and answer Question 0 in preparation for that lesson.

Answers to "Did You Get It?"

1. $y = 8$ and $y = -5$ **2.** $(x - 5)(x + 7)$

3. Zeros: $x = -2, 10$; x intercepts: $(-2, 0)$ and $(10, 0)$

4. a. $(x + 4)(x + 2)$ **b.** $(y - 6)(y + 5)$ **c.** $(2x + 1)(x - 9)$

Answers to "Prep Skills"

1. a. $x = 4$ **b.** $y = -3$ **c.** $x = -2.5$

2. a. $-12, 4, -24$ **b.** $-1, 13, 8$ **c.** $0, 72, 75, 0$

3. a. $x^2 + 9x + 18$ **b.** $t^2 - 16$ **c.** $6y^2 + 5y - 21$

4-9 Applications

Name ________________________________

A trendy nightclub has a \$10 cover charge, and gets 540 patrons on an average night. After surveying customers, they're able to estimate that for each additional \$1 they add to the cover charge, they'll lose an average of 20 patrons per night.

1. If they add x one-dollar increments to the cover charge, write an expression describing the new charge.

2. If they add x one-dollar increments, how many patrons will they lose on average? What will be the new average number of customers?

3. Write a function $R(x)$ that describes the average nightly revenue if they raise the current price by $\$x$. (Hint: You'll need your answers to Questions 1 and 2.)

4. Find $R(0)$ and describe what it means.

5. Find the x intercepts of your function and describe what each tells us about the nightclub's cover charge and revenue.

6. If all has gone well, the graph of your revenue function should be a parabola that opens down. Since parabolas are symmetric, the highest point on the graph is halfway between the x intercepts. Use that fact to find the highest point on the graph, and describe in detail what it tells you about the nightclub's cover charge and revenue.

4-9 Applications

Name ______________________________

As part of a stunt for a late-night TV show, an intern throws a golf ball upward from a Manhattan balcony at a speed of 32 feet per second. Any halfway competent physicist would tell you that in that case, the height of the ball t seconds after it's released can be modeled by the function $h(t) = -16(t^2 - 2t - 3)$. For what it's worth, this is only true if you're willing to ignore friction and wind resistance (which I do as often as I can—I hate those guys), but for a golf ball those are reasonably negligible.

7. From what height is the ball thrown? (Hint: How many seconds after it was released is the time when it was released?)

8. Factor the polynomial $t^2 - 2t - 3$.

9. Using your answer to Question 8, find the zeros of the height function, and describe what each of them tells us about the golf ball.

Lesson 4-10 Prep Skills

SKILL 1: EVALUATE FUNCTIONS

Evaluating an algebraic expression for a certain input is probably one of the two or three skills that we've used most in this entire course, so I have to think you're pretty comfortable with that by now. The only thing that's a little bit different now is the notation, since we've introduced functions. Before, we may have given you an expression like $3x^2 - 10$ and asked you to evaluate it for an input like $x = 2$. So you'd replace x with 2, do some arithmetic, and life is good. Now, we'll give you a function $f(x) = 3x^2 - 10$ and ask you to find $f(2)$. *This means exactly the same thing!* When you see the symbol $f(2)$, you just think of it as finding the value of the function f when the input variable is replaced by 2.

SKILL 2: PLUG NUMBERS INTO A COMPLICATED FORMULA

Wait, we just got done talking about evaluating a function for a given input. Isn't this the same skill? It is, but most of the formulas that we've plugged numbers into have been relatively simple. But in this lesson, you'll need to evaluate a formula that has a lot going on, so a little practice at that would be a nice preparation for the lesson.

- For $f(x) = \dfrac{\sqrt{x^2 - 2x} + x}{2x}$, find $f(5)$ and $f(-4)$.

  As always, the goal is to first replace the variable (x in this case) with the given input. Then we'll have to be extra careful with the arithmetic.

$$f(5) = \frac{\sqrt{5^2 - 2(5)} + 5}{2(5)} = \frac{\sqrt{25 - 10} + 5}{10} = \frac{\sqrt{15} + 5}{10} \approx \frac{3.87 + 5}{10} = \frac{8.87}{10} = 0.887$$

$$f(-4) = \frac{\sqrt{(-4)^2 - 2(-4)} + (-4)}{2(-4)} = \frac{\sqrt{16 - (-8)} + (-4)}{-8} = \frac{\sqrt{24} - 4}{-8} \approx \frac{4.90 - 4}{-8} = \frac{0.9}{-8} \approx -0.11$$

SKILL 3: REVIEW ECONOMIC TERMS

In Lesson 4-8, we used the economics of operating a business to study quadratic functions. This is just a quick review of the econ concepts.

The **price** p that a company charges for a certain item is often described as a function of the number of items that sell. An example is something like $p = -0.02x + 18$, where p is price in dollars, and x is units sold.

The **revenue** is the total amount of money brought in by the company for selling x items. To find revenue, multiply the number of items sold x by the price function p. In this case, the revenue is $R = x(-0.02x + 18) = -0.02x^2 + 18x$.

The **cost** function describes the cost to the company of making and/or selling x items. It's also a function with variable x.

The **profit** function describes the amount of money left over from the revenue after costs are subtracted. So a formula for profit is $P = R - C$, where R is the revenue and C is cost.

- If the cost function for the product above is $C = 850 + 3.5x$, the profit is

$$\begin{aligned} P = R - C &= -0.02x^2 + 18x - (850 + 3.5x) \\ &= -0.02x^2 + 18x - 850 - 3.5x \\ &= -0.02x^2 + 14.5x - 850 \end{aligned}$$

SKILL 4: FIND AND INTERPRET INTERCEPTS FOR A GRAPH

This skill was reviewed in the Prep Skills for Lesson 4-7 if you need a refresher.

PREP SKILLS QUESTIONS

1. For $f(x) = 1.3x^2 + 5x - 10$, find $f(0)$, $f(60)$, and $f(330)$.

2. Evaluate the formula for $a = -6.2$ and $b = 12.1$:

$$\frac{7a - \sqrt{b^2 - 4(3a)}}{5b}$$

3. When selling x units of a certain product, a company can charge $p = 8.5 - 0.035x$ dollars per item. The cost of producing and shipping the items is $C = 1{,}462 + 1.8x$. Find functions that describe the revenue and the profit for this product.

4. For the price function in Question 3, find the intercepts and describe what each means in the situation.

Lesson 4-10 Going. . . Going. . . GONE!

LEARNING OBJECTIVES

- ☐ 1. Solve a quadratic equation using the quadratic formula.
- ☐ 2. Find the vertex of a parabola using $x = \frac{-b}{2a}$.
- ☐ 3. Solve application problems using the quadratic formula.

©Akihiro Sugimoto/agefotostock RF

Every strike brings me closer to the next home run.
— George Herman "Babe" Ruth

Baseball purists will tell you that a well-pitched game is the pinnacle of the sport, but the average fan goes to the park hoping to see some home runs—and the longer, the better. There's just something cool about seeing a ball hit back back back, and gone. And in the information age, fans don't just want to speculate on how far a home run was hit: They want a measurement, and want it RIGHT NOW. So most ballparks and TV broadcasts now supply a distance after a home run is hit. But how do they calculate the distances? If you guessed "math," you're pretty smart. We've already seen that objects flying through the air tend to follow a parabolic path. In this lesson, we'll learn more about studying parabolas from an algebraic standpoint. This allows us to get more information about things like—you guessed it—how far a home run travels.

0. What information do you think we would need about the flight of an object in order to find an equation that models its path?

4-10 Class

As of this writing, it's been over 8 years since a major league baseball player hit a home run longer than 500 feet. There have been a few close calls, the longest of which was hit on August 6, 2016, by Giancarlo Stanton of the Miami Marlins. Based on information provided by a cool website called "Home Run Tracker," we can approximate the flight of that ball with this function: $f(x) = -0.00146x^2 + 0.718x + 2.5$. The input x represents the number of horizontal feet from where the ball was hit, while the output $f(x)$ is the height above the ground after the ball had traveled x feet horizontally. In order to find the total distance the ball was hit, we'd need to find the horizontal distance (x) that corresponds to a height of zero feet (when the ball would have returned to ground level if it hadn't been stopped by the stands).

So we'd need to solve this equation:

$$-0.00146x^2 + 0.718x + 2.5 = 0$$

In Lesson 4-9, we learned how to solve a quadratic equation if the side with the variable is factored. But that's not the case here, and we don't know how to factor something with decimals. So we need a method to solve a quadratic equation that isn't in factored form.

Math Note

If you want to learn more about the math behind estimating home run distance, do a Google search for "how hit tracker works."

Fortunately, there's a formula that provides solutions to any quadratic equation, as long as it's written in the **standard form** $ax^2 + bx + c = 0$, where a is not zero.

The Quadratic Formula

The solutions to a quadratic equation in the form $ax^2 + bx + c = 0$ are

$$x = \frac{-b + \sqrt{b^2 - 4ac}}{2a} \text{ and } x = \frac{-b - \sqrt{b^2 - 4ac}}{2a}$$

This is sometimes written more concisely as

$$x = \frac{-b \pm \sqrt{b^2 - 4ac}}{2a}$$

Important Note
a is the coefficient of the term with the variable squared, b is the coefficient of the variable to the first power, and c is the constant term.

where the symbol "$\pm$ " is read "plus or minus" and means one solution comes from adding and the other from subtracting.

1. For the equation describing Giancarlo Stanton's home run on the previous page, what numbers correspond to a, b, and c in the quadratic formula?

2. Use the quadratic formula to find the two solutions to that equation. Round to one decimal place if necessary.

3. One of the solutions provides the length of the home run. What was it?

4. What is the significance of the other solution in the physical situation?

Did You Get It

Try this problem to see if you understand the concepts we just studied. The answer can be found at the end of the Portfolio section.

1. In Did You Get It 1 and 2 in Lesson 4-8, we found that the revenue for a company that makes speakers was $R = 32x - 0.04x^2$ and the profit was $P = -0.04x^2 + 20.8x - 785$, where x in each case is number of speakers sold. Find the number of units that will make each of the revenue and profit equal to zero.

If we rewrite the two solutions provided by the quadratic formula just a bit, something interesting happens:

$$x = -\frac{b}{2a} + \frac{\sqrt{b^2 - 4ac}}{2a} \text{ and } x = -\frac{b}{2a} - \frac{\sqrt{b^2 - 4ac}}{2a}$$

Let's visualize that a bit differently:

$$x = -\frac{b}{2a} + \textit{some number} \text{ and } x = -\frac{b}{2a} - \textit{that same number}$$

This allows us to see that the x value $-\frac{b}{2a}$ is exactly halfway between the two solutions.

5. What's the connection between the solutions of the equation $ax^2 + bx + c = 0$ and the x intercepts of the graph of $f(x) = ax^2 + bx + c$?

6. The point where a parabola changes direction is called the **vertex** of the parabola. How do we know that the vertex has to be halfway between the x intercepts?

Combining Question 6 with the observation at the top of the page, we get a simple way to find the vertex of a parabola when we know the equation:

The Vertex Formula

For a quadratic function $f(x) = ax^2 + bx + c$, the x coordinate of the vertex can be found using the formula

$$x = -\frac{b}{2a}$$

Once you know the x coordinate, you can find the y coordinate by substituting the x coordinate in for x in the function and simplifying.

7. Use the vertex formula to find the first coordinate of the vertex for the function describing Stanton's home run: $f(x) = -0.00146x^2 + 0.718x + 2.5$.

8. What does your answer describe about the home run?

9. Now evaluate $f(x)$ for the input you found in Question 7. What does that tell you about the home run?

Did You Get It

2. Find the vertex of the parabola $f(x) = 4x^2 - 12x + 65$. Remember that the vertex is a POINT, not a number.
3. Find $f(0)$. Based on your results, is the vertex the highest or lowest point on the graph? How can you tell?

4-10 Group

In Lesson 4-8, we learned a lot about the economics of a boutique that sold a popular brand of bath beads. This was because we had enough information to model their revenue and profit with quadratic functions. But when it came down to analyzing those functions to find things like the maximum profit and the number of items they'd need to sell to break even, we were limited by having to estimate values from the graph.

There's nothing *wrong* with doing that—one of the central themes of this course is the value of visualizing data graphically. But now that we know the quadratic and vertex formulas, we can find exact answers to questions that we had to estimate before.

1. The price function for bath beads was $p = -0.025x + 22$, the revenue function was $R = -0.025x^2 + 22x$, and the profit function was $P = -0.025x^2 + 18.5x - 325$. Do you think the same number of items would make BOTH the revenue and profit have output zero? Don't use the formulas: Think about what revenue and profit mean, and explain your answer.

2. Use the quadratic formula to find the zeros of the revenue and profit functions and discuss whether what you find matches your thoughts in Question 1.

3. Do you think the number of units that provides the most revenue is the same as the number of units that provides the most profit? Again, this isn't based on the formulas. Think about the situation.

4. Find the number of units that provides the maximum revenue and the maximum profit and discuss whether what you found matches your thoughts in Question 3.

5. Find the exact maximum profit that the boutique can make from these bath beads.

6. What should they charge for each jar to make the most profit?

7. What price provides the largest revenue? What would the profit be at that price?

8. Superficially, you might think that charging more for a product will lead to making more money; that selling more items will lead to making more money; and that taking in the most revenue would lead to making the most money. Discuss what you've learned in studying the boutique business about all of those statements. More detail means you learned more!

Did You Get It

4. For the company in Did You Get It 1, the price function was $p = 32 - 0.04x$. Find the number of items and the price that would lead to the largest possible profit. How much would the company make in that case?

4-10 Portfolio

Name ______________________________

Check each box when you've completed the task. Remember that your instructor will want you to turn in the portfolio pages you create.

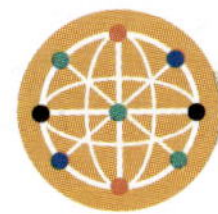

Technology

1. ☐ Build a spreadsheet that finds the solutions of any quadratic equation of the form $ax^2 + bx + c = 0$ when values for a, b, and c are entered. Then use your spreadsheet to confirm the distances you found for the home run described on the first page of this lesson. Do the results agree with your answers to Question 2 in the Class portion of this lesson? A template to help you get started can be found in the online resources for this lesson.

	A	B	C	D	E
1	a	b	c	Solution 1	Solution 2
2					

Online Practice

1. ☐ Include any written work from the online assignment along with any notes or questions about this lesson's content.

Applications

1. ☐ Complete the Applications problems.

Reflections

Type a short answer to each question.

1. ☐ What is the quadratic formula used for? Why is that so useful? Discuss.

2. ☐ What types of applied problems can be solved using the vertex formula?

3. ☐ Name one thing you learned or discovered in this lesson that you found particularly interesting.

4. ☐ What questions do you have about this lesson?

Looking Ahead

1. ☐ Complete the Prep Skills for Lesson 4-11.

2. ☐ Read the opening paragraph in Lesson 4-11 carefully and answer Question 0 in preparation for that lesson.

Answers to "Did You Get It?"

1. Revenue: 0 and 800 units; Profit: About 41 and 479 units
2. (1.5, 56)
3. $f(0) = 65$; since there's a point higher than 56, the vertex must be the lowest point.
4. 260 items at $21.60 profit is $1,919

Answers to "Prep Skills"

1. $f(0) = -10; f(60) = 4{,}970; f(330) = 143{,}210$
2. −0.96
3. $R = 8.5x - 0.035x^2$
   $P = -0.035x^2 + 6.7x - 1{,}462$
4. p intercept: (0, 8.5); This means if the company sells no product, they can charge $8.50 per item (this doesn't really make sense in real life).

   x intercept: (243, 0); This means if the company gives the items away (price is $0), they would give away about 243 items.

4-10 Applications

Name ______________________________

The height of a golf ball in meters can be described by the equation $h = -4.9t^2 + 23.5t$, where t is the number of seconds after it was hit. (Note: This is different from our home run equations, where the input x represented the number of horizontal feet traveled.)

1. Use a calculator or spreadsheet to make a table of inputs and outputs for this function. The variable t should start out at zero and increment by 0.2. Then use the result to draw a detailed graph of the function.

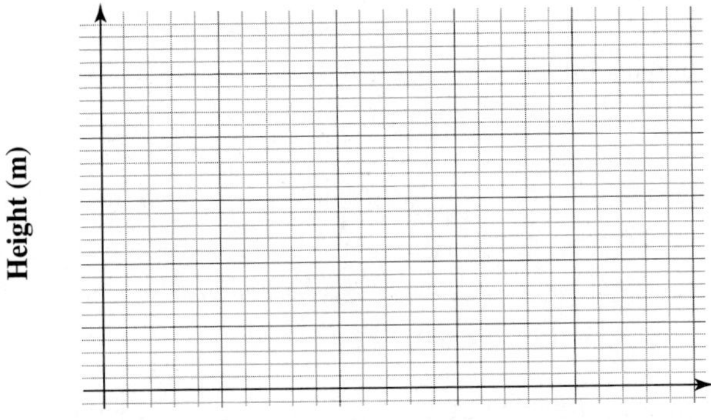

2. Find the vertex of the parabola using the formula we developed in this lesson, rounding each coordinate to the nearest tenth.

3. Explain what each coordinate of the vertex means.

4. Find the intercepts for the function.

4-10 Applications

Name ____

5. Explain what each intercept means.

6. Describe when the ball is headed upward, and when it's coming back down.

7. The function $d(t) = 49.2t$ describes the horizontal distance (in meters) traveled by the ball after t seconds. How far did it go?

8. The distance for golf shots is traditionally measured in yards. Given that there are 3 feet in a yard and that 1 meter is 3.28 feet, how far in yards did the shot go?

On November 2, 2016, Rajai Davis of the Cleveland Indians hit a 2-run home run that tied game 7 of the World Series in the bottom of the eighth inning. Using Home Run Tracker data, I was able to find a function that models the flight of the ball:

$$f(x) = -0.00175x^2 + 0.6315x + 1.9$$

where x is the horizontal distance in feet from where the ball was hit and $f(x)$ is the height in feet.

4-10 Applications

Name ______________________________

9. How high was the ball at the time it was hit?

10. Find the solutions to the equation $-0.00175x^2 + 0.6315x + 1.9 = 0$ and describe what each tells us about the home run.

11. The home run fence is 19 feet high at the point where the ball went over it, which is about 325 feet from where the ball was hit. By how much did the ball clear the fence?

Lesson 4-11 Prep Skills

SKILL 1: INPUT DATA AND FIND A REGRESSION EQUATION

This skill was the topic of Lesson 3-9, where we learned how to use both calculators and spreadsheets for finding regression equations. Refer to the Tech boxes in that lesson along with the accompanying videos in online resources if you need a refresher. We also practiced this skill in Lesson 4-6, where we found exponential equations of best fit. You should probably review the Tech box in that lesson as well.

SKILL 2: FIND THE CORRELATION COEFFICIENT AND RESIDUALS

Each of the correlation coefficient and residuals are means for deciding how accurately a given regression equation matches the data that it came from. In the case of correlation coefficient, this is provided by the technology used to find the equation. But you need to know that in some cases the technology provides r^2, which is the square of the correlation coefficient r. In that case, you'll need to find the square root.

The residual of a data point is defined to be the difference between the actual value and the value predicted by the regression equation. So the way you find a residual is

1. Evaluate the regression equation for the input of a known data point. This is the value predicted by the equation.
2. Compute Actual value – Predicted value.

SKILL 3: USE THE QUADRATIC FORMULA

This was the main skill that we studied in Lesson 4-10. The quadratic formula is good for exactly one thing: solving quadratic equations in the form $ax^2 + bx + c = 0$. The order is really important here: a always stands for the coefficient of the x^2 term, b is always the coefficient of the x term, and c is always the constant term. Once you have an equation in that form, you should identify and WRITE DOWN a, b, and c. Then substitute those values into the quadratic formula

$$x = \frac{-b \pm \sqrt{b^2 - 4ac}}{2a}$$

and do the arithmetic to find the solutions. Remember that there are TWO solutions. The symbol "$\pm$" is read as "plus or minus" and it means BOTH add and subtract. So another way to look at the quadratic formula is that it supplies two solutions:

$$x = \frac{-b + \sqrt{b^2 - 4ac}}{2a} \text{ and } x = \frac{-b - \sqrt{b^2 - 4ac}}{2a}$$

- The solutions of $4x^2 - 8x - 9 = 0$ are

$$x = \frac{-b \pm \sqrt{b^2 - 4ac}}{2a} = \frac{-(-8) \pm \sqrt{(-8)^2 - 4(4)(-9)}}{2(4)} \qquad a = 4, b = -8, c = -9$$

$$= \frac{8 \pm \sqrt{64 - (-144)}}{8} = \frac{8 \pm \sqrt{208}}{8} \approx \frac{8 \pm 14.4}{8}$$

This provides two solutions: $\frac{8 + 14.4}{8} = 2.8$ and $\frac{8 - 14.4}{8} = -0.8$.

SKILL 4: FIND THE VERTEX OF A PARABOLA

The vertex of a parabola is the key point on the graph: It's where a parabola changes direction, making it either the highest or lowest point on the graph. There's a simple procedure for finding the vertex, and it uses the same symbols as the quadratic formula. The first coordinate of the vertex is given by the formula

$$x = -\frac{b}{2a}$$

where b is again the coefficient of the x term and a is the coefficient of the x^2 term. Once you have the first coordinate from this formula, you find the second coordinate the same way you would find the second coordinate for any input: by plugging back into the equation.

- The graph of the function $f(x) = -3x^2 + 12x - 4$ is a parabola. The first coordinate of the vertex is $x = -\frac{b}{2a} = -\frac{12}{2(-3)} = -\frac{12}{-6} = 2$. The second coordinate is $f(2) = -3(2)^2 + 12(2) - 4 = -12 + 24 - 4 = 8$. So the vertex is the point (2, 8).

PREP SKILLS QUESTIONS

1. Find an exponential function of best fit for the data in the table.

x	0	5	10	15	20	25
y	12,406	9,430	6,023	4,803	3,411	1,000

2. Find the correlation coefficient for the exponential model in Question 1, then find the residuals for each data point.

3. Find all solutions to each equation.

   a. $x^2 + 15x - 3 = 0$ b. $4y^2 - 3y - 10 = 0$ c. $-0.03x^2 + 0.276x + 12.6 = 0$

4. Find the vertex of the parabola for each quadratic function that makes up the left side of the equations in Question 3.

Lesson 4-11 Down the Drain

LEARNING OBJECTIVES

- ☐ 1. Calculate the equation of best fit for a set of quadratic data using technology.
- ☐ 2. Solve application problems that involve a quadratic set of data.

©Shutterstock/Pavel Ilyukhin RF

A great accomplishment shouldn't be the end of the road, just the starting point for the next leap forward.

—Harvey Mackay

So this is it, friends: the end of our journey. But as the quote indicates, we're hoping that you won't see this as the end of math literacy, but the beginning of the rest of your life, a life that you can proudly live as someone who understands the value of using math in our world. We've modeled so many different things using math that it seems appropriate to finish up with one more lesson on modeling. You may have wondered where exactly the functions we used to model home runs in the last lesson came from. Does the Home Run Tracker website give formulas? The answer is no. But it does give more than enough information for us to use the regression techniques that we've learned to find a parabola of best fit. In this lesson, we'll get you up out of your seats for one more shot at active learning, and develop some data that might be modeled well by a quadratic function.

0. Do you think that data can be modeled by a quadratic function even if it doesn't change direction? That is, if it always increases or decreases?

4-11 Group

Experiment: The Time Needed to Drain a Bottle

Supplies needed:

- A clear 2-liter plastic bottle (or larger) with a hole drilled near the bottom of the straight-walled portion. The heights in the table on the next page should be marked on the bottle either with a permanent marker, or by attaching a paper strip with height markings. Height zero should be marked at the height of the hole. Experiment with the hole size so that it takes roughly 3 minutes for the bottle to drain. About 1/4 inch is a good place to start. You might also consider different size holes for different groups.
- Water and a bucket to drain water into.
- A watch that displays seconds. The stop watch feature on a smartphone works well.

1. Cover the drain hole, then fill the bottle up to the 20-cm mark. Set the time to zero seconds. Then uncover the drain hole and take the cap off the bottle. As the water drains out, record the time for each height in the table.

x Time (sec)	y Height (cm)
0	20
	18
	16
	14
	12
	10

x Time (sec)	y Height (cm)
	8
	6
	4
	2
	0

2. Use your data to create a scatter plot on the grid provided.

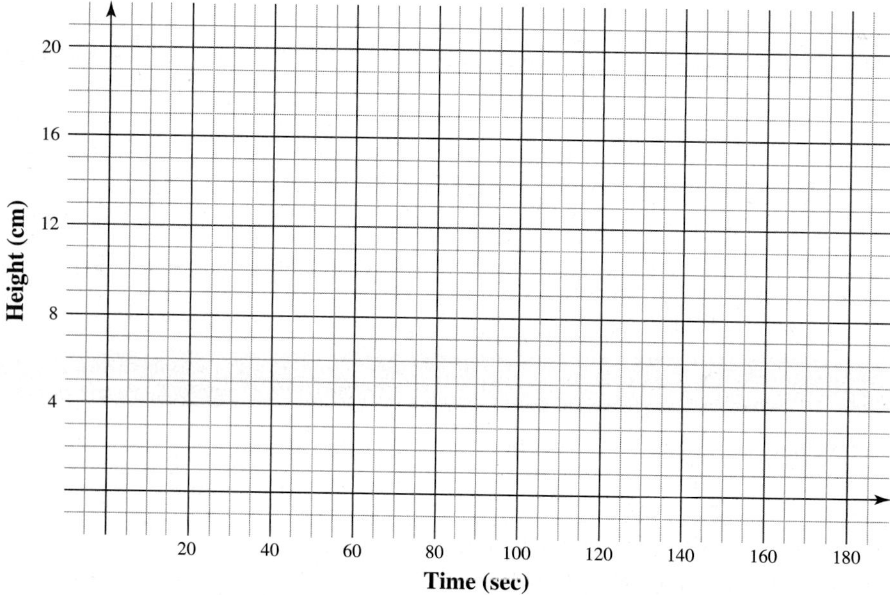

3. Do you think a linear or quadratic model is best suited to fit the data from this experiment?

4. What information in the table or graph supports your choice in Question 3? Explain why you think your choice makes sense.

5. Now try your best to ignore the scatter plot and data. Was there anything about what physically happened during the experiment that you think supports your choice in Question 3? Explain.

6. Find either the line or parabola of best fit, depending on which you chose in Question 3. You might encounter coefficients in scientific notation, so watch out for that. Round to four decimal places. Find the correlation coefficient if you choose a linear model, and the value of r^2 if you choose a quadratic model. (Calculators and spreadsheets will most likely show only r^2 if the model isn't linear: the correlation coefficient r is designed to measure how close to linear a model is.) The closer r^2 is to one, the closer a quadratic model fits the actual data.

7. Use your model to predict the time when the water in the bottle reaches 2 cm above the drain hole. How does this compare to your experimental data?

8. What's the value of y when x is 50? Explain the meaning of these values.

9. Find the value of x when y is 11. Explain what these values mean, and compare to what you would predict based strictly on the table.

Did You Get It

Try this problem to see if you understand the concepts we just studied. The answer can be found at the end of the Portfolio section.

1. Find the quadratic equation of best fit for the data shown, which represent the profit made by one company when they produce and sell x items. Find the value of r^2 as well.

x (Items)	0	100	200	300	400	500
y (Profit in $)	−1,243	−12	1,161	2,503	3,211	3,617

2. How much does the model predict the company would make if they sell 700 items?

10. In case you're not wet enough already, we're going to repeat the experiment. This time, instead of noting the time every 2 centimeters, we're going to note the height every 10 seconds. Fill in the table with your group's results. If the bottle isn't empty by the time you get to the end of the table, that's okay. You should still have plenty of data.

x Time (sec)	y Height (cm)
0	
10	
20	
30	

x Time (sec)	y Height (cm)
40	
50	
60	
70	

x Time (sec)	y Height (cm)
80	
90	
100	
110	

11. Find a best fit model for the new data, and write the correlation coefficient or r^2, whichever is appropriate.

12. Compare your two models, both algebraically and graphically. Discuss similarities and differences. Which is the better fit for the data you found? How do you know?

13. Why do you think the bottle didn't drain at a constant rate? Think about what's happening physically.

14. In the table, use your equation of best fit to find the residual for each data point. Are there any points that jump out as being very unusual? Why do you think that is?

x Time (sec)	y Height (cm)	Predicted height from equation	Residual
10			
20			
30			
40			
50			
60			
70			
80			
90			
100			
110			
120			

Did You Get It

3. Find the residuals for your quadratic model in Did You Get It 1. Are you surprised by the results, based on the value of r^2?

4-11 Portfolio

Name ____

Check each box when you've completed the task. Remember that your instructor will want you to turn in the portfolio pages you create.

Online Practice

1. ☐ Include any written work from the online assignment along with any notes or questions about this lesson's content.

Applications

1. ☐ Complete the Applications problems.

Reflections

Type a short answer to each question.

1. Would you change your answer to Question 0 at the beginning of the lesson now? Explain.
2. Why is it useful to first draw a scatter plot for data before deciding on what type of model to use?
3. Name one thing you learned or discovered in this lesson that you found particularly interesting.
4. What questions do you have about this lesson?

Looking Ahead

1. Start preparing for your final!
2. Consider sending us some feedback on the book. We love to hear from students! Our email addresses can be found in the front of the book.

Answers to "Did You Get It?"

1. $y = -0.0107x^2 + 15.433x - 1{,}339 \quad r^2 = 0.99497$
2. \$4,221.10
3.

x (Items)	y (Profit in \$)	Predicted Profit	Residual
0	-1243	-1339	96
100	-12	97.3	-109.3
200	1161	1319.6	-158.6
300	2503	2327.9	175.1
400	3211	3122.2	88.8
500	3617	3702.5	-85.5

Answers to "Prep Skills"

1. $y = 14{,}767.183(0.913)^x$
2. $r = -0.9500$; residuals: −2,361.18, 61.89, 80.00, 1,032.84, 1,019.26, −517.29
3. **a.** $x = 0.20, -15.20$ **b.** $y = 2, -1.25$ **c.** $x = 25.60, -16.40$
4. **a.** (−7.5, −59.25) **b.** (0.375, −10.5625) **c.** (4.6, 13.2348)

4-11 Applications

Name ______________________________

1. The table shows data describing a batted ball in a baseball game. Find a quadratic function of best fit for the data.

Feet from home plate	Feet in the air
0	4
100	75
200	95
300	46

2. How far away from home plate did the ball land?

3. How high did it go?

4-11 Applications

Name ______________________________

4. Did the ball reach its highest point exactly halfway in between where it was hit and where it landed? Should it have? Discuss, and keep in mind that the model is a parabola!

5. The Home Run Tracker data actually describes where the ball WOULD land if nothing stopped it from reaching the ground, like a fielder, the stands, or home run fence. In the direction this particular ball was hit, the fence is 344 feet from home plate, and is 8 feet high. Did the ball clear the fence?

6. Find the value of r^2 and the residuals for your model, then use both to discuss how well you think the model fits the data.

Unit 4 Language and Symbolism Review

Carefully read through the list of terminology we've used in Unit 4. Consider circling the terms you aren't familiar with and looking them up. Then test your understanding by using the list to fill in the appropriate blank in each sentence.

$d = \sqrt{(x_2 - x_1)^2 + (y_2 - y_1)^2}$

$x = \dfrac{-b \pm \sqrt{b^2 - 4ac}}{2a}$

$x = -\dfrac{b}{2a}$

arbitrary
binomial
coefficient
conjecture
counterexample
deductive reasoning
equivalent
expanded form
exponential decay
exponential function
exponential growth
$f(x)$
factored form
factoring
factors
function
growth factor
hypotenuse
inductive reasoning
inverse variation
isosceles
margin of error
parabola
parameters
perfect squares
polynomial
prime polynomial
profit
quadratic function
revenue
right triangle
standard form
symmetry
terms
trinomial
vertex
zero

1. When an expression has several pieces that are added or subtracted, the individual pieces are called ______________.
2. For a term that has a number multiplied by some power of a variable, the number part is called the ______________.
3. ______________ is the process of reaching a conclusion based on specific examples.
4. ______________ is another name for an educated guess based on available evidence.
5. A ______________ is an example that proves your conjecture was false.
6. ______________ is the process of reasoning that arrives at a conclusion based on previously accepted general statements.
7. An ______________ number is a number that is able to represent ALL possible numbers.
8. Each of 49, 25, and 144 are called ______________ because each is the square of a whole number.
9. A triangle with two sides that are perpendicular is called a ______________.
10. The longest side of a right triangle is called the ______________.
11. The distance between two points (x_1, y_1) and (x_2, y_2) is given by ______________.
12. The ______________ for a poll is a statistical measure that provides a range of values that the true outcome of the poll is likely to be inside.
13. An ______________ relationship exists between two variable quantities when that relationship can be described by the equation $y = \frac{k}{x}$, where k is a constant.
14. When a quantity grows according to exponential growth, the total is repeatedly MULTIPLIED by the same number. We call this multiplier the ______________.
15. A ______________ is a relationship between two quantities where each input produces a unique output.
16. The function notation used to indicate the output of a function named f for an input x is ______________.
17. In an ______________ equation like $y = a \cdot b^x$, a will be positive and the value of b will be greater than 1.
18. In an ______________ equation like $y = a \cdot b^x$, a will be positive and the value of b will be between 0 and 1.
19. A function where the input is an exponent is called an ______________.

20. ______________ are constants in an equation of a certain form that are determined by the data for a particular model.

21. The shape of the path that an object moving through the air under the influence of gravity follows is called a ______________.

22. When a graph can be cut in half so that one half is the mirror image of the other, this is known as ______________.

23. A ______________ can be written in the form $f(x) = ax^2 + bx + c$.

24. In economics, the word ______________ refers to the amount of money that a business takes in as a result of selling a product or service.

25. The difference between revenue and cost is the ______________ made by the business.

26. ______________ is a fancy math term for an expression with two terms.

27. The ______________ of an expression is written as a product.

28. Multiplying factors in an expression will give what is called the ______________.

29. If you have two ______________ forms of the same expression you should be able to input ANY value for the variable quantity x and get the same result from either expression.

30. An algebraic expression consisting of one or more terms that look like either a real number or a real number times a variable raised to a whole number power is called a ______________.

31. ______________ is simply the process of writing a number or expression as a product.

32. When the only way to write a polynomial as a product is one times the polynomial, the polynomial is called a ______________.

33. If $(c, 0)$ is an x intercept of the graph of $P(x)$, then c is a ______________ of $P(x)$.

34. If $(c, 0)$ is an x intercept of the graph of $P(x)$, then $x - c$ is one of the ______________ of $P(x)$.

35. A polynomial expression with three terms is called a ______________.

36. ______________ of a quadratic equation is $ax^2 + bx + c = 0$, where a is not zero.

37. The quadratic formula is ______________.

38. The point where a parabola changes direction is called the ______________ of the parabola.

39. The x coordinate of the vertex of a parabola defined by the function $f(x) = ax^2 + bx + c$ can be found using ______________.

Unit 4 Learning Objective Review

This is a short review of the learning objectives we've covered in Unit 4. In each case, rate your confidence level by checking one of the boxes. If you feel like you're struggling with these skills, consult the lesson referenced next to the objective and see the online resources for extra practice.

1. Apply inductive reasoning to make a conjecture. (Lesson 4-1)
2. Disprove a conjecture by finding a counterexample. (Lesson 4-1)
3. Apply deductive reasoning to solve a problem. (Lesson 4-1)
4. Solve application problems using the Pythagorean Theorem. (Lesson 4-2)
5. Solve application problems involving the distance formula. (Lesson 4-2)
6. Determine the margin of error in a given poll. (Lesson 4-3)
7. Explain the meaning of the margin of error in a given poll. (Lesson 4-3)
8. Calculate the number of poll respondents needed for a given margin of error. (Lesson 4-3)
9. Identify situations where inverse variation occurs. (Lesson 4-4)
10. Solve problems involving direct and inverse variation. (Lesson 4-4)
11. Define function and use function notation. (Lesson 4-5)
12. Identify the significance of a and b in an equation of the form $y = ab^x$. (Lesson 4-5)
13. Find exponential models. (Lesson 4-5)
14. Compare exponential models using graphs, tables, and formulas. (Lesson 4-5)
15. Gather and organize data from an experiment. (Lesson 4-6)
16. Find an exponential curve of best fit for an experimental data set. (Lesson 4-6)
17. Study the decay rate for exponential decay. (Lesson 4-6)
18. Identify a parabolic graph. (Lesson 4-7)
19. Solve problems using the graph of a quadratic equation. (Lesson 4-7)
20. Combine algebraic expressions using addition or subtraction. (Lesson 4-8)
21. Demonstrate the relationships between revenue, cost, and profit functions. (Lesson 4-8)
22. Combine algebraic expressions using multiplication. (Lesson 4-8)
23. Write a revenue function from a demand function by multiplying algebraic expressions. (Lesson 4-8)
24. Explain why factoring is useful in algebra. (Lesson 4-9)
25. Explain the connection between zeros and x-intercepts. (Lesson 4-9)
26. Factor a trinomial. (Lesson 4-9)
27. Solve a quadratic equation using the quadratic formula. (Lesson 4-10)
28. Find the vertex of a parabola using $x = -b/2a$. (Lesson 4-10)
29. Solve application problems using the quadratic formula. (Lesson 4-10)
30. Calculate the equation of best fit for a set of quadratic data using technology. (Lesson 4-11)
31. Solve application problems that involve a quadratic set of data. (Lesson 4-11)

After you've evaluated your confidence level with each objective and gone back to review the objectives you weren't sure about, use the problem set as an additional review.

In Questions 1 and 2, determine whether the type of reasoning used is inductive or deductive. Explain how you decided.

1. Eric knew Jamal was late the last three times they met for lunch, so he decided to lie and tell Jamal to meet 15 minutes before he really planned to show up.

2. Beth was worried that if she didn't do enough homework, she might fail her class.

3. Find a counterexample to the conjecture: This unit review only has three questions.

4. A roller coaster is being built to include a section that is set at 40% grade. How much vertical distance will this roller coaster rise along this section that covers a horizontal distance of 160 ft? Draw a picture to help illustrate your work.

5. What would the length of the actual track be in this section? Round to the nearest tenth of a foot if needed.

6. With the origin at Champaign, the coordinates for Springfield and Bloomington on the map shown are (−74, −21) and (−29, 27), respectively. Assuming the coordinates are given in miles, find the distance between the two cities. Round to the nearest tenth of a mile. Show your work!

7. A recent news article reported that according to a poll, only 29% of Americans support a proposed infrastructure plan. At the bottom of the article was the following sentence: "The Post-ABC poll was conducted by telephone Jan. 12–15, 2017, including landline and cellphone respondents. Overall results have a margin of sampling error of plus or minus 3.5 percentage points." The confidence level was not given. Supposing the confidence level was 95%, write a sentence or two explaining the results of the poll.

For Questions 8 and 9: In conducting a survey of 200 students, a researcher found that 22% had significant (more than $1,000) credit card debt.

8. Calculate the margin of error with a 95% confidence level for this survey.

9. If the researcher wanted to get a margin of error of 4%, how many total students would she need to survey? Assume the number of students with significant credit card debt stays at 22%.

For Questions 10–13: While running for city council, Erika plans to hand out fliers to every house or apartment in the largest neighborhood in her district. The time it takes to hand out fliers to the entire neighborhood varies inversely with the number of people who help hand out the fliers. From her previous election, she knows it takes 25 hours for 2 people to hand out fliers to the neighborhood.

10. Write an equation that relates the time required to the number of people.

11. Use the equation to complete the table. Round to the nearest tenth of an hour if needed.

People	Time
1	
2	
3	
4	
5	
6	
7	
8	
9	
10	

12. How long does it take for 10 people to hand out these fliers?

13. Does this result make sense with what you know about inverse variation? Explain.

14. Thinking in terms of exponential growth, what factor would you want to multiply by in order for a quantity to grow by 15% (that is, gain 15% of its value)?

15. Write an exponential growth function that describes an investment that starts at $500 and grows exponentially by 15% of its value each year x.

16. For the function in Question 15, find $f(4)$ and describe what it means.

17. Thinking in terms of exponential decay, what factor would you want to multiply by in order for a quantity to decay by 15% (that is, lose 15% of its value)?

18. Write an exponential decay function that describes a quantity of bacteria that have been treated with an experimental antibiotic that starts at 800,000 and decays exponentially by 15% of its value each day x.

19. For the function in Question 18, find $f(4)$ and describe what it means.

20. The graph shows the number of bacteria over a 20-day period based on your equation in Question 18. Use the graph to estimate how many days it takes for the number of bacteria to drop below 100,000.

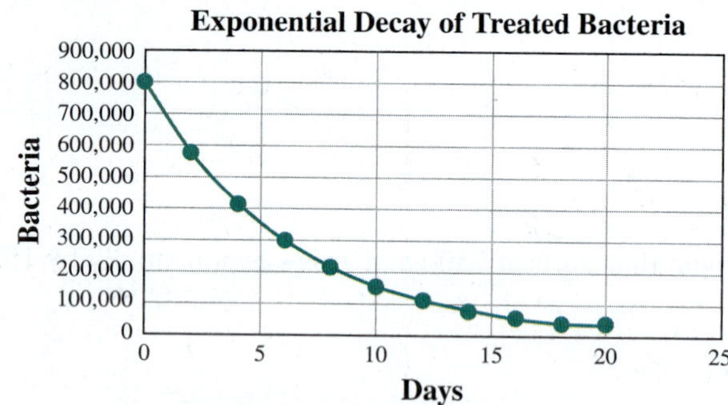

The amount of money in an investment account has been recorded in the table. Use the table to answer Questions 21–23.

Years	Value
0	$10,000
1	$12,320
2	$15,120
3	$18,610
4	$22,880

21. Use a spreadsheet or your calculator to find the exponential equation of best fit for this data. Write your equation in the form $y = a(b)^x$.

22. Based on your equation, describe the rate at which this investment has been growing over this 4-year period.

23. A common legalish statement in advertisements for investment opportunities is, "Past performance is not indicative of future results." If we assume this trend will continue, what would this investment be worth after 6 years?

The graph shows the daily profit y in dollars of an auto parts maker where x parts are produced in a day. Use the graph to answer questions 24–27. Answer each question with a complete sentence.

24. Estimate the y-intercept of this graph. Explain what it means.

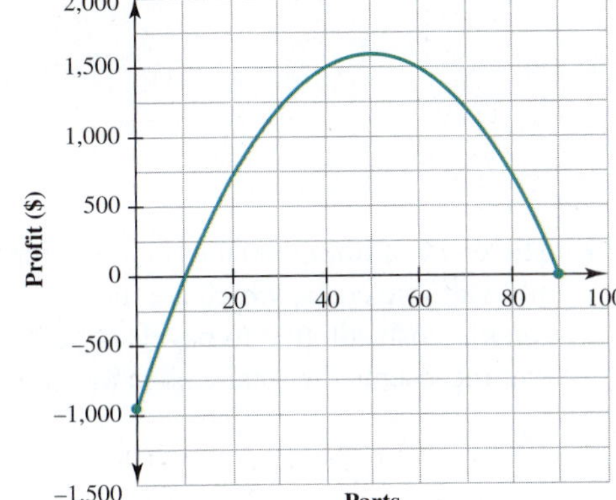

25. Estimate the x-intercepts of this graph. Explain what they mean.

26. Estimate the number of parts that will need to be produced in order to make a $1,500 profit.

27. Estimate the maximum profit the business can generate. How many parts need to be produced to reach that profit?

For Questions 28–37, refer to the following: From years of experience, a bagel shop has determined that at a price of $p = \$2.50$, they will sell $x = 450$ bagels each week. If they lower the price to $p = \$2.00$, they will sell $x = 550$ bagels each week.

28. Assuming this relationship is linear, write an equation that relates the price, p, to the number of bagels, x, that will be sold. Your equation should look like $p = mx + b$.

29. The revenue generated by selling these bagels is the product of the price p times the number of bagels sold, x. Write an equation that represents revenue by substituting your result from above into the equation $R = xp$.

30. The costs of doing business for a company can be found by adding fixed costs, such as rent, insurance, and wages, and variable costs, which are the costs to purchase the product you are selling. The portion of the bagel shop's fixed weekly costs allotted to bagels is \$200, and the supplies to actually produce each bagel cost \$0.75. Write an equation that represents the total cost to the shop of producing x bagels. Your equation should look like $C = mx + b$.

31. The profit made by the sale of these bagels is found by subtracting the costs from the revenue: $P = R - C$. Find an equation that represents profit by substituting your expressions for R and C into the profit formula. Simplify your answer.

32. What is the profit made by selling 300 bagels each week?

33. What is the profit made by selling 450 bagels each week?

34. What is the profit made by selling 0 bagels each week?

35. Find the number of bagels sold per week that will generate the highest profit.

36. What is the maximum profit that can be generated by this business?

37. At what price would you sell the bagels to realize this profit?

According to the baseball tracking system installed at a ball park, a home run was at the following horizontal and vertical positions along its path. Assume x represents the horizontal distance in feet from home plate and y represents the height in feet.

Horizontal distance	Height
0	3
100	128
200	161
300	75

38. Use a calculator or spreadsheet to find the quadratic equation of best fit for this data. (Do not round your results.)

39. How far did the home run travel?

40. What is the highest point reached by the ball?

In Questions 41–48, simplify each expression by performing the indicated operations.

41. $(x + 8) + (3x - 5)$

42. $2(x + 3) - 3(x - 4)$

43. $(x^2 - 7x - 9) + 2(5x^2 - 11)$

44. $-(3x^2 + 4x - 5) - 2x(2x + 5)$

45. $-5x(3x - 6)$

46. $(x + 4)(x - 7)$

47. $(x - 3)(x - 8)$

48. $(2x - 5)(3x + 7)$

In Questions 49–52, factor each trinomial.

49. $x^2 - 7x + 10$

50. $x^2 + 11x + 18$

51. $2x^2 + x - 6$

52. $3x^2 - 4x - 15$

53. Solve the equation $x^2 - 7x + 10 = 0$ using the quadratic formula. Compare your answers to those for Question 49. What do you notice?

Index

X

Y

Z